Long-term hazard to drinking water resources from landfills

nts

Preface

'THINKING IS NOT ENOUGH. IDEAS MUST BECOME WORDS AND WORDS DEEDS'

– Once said by Asaf Youbiner, a New York event manager and his words are still appropriate to describe the contents of this book and of Professor Spillmann's and his many fellow collaborators' lives' work. The ideas of 30 years of research accompanied by practical large-scale application of the results have already been converted into words and practice. The term 'combatant' has often been on the agenda in the real sense of the word because the results documented in this book inevitably have not always coincided with everyone's economic interests.

I have been collaborating with Peter Spillmann since 1974 and still remember his strict guidelines with regard to driving a car: 'The risk increases with the square of velocity.' If one transfers this idea to the history of this book, then no risks remain in the results. The deeds can now follow the words!

From my personal experience in the field of international environmental protection, I see that there is urgent need for these deeds because 90% of mankind on 'landfill earth' cannot afford a waste management system based on the example of the industrial states. The monthly per-capita income of some third world countries is equivalent to the cost of a ton of waste in Germany. So there are no economic interests to establish waste management, so alternatives have to be found. The willingness for environmental reclamation in the countries is there, but the money is not available.

Based on the results of this book and a practical approach taking into account the socio-economic conditions in these countries, we have developed affordable projects which only require start-up financing and then very soon will show their own economic advantages and offer a large potential for jobs and a clean environment. The cost equivalent is around one or two cigarettes a month in Germany, whereas 50–60% of these expenses flow back into the wages of employees.

I would like to enumerate the 'ideas' of the last 30 years again for all interested parties because who would have thought that:

Waste would be stored in high-quality structures with a geological barrier, base liner and cover system.

Waste must be treated according to a multibarrier concept.

Waste may not be deposited without any pre-treatment.

Waste in the form of slags from metallurgical industries burns spontaneously and produces high temperatures, sometimes reaching the red hot state over decades.

Waste in the form of ash, when deposited, goes through a second exothermic process.

Waste is ignited by thermophylic and hyperthermophilic bacteria.

Waste may produce pH values as low as 0.5 when coming into contact with *Thiobacillus ferrooxidans*.

Waste from plastic is attacked by thermophylic bacteria in their search for food.

Waste produces leachate over more than 100 years in landfills.

Waste produces toxic, carcinogenic and explosive gases in landfills over several decades.

Waste in landfills may be redirected to a material-flow specific utilisation after an aerobic stabilisation, where the gas and leachate problems can be completely eliminated. Certificates as under the CDM (Clean Development Mechanism) Act can be issued.

Waste in landfills may become a repository of raw materials or valuable materials for future generations.

Waste from plastics may be converted into diesel oil, e.g. using catalytic depolymerisation.

Waste represents economic assets and valuable resources when a material flow-specific treatment is used.

The investments in research and science are absolutely necessary and show striking initial results in this book: True knowledge is knowledge by causes (F. Bacon). There is enough knowledge, ideas and words, the deeds must now follow, i.e. engineers must implement the 'cooking recipes' of this book in a detailed and conscientious way.

Braunschweig, April 2008
Dipl.-Ing. Michael Struve

Foreword

Elected DFG (German Research Foundation) expert reviewer for hydrogeology and engineering geology for two periods (1976–1979), member of two DFG senate committees among others for environmental research (1990–1996) and two DFG senate committees among others for water research (1990–1996). He has produced numerous expert opinions on single projects, integrated projects, key research programmes and specialist research areas, as well as graduate colleges.

Summary

With all the required scientific care and reliability, the findings described in this book support the view that our current understanding about landfill stability must be fundamentally corrected, particularly with regard to the type and duration of emissions. Examples prove that detailed problems can be solved using state of the art techniques. This book therefore represents the foundations from which the necessary changes in waste management have to be developed.

Objective of the expert opinion

This book is the final report of an interdisciplinary research programme for the determination of long-term contamination of drinking water resources by landfills. The author of this expert opinion has been familiar with the research programme developed and managed by Peter Spillmann, from its early planning stages in 1972 up to its completion in 2007, from the viewpoint of an expert reviewer and advisor. In addition, the author and his working group at the Christian Albrechts University in Kiel researched into the field of hydrogeology in Part A of the DFG key programme 'Pollutants in Groundwater'. He has therefore been asked by the publishers to assess the research programme and its results from the viewpoint of an expert reviewer.

Completed expert opinions of the research programme

The research programme sponsored by DFG from 1976–1991 was annually checked for scientific quality and proper execution by independent experts. During the first 5 years, two specialists provided expert opinions about each field and three more specialists assessed the entire programme. While the specialists remained anonymous in this phase, the results and the scientific and technical execution of the subsequent key research programme over 10 years was adjudicated by a team of named experts. It consisted of Mr H. Bernhardt (Chairman), H. Kobus, W. Kölle, H.-G. Schlegel, C.-J. Soeder, E. Thofern, G. Tölg, L. Weil and the author of this preface. The subsequent programmes sponsored by the Volkswagen Foundation, the European Union and the Federal Ministry for Economics were anonymously adjudicated at least twice for each subject, so that the complete programme and its results were available to numerous specialist scientists. Sixteen theses and two German habilitation papers prepared in this field and adjudicated in each case by at least two specialist scholars contributed to a thorough evaluation of the basic and applied research. These intensive checks allow one to concede that this research programme is one of the most thoroughly examined research programmes in the field of landfill research.

The objective of this preface is the assessment of new findings about long-term groundwater protection in view of the research reports already published.

Methodology and results of reviewed research

Concept and structure of the programme

The research programme was designed to determine the expected long-term groundwater contamination from municipal waste deposits. For the first time ever in landfill research, all relevant causes, effects and relationships were to be determined within a programme at a central test facility and looked at:

- the effects of different deposition conditions and industrial additives on leachate contamination from municipal solid wastes;
- the effects of differently contaminated leachates on surface waters, the soil between landfill and groundwater and on porous aquifers.

For the first time ever in landfill research, the investigation procedures and test facilities matched the exacting requirements necessary to produce the high-quality results in an experimental project in engineering sciences. The tests took place in specially designed lysimeters, each of which had been precisely specified, were easy to access from all sides, and the results were both reproducible and transferable because the equipment not only fulfilled all landfill engineering and groundwater–hydraulic boundary conditions, but also all six relevant physical model laws. The previously unknown but very important model law concerning the progression of aerobic biological consolidation as a function of the diffusion path within the waste body has been established by Peter Spillmann from results of his preliminary tests analogous to Terzaghi's criterion (see Section 2.8).

By consistently using the integrating measurement principle 'from large toward small' and 'from outside inwards' a clear interpretation of the results of the differentiated measurements was possible even where the waste body exhibited large inhomogeneities. In addition, sources of serious error were detected in common groundwater sampling techniques and then eliminated using improved methods (see Section 11).

Simultaneous to basic research, the results were used in practical groundwater investigation, landfill engineering and landfill mining. A direct feedback with the actual landfill and groundwater regimes indicates that the results determined in the tests can be readily used in practice.

The scientists from the 17 groups within Germany that participated in the research project were all specialists of their own discipline, each having their own responsibility within the project as a whole. These disciplines covered all of the technical aspects of the emerging complex issues. Overlaps in the respective disciplines enabled error-free communication among the specialists and an active exchange of experience.

The German Research Foundation Senate Committee for Water Research emphasised in its decision that the interdisciplinary research project was successfully completed despite its broad scope. This was ensured by the long-term support from DFG owing much to the efforts of Dr Ulrich de Haar, expert of the Senate Committee for Water Research who unfortunately passed away prior to completion of the project. The well-structured and thoughtful research co-ordination by Professor Hans-Jürgen Collins of Braunschweig Technical University proved to be a substantial contribution to a successful interdisciplinary co-operation both in terms of scientific issues and schedule.

The supplementary studies, sponsored by the Volkswagen Foundation, the European Union and the Federal Ministry for Economics were carried out at a smaller scale but based on the same principles.

Coverage of past publications

The results of the project including the related supplementary procedural research has so far been published in six books, two German habilitation papers, 16 theses and approximately 150 individual papers or text contributions in Germany and abroad during the research project. Summary research reports including interdisciplinary interpretation of the results, were published by the German Research Foundation in the volumes *Wasser- und Stoffhaushalt von Abfalldeponien und deren Wirkungen auf Gewässer* (*Water and Material Balance of Landfills and their Effects on Waters*) (approx. 340 pages) in 1986 and *Langzeitverhalten von Umweltchemikalien und Mikroorganismen aus Abfalldeponien im Grundwasser; Schadstoffe im Grundwasser, Band 2* (*Long-term Behaviour of Environmental Chemicals and Microorganisms from Landfills in Groundwater; Pollutants in Groundwater, Volume 2*) (approx. 740 pages) in 1995 by the Weinheim-based VCH Publishing House (Publishing House Chemistry). Both volumes contain an English translation of the interdisciplinary interpretations. Additionally, in the report *Schadstoffe im Grundwasser, Band 2* (*Pollutants in Groundwater, Volume 2*) all figure and table captions were translated into English. The most important results are therefore available in English. The two books are now out of print but they can be borrowed from German university libraries by way of interlending (see details in the 'Literature' section).

Summary of the results published by DFG

The first volume (1985) contains results from short- and medium-term reactions in the waste body at rest, the subsequent emissions under different deposition and contaminant conditions and the results of biological leachate purification. These unparalleled results were obtained under defined and dimensionally correct conditions confirmed by landfill construction practice and could not be disproved by competitive research groups. It was shown that leachate from landfills will contaminate groundwater and surface water in each case with organic and mineral materials. Pathogenic organisms, e.g. from settlement lagoons, are once more discharged within a relatively short period of time. Chemicals are only temporarily sorbed. An intensive biological stabilisation before deposition reduces the initial elevated emissions, however it does not eliminate partially degradable materials or the risk of chemical contamination.

In countries with properly developed environmental technology, these findings and their consequences for landfill engineering are considered the international state of the art. This book will only deal with them to such an extent as they influence the long-term effect.

The research report on groundwater contamination published in 1995 deals with the long-term effect of different leachates and/or their components on relatively shallow porous aquifers. The excellent self-cleaning capacity of such systems is well-known and has been confirmed. In addition, it has been found that self-cleaning can only be used reliably if the process is completely manageable and can be controlled with certainty. The degradation process ceases if the concentration of a degradable chemical becomes too low or the necessary supplementary nutrients are missing. In each case the groundwater is unfavourably changed if it is contaminated by leachates from landfills. These results are also today's state of the art and represent the basis of construction regulations for landfills. Chapter 11 of this book illustrates some very important engineering consequences.

Assessment of the new findings

Extent of innovative results

This book covers the long-term behaviour of landfills and answers the still open questions: when and for how long can emissions from landfills be expected and actually how stable are those materials that are considered landfillable under today's standards? The essential new knowledge, in my view, is that current concepts about the stability of waste bodies are far less certain and it is absolutely necessary to correct them due to the clearly proven relationships.

The following new results fundamentally contradict current assumptions about the behaviour of waste bodies:

- An acidic conservation which can be reactivated may be limited to certain zones or components, which normal inspection techniques fail to recognise.
- An anaerobic deposit, which is normally classified as stable based on both common landfill engineering tests and careful biological investigation of the resting material is only dormant and can be reactivated at any time.
- Biological reactivation of dormant wastes is accompanied by mobilisation of the inventory of polluting chemicals, whose mass and toxicity represent a hazard potential which cannot be recognised in a dormant state.
- The conservation time can exceed by far the guaranteed lifetime of technical barriers.
- Conservation can only be avoided when the waste body is built as a permanently aerobic deposit using a pre-arranged intensive biological treatment.
- Plastics in plastic mono landfills, which are usually classified as chemically stable, can be exposed to thermogenic biochemical processes which can result in fires.

From these new findings the following consequences can be obtained:

- Stability criteria and landfill engineering and operational regulations as implemented in countries with highly developed environmental technology are necessary and effective, but they are not sufficient to ensure long-term protection for drinking water. They must be corrected and supplemented.
- Uncontained landfills, particularly in arid areas, endanger drinking water resources to a much larger extent than assumed in these countries, therefore urgent action is needed.
- Long-term protection of drinking water resources can only be ensured by disposing of materials that can be integrated into the environment in a way that is suitable to the specific location.

Reliability and transferability of the results

The supplementary tests also constituted basic research which took into account the precise test conditions of natural science including model laws. Relying on model theory based on these laws, the results can be converted to other climatic areas and other waste compositions. There was no doubt about this in the numerous expert opinions. The working group of Peter Hartmann and Günter Ballin, University of

Rostock, has managed to improve the agreement of the dynamics of thermal processes between landfill and laboratory test to an accuracy of 0.01°C. This has established new standards in the field of landfill research and clarified the causes of fires in plastic mono landfills.

Reliability and transferability of the results have been verified by findings obtained in landfill engineering, land reclamation and landfill mining in completely different climatic areas.

The limiting values of German and European regulations used for assessing the results are derived from the administratively specified minimum requirements of the region without claiming transferability. This research project has proved that adherence to these limiting values is necessary, but is not in itself sufficient to identify materials which can be integrated into the environment. It is therefore left to the reader to consult currently valid criteria of another region for comparison. Measured values and scientifically justified requirements represent the scientific standards for residual substances to be integrated into the environment in a way suitable to the specific location. These must be specified for each location.

Assessing the practical application of the results, Chapter 11

The practical solutions illustrated in Chapter 11 use previously reviewed results and represent the state of the art. They also solve the newly identified problems of the long-term hazards to drinking water resources. The consultants and engineering companies referred to in the appendix have already proved that they are capable of implementing these solutions using economically justifiable methods. This book not only deals with real hazards to drinking water, but also sets out economically feasible solutions based on the state of the art. Current practice falls short of this standard in as much as certain risks are not recognised and important goals are not achieved. Therefore two new solutions should be emphasised:

(a) the spatial distribution of emissions should be determined;
(b) wastes should be transformed into materials which can be integrated into the environment.

As for (a): It has been proved in the research programme 'Pollutants in Groundwater' that chemical contaminants can spread in high concentrations yet exhibit small transverse dispersion on narrow paths. Profile measurement and layer sampling (Section 11.2.1.2) can determine these paths in a cost-efficient way.

The current method used for extracting and monitoring drinking water only provides information about the expected contaminants in the water of a planned production well. It is not at all suitable for the investigation of emissions from contaminated sites and is only of use as a supplement to sampling.

Where (b) is concerned: Based on proven biological stabilisation, Section 11.4.2 shows by practical example that mixed municipal waste can be separated into chemically defined material flows using a simple wet technology process which does not generate sewage. This separation process can also be applied to landfill mining (Section 11.3.2). Industrial utilisation of chemically defined materials or their transformation into materials which can be integrated into the environment is state of the art in industry. The results of this programme that stipulate only materials which can be integrated into the environment should be landfilled are not restricted to first world countries, this condition can also be achieved in countries with a far lower gross domestic product.

This technology also meets the requirements for incineration residues established by EAWAG/ETH Zurich (Head: Professor P. Baccini) and based on detailed investigations. It has been proven that incineration residues from mixed municipal waste are too dissimilar to naturally occurring materials and minerals for them to be integrated into the environment without causing any long-term damage (Baccini and Gamper, 1994), even after complete incineration. Efficient treatment therefore demands a controlled waste materials flow process. Since this can be achieved in a cost-efficient way using the techniques illustrated in the book, undifferentiated thermal treatment of mixed municipal wastes no longer represents state of the art. (Important: separating an energy-rich fraction, as still being practiced today, does not provide a chemically defined material flow!)

Overall assessment

Based on scientific methods, Chapters 1 to 10 have proved that landfills represent considerable long-term hazards to drinking water resources which has so far not been recognised, necessitating an urgent rethinking in waste management. Solutions to actual problems are described in Chapter 11, using practical examples, and experienced companies are referenced in the Appendix that have either implemented these solutions or are definitely capable of performing them. This book therefore provides a reliable basis for the reduction and removal of current long-term hazards to drinking water resources.

Literature

Baccini, P. and B. Gamper (eds) 1994 *Deponierung fester Rückstände aus der Abfallwirtschaft* (*Landfilling of Solid Residues from Waste Management*) vdf Hochschulverlag AG a.d. ETH Zürich.

Spillmann, P. (ed.) 1986 *Wasser- und Stoffhaushalt von Abfalldeponien und deren Wirkungen auf Gewässer* (*Water and Material Balance of Landfills and their Effects on Waters*). (Forschungsberichte Deutsche Forschungsgemeinschaft – Research Reports of the German Research Foundation) VCH Verlagsgesellschaft Weinheim.

Spillmann, P., H.-J. Collins, G. Matthess and W. Schneider (eds) 1995 *Schadstoffe im Grundwasser, Band 2: Langzeitverhalten von Umweltchemikalien und Mikroorganismen aus Abfalldeponien im Grundwasser* (*Pollutants in Groundwater, Volume 2: Long-term Behaviour of Environmental Chemicals and Microorganisms from Landfills in Groundwater*). (Forschungsberichte Deutsche Forschungsgemeinschaft – Research Reports of the German Research Foundation) VCH Verlagsgesellschaft Weinheim.

Professor Georg Matthess, Darmstadt

Acknowledgements

The results of the long-term research have already been used in practice by experts for planning, monitoring and construction for more than 15 years. Competent companies have successfully implemented these design methods in the industry and manufactured and supplied the facilities required. These consultants and companies will introduce themselves in the last section.

The authors also express their utmost gratitude to the scientific translator, Mr. Nigel Pye, NP Services, Foley View, Farleigh Lane, Maidstone, England, ME16 9LX (npservices4u@gmail.com) for his technical competence and care in the translation and personal commitment in the work leading to its publication and to Prof. Tamás Meggyes, University of Wolverhampton, England and ICP Hungária, Budapest, Hungary for his scientific thoroughness in editing the translation.

The editors express their sincere gratitude for the support received to the research and this final publication.

Rostock, April 2008

Peter Spillmann
Timo Dörrie
Michael Struve

Scientific institutions and their research groups to investigate waste bodies

(1) DFG Integrated Programme, 'Pollutants in Groundwater' Part A

Leichtweiß Institute of Hydraulic Engineering and Water Resources, Dept. Agricultural Hydraulic Engineering and Waste Management, Technical University of Braunschweig
Prof. H.-J. Collins

Coordination of the Integrated Programme Part A	Hans-Jürgen Collins
Scientific concept; model design, building and operational management of the central test facility; water and solids balance; interdisciplinary evaluation and publication	Peter Spillmann
Construction and operation of the central test facility and the additional facilities	Adib Chammah Bernd Kaluza Detlef Duncker Wolfgang Stenzel Michael Röckelein

Institute of Hydrology of GSF Research Centre for Environment and Health
Prof. H. Moser

Investigation of water movements, evaporation processes, storage and water regeneration using the environmental isotopes ^{2}H and ^{18}O	Piotr Maloszewski Heribert Moser Willibald Stichler Peter Trimborn

Institute of Sanitary and Environmental Engineering, Technical University of Braunschweig
Prof. R. Kayser

Characterisation of flow path emissions using waste water parameters	Klaus Kruse

xxx

Degradability of leachate components   Hans-Jürgen Ehrig
(in Spillmann (ed.), 1986)   Rainer Stegmann

Institute Fresenius, Chemical and Biological Laboratories, Taunusstein-Neuhof
Prof. W. Fresenius

Waste analysis and analysis of transport of typical   Hans-Hermann Rump
residues from industrial production   Wilhelm Schneider
                                                               Heinz Gorbauch
                                                               Key Herklotz

**Institute for Weed Research of the Biological Federal Institute for Agriculture and
Forestry, Braunschweig**
Prof. W. Pestemer

Analysis of transportation and degradation of the   Henning Nordmeyer
pesticides simazin and lindane – examples of toxic   Wilfried Pestemer
industrial products   Key Herklotz

**Institute for Microbiology and Agriculture of the Justus-Liebig University in
Giessen**
Prof. E. Küster

Characterisation of stabilisation processes based on   Wolfgang Neumeier
microbiological criteria   Eberhard Küster
                                                               Zdenek Filip

**Medical Centre for Hygiene and Medical Microbiology of the Philipps University
of Marburg**
Prof. K.-H. Knoll

Transportation and survival period of indicator organisms   Karl-Heinz Knoll
(in Spillmann (ed.), 1986)   Klaus-Dieter Jung

Institute for Agricultural Chemistry of the University of Bonn
Prof. Kick

Landbehandlung von Sickerwässern   Michael Lohse
(in Spillmann (ed.), 1986)

Geological-paleontological Institute and Museum of the University of Kiel
Prof. G. Mattheß

Geochemical investigations and statistical assessment   Margot Isenbeck
(in Spillmann (ed.), 1986)   Walter Kretschmer
                                                               Asaf Pekdeger
                                                               Jürgen Schröter

Hesse State Institute for Geology
Geochemical consultation   Arthur Golwer

(2) Investigation into the reactivation of biochemical degradation processes in the waste, performed in the research project 'Mining – re-using – depositing landfill wastes from old sites', sponsored by the Volkswagen foundation

Leichtweiß Institute of Hydraulic Engineering and Water Resources, Dept. Agricultural Hydraulic Engineering and Waste Management, Technical University of Braunschweig
Prof. H.-J. Collins

Coordination of the project	Hans-Jürgen Collins
Waste management investigations	Friederike Brammer
	Hans-Jürgen Collins

Institute for Ecological Chemistry and Waste Analysis, Technical University of Braunschweig
Prof. M. Bahadir

Chemical analysis of toxic residues	Jan Gunschera
	Jörg Fischer
	Wilhelm Lorenz
	Müfit Bahadir

Institute for Microbiology, Working Group Technical Environmental Microbiology, Technical University of Braunschweig
Prof. Hans Helmut Hanert

Ecotoxicological and ecophysiological monitoring, control of in-situ activities and optimisation of decomposition processes	Martin Kucklick
	Peter Harborth
	Hans-Helmut Hanert

(3) Special investigations into landfill behaviour of plastics and biological stabilisation of natural organic substances – sponsored by the EU Commission DG and BMWi Federal Ministry of Economics and Technology

Institute of Landscape Construction and Waste Management, Waste Management Section, University of Rostock
Prof. Peter Spillmann

Design and construction of large-volume laboratory lysimeters enabling exact simulation of heat accumulation as well as material compaction and pore pressure; investigation of landfill behaviour of plastic wastes, in particular spontaneous ignition	Günther Ballin
	Peter Hartmann
	Peter Spillmann
	Frank Scholwin

Design and construction of a large-volume laboratory lysimeter facility enabling the exact simulation of heat accumulation for simultaneous investigation of the stabilisation of different wastes; investigation of changes in material from natural organic substances by extensive aerobic degradation up to 'forest soil stability'

Matthias Franke
Peter Degener

Institute of Soil Science and Plant Nutrition, University of Rostock
Prof. Peter Leinweber

Pyrolysis field ionisation mass spectrometry

Rolf Beese
Kai-Uwe Eckhardt

Pyrolysis gas chromatography/mass spectrometry

Gerald Jandl

(4) Interdisciplinary interpretation of the results, implementation of the results in practice, editing the final publication

Institute of Landscape Construction and Waste Management, Section Waste Management, University of Rostock
Prof. Peter Spillmann

Interdisciplinary interpretation of the results

Peter Spillmann

Concept of large-scale application, remediation of available old landfills, material-differentiated waste treatment

Timo Dörrie
Helmut Eschkötter

Scientific editing

Timo Dörrie

Production of the manuscript

Gisela Beckmann
Timo Dörrie

Dimensions, symbols and abbreviations

Physical dimensions
The 'SI' units apply (SI = Systeme International d'Unites).

Chemical symbols
The international symbols of the periodical system, including isotope identification, apply.

Abbreviations for the characterisation of contamination quantities of waste and waste water management
The abbreviations of the Waste Water Engineering Association (Abwassertechnische Vereinigung ATV, now German Association for Water, Wastewater and Waste (Deutsche Vereinigung für Wasserwirtschaft, Abwasser und Abfall DWA)), the Waste Deposition Ordinance (Abfallablagerungsverordnung) 2001, the Waste Water Ordinance (Abwasserverordnung) 2002 and the Sewage Sludge Ordinance (Klärschlammverordnung) 2002 apply.

Analogous notations have been translated into English.

Specific abbreviations and symbols outside of waste and sewage engineering are initially written out in full by the authors at their first occurrence.

Introduction

Objective and extent of the investigation

Calculations by German water management experts indicate that water may become a limiting factor for the development of settlements and industrial communities even in temperate areas such as Germany. Based on this finding, the Federal Water Act (Wasserhaushaltsgesetz) was issued requiring groundwater to be protected under the principle of 'cause for concern'. Because the measurements of groundwater contamination downstream of landfills indicated a problem, the Water Research Commission of the German Research Foundation DFG and the Volkswagen Foundation sponsored an interdisciplinary long-term research programme in which 15 institutes with 36 scientists investigated the long-term contamination of groundwater by leachates from waste deposits from 1976 to 1996. Unanswered questions about the stability of plastics and apparently stable earth-like substances were studied in another project sponsored by the European Union and industry from 1996. The research was fully completed by 2005. The basic relationships of long-term water contamination from local waste deposits with and without industrial wastes have been clarified.

Test method

Leachate emissions from 15 landfill types were tested in 1:1 scale models which, being fully manageable from all sides, provided reproducible results. Natural cleaning of two characteristic leachates was tested in four 100-m-long aquifer models, also manageable from all sides. The landfill models were built as cylindrical sectional cores from actual wastes (about 60 tonne content each) and the aquifer models were designed as sections of a real aquifer and were provided with two different covering layers. The results of the groundwater investigations have already been published (*Schadstoffe im Grundwasser* (*Pollutants in Groundwater*) Volume 2, VCH Publishing House, Weinheim, 1995).

The investigation on landfills whose results are published in this book cover all main stages of landfill decomposition: from the long-term acidic phase through a rapid anaerobic degradation phase within the alkaline range to aerobically stabilised forest-like soil materials; from population equivalent deposition of waste and sewage sludge to residual waste after an extensive separation of 'bio waste' and valuable materials; from municipal solid waste with almost no industrial contaminants to three-stage industrial contamination with typical industrial residues and/or environment-polluting industrial products:

- galvanic sludge (non-degradable, immobilised trace elements)
- cyanides (hardening salts, potentially degradable)
- phenol sludge (non-chlorinated organic, potentially degradable)
- lindane (= wide-spread chlorinated cyclic compound, potentially degradable)
- simazin (= simply chlorinated atrazine ring, potentially degradable).

Direct comparison of different stabilisation processes, in the same initial material and at the same location, clearly indicated small differences. The results can be directly transferred to flat landfills and model laws have been established for deep landfills. The results have been obtained under real, directly transferable field conditions with the additional benefit of laboratory accuracy.

Plastics and extensively biologically stabilised materials, which were believed to be very inert after the completion of the large-scale open air tests, were tested in large-scale laboratory lysimeters (D = 30 cm, H = 150 cm) which can simulate landfill-typical heat accumulation precisely to 0.1°C and develop a distribution of both stresses within the solids and internal pore pressure for special analysis.

The organic soil-like substance was analysed using both pyrolysis gas-chromatography/mass spectrometry and simultaneously by pyrolysis field ionisation mass spectrometry. The research programmes and their results were checked annually by experts.

Results of the research project

Water management
- The waste also acts as mulch in a highly compacted state. If the surface of the landfill is not covered by a sufficiently thick plant

cover, no more than 20–25 mm of rainfall evaporates even in arid areas. Therefore, leachates will have the potential to contaminate the groundwater due to rare but intensive rainfalls, even in desert-like territories. If the waste predominantly consists of aqueous vegetable residues (typical for southern countries), this water does not evaporate but flows off as leachate.

- Water storage and leachate discharges show simultaneous profiles. The key emissions of water-soluble pollutants will only be emitted if storage capacity is saturated and a substantial part of the storage material is degraded by biological decomposition. Depending upon storage conditions the storage process can drag on over several decades or even centuries.

Biological stabilisation

- The uncontrolled biological degradation fails to run completely in conventional landfills. It is interrupted by conservation processes which may last incalculably long but can be re-activated at any time, so transferring the problem to future generations.
- Biological stabilisation before deposition, which meets the current stability conditions in Germany, produces an extensively degassed anaerobic waste body which is then conserved for an unforeseeable long time.
- Extensive aerobic stabilisation produces a forest-soil-like low-reaction material – being the state-of-the-art 'Giessen model' since 1968 – which meets the eluate limiting values of the TASi document (Technical Instructions on Recycling, Treatment and Disposal of Municipal Waste) for thermal treatment without any exception. This aerobic stabilisation also enables a permanently aerobic waste body to develop where no conservation takes place. However, biological stabilisation on its own cannot produce a material which could be integrated into the environment due to the anthropogenically determined composition of the initial material, in particular the excessive heavy metal content.

Emission behaviour

- Anaerobic deposits from natural wastes contaminate the ground-water with an excessively elevated salt input and, to a similar extent, with organic substances.
- An intentional aerobic stabilisation of the waste before deposition and the production of a soil-like material up to the point where a

permanently aerobic waste body is achieved can reduce the ground-water contamination by natural materials not permitted in drinking water by an order of magnitude but cannot completely eliminate them.

- The placement of toxic industrial wastes in MSW landfills disturbs and interrupts the natural stabilization processes. It prevents the production of soil-like stable materials by intentional aerobic degradation and delays the normally slow biological stabilisation under anaerobic conditions for an incalculably long time.

- Although the natural potential for the degradation or immobilisation of industrial wastes can be recognised in co-disposed industrial and municipal solid wastes, it can only be fully used for phenol even if targeted aerobic degradation is applied. All further substances will be emitted over an incalculably long period of time if the materials are not subjected to a focused conversion.

- In waste mixtures made up of plastics which are chemically stable when used properly, e.g. resistant to acids, contamination with soil organisms can produce conversions including spontaneous ignition.

- Current assumptions of approximately 30 years for the aftercare are too short by about one order of magnitude for flat landfills with a low contaminant level and by more than two orders of magnitude for deep landfills with considerable industrial contamination.

Application of the results

- Industrial removal has been carried out since 1992 in landfills containing industrial residues in the Vienna Basin in Austria, in an area above one of Europe's largest groundwater reservoirs. This is based on the prediction for the reactivation of degradation and emission processes in landfills which has been confirmed in practice. In addition, landfill mining is being practised in large deposits outside this area.

- Since it has been proved that an aerobic stabilisation of the waste before its removal from an old deposit provides an efficient protection for personnel and people living nearby, this stabilisation method has become state of the art in Austria.

- An extensive aerobic stabilisation before deposition is permitted in Germany as an alternative to thermal treatment and is also used as a pre-treatment in capital-deprived countries.

Forecast for existing deposits

- Those waste deposits whose only protection consists of a capping system will contaminate the groundwater over the long term, even if no groundwater contamination can currently be observed.
- The extent and length of groundwater contamination depends on the depth of landfills and industrial pollutants contained within.
- Contamination from a large landfill lasts longer than the possible warranty life of a technical/mechanical barrier.
- The hazard of spontaneous ignition in plastic mono landfills must be excluded by specific material tests which precisely determine landfill conditions including contamination with soil micro-organisms and heat accumulation.
- The results of this research project indicate that the requirements of the Federal Water Act can only be fulfilled over the long term by converting the wastes into materials that can be integrated into the environment.

Informative power of the investigation

In view of the time scale and the interdisciplinary completeness, accuracy and simultaneous direct transferability of the measurements, the research results presented here belong to the most comprehensive technical research activities which have been performed so far on long-term emissions from landfills. The conservation and retardation effects of stabilisation which create the basis for the assessment of landfills have been confirmed in large landfills in the Vienna Basin within the framework of landfill mining while spontaneous heating of plastics has been investigated and verified in mono landfills in northern Germany. The authors take the view that it has been proved beyond any doubt that materials in existing waste deposits must be remediated further in order to protect groundwater while only materials that can be integrated into the environment may be deposited in the future over an unlimited period of time.

1

Object and concept of the research project

Peter Spillmann and Hans-Jürgen Collins

1.1   Problem

In the past, municipal waste was deposited outside of the residential areas and usually used as compensation backfill for areas where natural materials had been removed from the landscape (e.g. backfilling of gravel and clay pits, quarries and open pit mines). As long as these wastes were of natural origin and it was possible to integrate them into the ecosystem regarding their composition and volume, the self-cleaning capability of the soil and the uncontrolled biochemical auto-stabilisation of the wastes were sufficient to protect following generations from adverse impacts.

Due to urban settlements getting bigger, the amount of waste disposed of these days can exceed the natural decomposition equilibrium with the consequence of affecting drinking water by salts and biochemically persistent humic substances. As a result of industrialisation, toxic industrial wastes were disposed of in municipal solid waste sites along with wastes of natural origin and sewage sludge in a diffuse way. In addition to dispersed disposal, industrial wastes were intentionally disposed of together with or on top of municipal solid waste (co-disposal), in order to benefit from the biochemical activity and sorption capacity of organic substances in order to degrade and immobilise the toxic wastes. Landfills of this kind are usually located around traditional industrial sites. Because of the expansion of residential areas today many of these deposits lie within the suburbs, some of them even having been built upon. About 200 to 300 deposits with the potential to contaminate groundwater are not uncommon in and around old industrial cities, and industrial contamination can also be extremely high in remote waste sites, especially in old mines, as far as the authors are aware. The total number of deposits which are not contained by base liners and leachate treatment is estimated by the Federal Environment Agency (UBA) to exceed 100 000 contaminated sites (UBA, as of 1998) in Germany alone.

Due to our increased understanding over the past 20 years of how the sum of isolated pollutants over a wide area posed a contamination potential for groundwater, the standards for barriers beneath waste deposits became more and more stringent. Currently, an engineered composite liner (HDPE on a 3×25 cm mineral barrier) on a high-sorption geological barrier is stipulated. Leachate must be collected and clarified, and the surface of the waste body must be completely sealed against rainwater (composite capping on non-neutralised municipal waste) according to current German guidelines. Burning the extracted gases enables widespread control of emissions from new landfills. However, the effect of manmade barriers and treatment plants is time limited and the storage capacity of geological barriers is also finite and not equally effective for all pollutants. Therefore, new landfills for non-neutralised wastes have to be regarded as entombments of limited stability over the long term which contain potential environmentally polluting substances.

The following basic aspects must be considered for the assessment of potential groundwater contamination from waste deposits:

- Drinking water – and thus groundwater as its main source – is the only foodstuff that is absolutely vital and cannot be replaced by any other substance.
- Extensive groundwater contaminants, which are highly resistant to the self-cleaning effect of soil, cannot be reduced to the initial non-contaminated level of drinking water using technological processing methods.
- Persistent pollutants accumulated in the biosphere might also reach humans through the natural food chain even when contaminated groundwater is not used directly as drinking water.

The prerequisite of optimising the protection measures is reliable understanding of the potential for groundwater contamination by wastes.

The following research objectives result from this requirement:

- Which long-term contaminants can be expected from wastes of natural origin?
- To what extent is a diffuse substance from an industrial origin retained and/or degraded?
- Does the interaction between spiked industrial wastes with municipal solid wastes of natural origin result in an actual reduction of

emissions over the long term as measured in situ using short-term tests?

- What influence does landfill technology have on the extent and duration of emissions?
- What is the effect of climatic water balance on emissions?
- When and to what extent can emissions be expected from remobilisation and reactivation processes?
- How does extensive aerobic stabilisation (decomposition) affect the emissions from the waste after being placed in a highly compacted state?
- Are soil-like substances really soil-like after aerobic stabilisation?
- To what extent are apparently chemically and biologically stable plastics actually inert in landfills?

1.2   Basic experimental concept

A reliable statement on the long-term potential for groundwater contamination from wastes requires investigation of the entire leachate path from its emergence in the waste body to the final phase of its conversion in the aquifer. It should also be noted that, because of the variety of deposits, the investigations must be restricted to the basic types of material mixtures and characteristic landfill types as well as to typical aquifers. The investigations should therefore be aimed at making fundamental statements which can be transferred to other, non-tested deposits and aquifers.

In order to be able to establish statements on the long-term effects, additional models using accelerated testing must be applied. Furthermore, the precise conditions of correct scientific investigation must be observed:

- accessible from all sides
- controllable
- reproducible
- transferable.

In order to be able to undertake experimental investigations at acceptable costs, the following general concept has been chosen:

- The three most important anaerobic landfill types (without sewage sludge, with compact sewage sludge inclusions and highly compacted sewage sludge waste mixes) and the most important aerobic landfill type (permanently aerobic, soil-like sewage sludge waste mix) were modelled as generally controllable sectional landfill cores in the form of large-scale, compressible lysimeters (see details in Chapter 2).

- The two most important aquifers concerning leachate contamination (porous aquifers near the surface with cover layers of good or occasionally poor permeability) were built as controllable sectional cores from real aquifers for the two most important leachate types (high organic contamination from the initial anaerobic phase and low organic contamination from the aerobic final phase of a landfill) in the form of a large-scale trough channel open to direct influence from the weather (see Spillmann *et al.*, 1995).

The different landfill model types were built on the same site (Braunschweig-Watenbüttel main test site) with waste from the same collection area in the form of two simultaneous models for each landfill types, alternatively with and without intermediate earth layers (daily cover). Short- and medium-term processes were measured on all simultaneous models until hydraulic equilibrium was reached (after about 5 years) and then the simultaneous models with intermediate earth layers were dismantled for precise physical, chemical and biological investigation of the materials (see results in Spillmann, 1986b). The second simultaneous model was observed in each case for a further 10 years and tested extensively for long-term effects after 15 years, particularly looking for reversible conservation. The stabilisation process was tested by two independent, complementary investigation methods:

- Intensive tests on the material of two simultaneous models for each landfill type at different times provide fundamental data on the transformation of a landfill type material as a function of time.
- Tests on the same wastes at the same site for landfill types with very different stabilisation rates enabled the time scale for the models to be established using the laws of convection and diffusion. This can be used to estimate the order of magnitude of the time interval needed to convert natural organic and mineral wastes into soil-like substances in common deposits.

The influence of the change in waste composition, primarily the effect of separate collections for composting organic substances, was tested in a supplementary programme using the same principles (Wolfsburg landfill test site).

    The influence of industrial wastes was tested in such a way that municipal solid waste with low industrial contamination was selected as the base mass of the deposit and was contaminated in selected landfill types with typical industrial chemicals and/or wastes and pathogenic organisms (cyanides, galvanic sludge, phenol sludge, persistent chlorinated hydrocarbon compounds; indicator bacteria in various dosage and application) in three successively increasing contamination stages (low, medium and high contamination) (see details in Chapter 2).

1.3   Areas of expertise

Complex questions cannot be properly answered by scientists from only one discipline. The following tasks are therefore interconnected in the investigation of the impact on groundwater of long-term emissions from landfills.

Representative landfill types (waste management) and aquifers (hydrogeology) have to be investigated using models in such a way that the measurement and analysis accuracy enable the long-term effects to be forecast at low degradation rates (hydraulic engineering experiments, hydrometry). Typical initial contaminants in leachates have to be specified (sanitary water engineering and analytical chemistry). The changes in natural and synthetic materials must be determined quantitatively (different special fields of analytical chemistry, humic substance analysis and trace material analysis). Interactions between water and solids must be described (hydrogeology, water chemistry). In order to be able to evaluate degradation performance, colonisation with microorganisms and their activity has to be described (microbiology of different special fields). Also, the currently somewhat less appreciated question of the survival period and transport distances of pathogenic organisms has to be considered. Clarification of collected leachate in treatment plants (sanitary water engineering) and by land treatment (agricultural chemistry) has to be investigated. The working groups which have fulfilled these interconnected tasks in connection with waste bodies are listed following the table of contents (Working Groups of Groundwater Investigations, see Spillmann *et al.*, 1995, pages XIX–XX).

1.4   Sponsoring the investigations and validation of results

A group of researchers was sponsored by the German Research Foundation (DFG) to undertake the investigation of the landfill sectional cores up to the point of hydraulic equilibrium (end of the storage phase) and the dismantling of selected alternatives, as well as the building and inclusion of the artificial aquifers. On recommendation of the adjudicators, this research work was included in the DFG integrated research programme 'Pollutants in Groundwater' and sponsored from 1986–1991.

In addition to the financial support from DFG, the city of Wolfsburg and the State of Lower Saxony sponsored another project which was carried out in a similar facility on the Wolfsburg landfill site. The latter project examined the influence of intensive recycling of natural organic substances and valuable materials from industry on the space requirement and leachate contamination of municipal residual waste. After completing this investigation (Chammah *et al.*, 1987; Collins and Spillmann, 1990), the effect of industrial wastes on residual waste landfills was investigated at this facility within the DFG integrated research programme.

The specified landfill sectional cores at the main facility in Braunschweig-Watenbüttel were particularly suitable for investigating the effect of landfill mining on different types of landfill. This research project 'Mining – re-using – depositing landfill wastes from old sites' from a new interdisciplinary group of researchers was sponsored by the Volkswagen foundation (Brammer *et al.*, 1997). The results of this research project, combined with the results of the preceding DFG integrated programme, made an important contribution to forecasting the long-term effect of landfills on groundwater, in particular with respect to the length of time of this effect. The results from this research group were therefore integrated into this book.

The scale of tests in a real landfill sectional core does not enable detailed research. Therefore, to determine the long-term stability of apparently inert material groups such as chemically resistant plastics and humic-like organic stabilisation products, large-scale laboratory lysimeters were built at the Department of Waste Management of the Agricultural and Environmental Faculty of Rostock University, which were capable of correctly simulating both heat accumulation and pressure distribution within the waste body. The stability of plastic wastes was investigated with the support of the European Union's Commission DG XI.E3 and industry, while the tendency of material transformation in organic substances was also investigated, sponsored by the Federal Ministry of Economics and Technology. Thus, in principle, all key questions were identified to determine the long-term behaviour of wastes in landfills.

The research results have already been implemented in industrial applications. The results of the latest applications up to 2007 have been aggregated in Chapter 11.

2

Central test facility and test procedure

Peter Spillmann and Hans-Jürgen Collins

2.1   Scope of investigation

The waste bodies of the central test facility on the Braunschweig-Watenbüttel landfill site and the tests performed between 1976 and 1985 have already been described in the first publication on short- and medium-term processes in water and material balance of landfills (Spillmann, 1986b). Therefore, the test facility and test procedure for the long-term effects of landfills is only described in this book to such an extent as is necessary to understand the long-term investigations.

The details of the central test facility aquifers and the tests to study the long-term effect of landfill leachates on groundwater near the surface were published by Spillmann *et al.* (1995). In this book, the results of the groundwater investigation are described to illustrate the long-term emissions from waste bodies.

The observations from 1986 to 1994, the detailed investigations on the apparently stable material up to 2007 and the application of the results in engineering up to 2007 will be presented in detail.

2.2   Concept and location

Well-specified, controllable and, if possible, variable test conditions as well as adherence to the model laws, are prerequisite of fundamental research. In addition, the object of research must be identical, or at least very similar, in an interdisciplinary co-operation of several researchers. Since test objects such as municipal waste landfills and aquifers are dependent on the location, the research group decided to establish a central test facility for the investigation of different waste bodies and different aquifers and, if necessary, perform detailed investigations at smaller supplementary facilities at their own institute. The city of Braunschweig's landfill was selected as the location for the central test facility for waste bodies and the open-air area of the Leichtweiß Institute, Braunschweig was chosen to investigate the aquifers.

The fundamental processes in the waste body can only be properly investigated when an average municipal waste in which the organic components exhibit a constant composition is selected as an initial material for the deposition tests. This material has to be subjected to diverse landfill processes in representative landfill sectional cores and intentionally contaminated with selected chemicals and bacteria. The spiked chemicals and bacteria – properly marked to make them traceable – are selected as representatives of major pollutant groups and/or pathogenic bacteria.

The characteristics of the initial material and the deposition conditions to be tested identify the minimum dimensions and the minimum number of test specimens as well as the the model laws to be fulfilled. The transfer of the results to different sized landfills has to be performed based on the laws of the model.

2.3   Selection of landfill types and waste materials

2.3.1   *Waste without recycling influence – the Braunschweig-Watenbüttel central facility*

The composition and the structure of the deposited wastes affect the processes in the waste materials. The structure is primarily shaped by type and technique of waste placement. The most important difference in long-term behaviour among domestic landfills can be expected between a pure municipal waste landfill and a co-disposed landfill containing municipal waste (MW) and communal sewage sludge (SS). Within these groups, the most important differences can be expected between highly compacted fresh waste (anaerobic) and the decomposition preceding compaction (aerobic, extensive biochemical degradation in presence of atmospheric oxygen). Further differences result from the placement of sewage sludge in lenses (SSL) or the deposition of a sewage sludge waste mix (SSM), the different thickness of the lifts (50 cm, 2 m), whether intermediate cover (daily earth cover) is used and the extent of industrial contamination (none, low, medium (fair) and high). Because greater technical complications were expected in co-disposed municipal waste and sewage sludge as opposed to a pure municipal waste landfill, special emphasis was placed on deposits containing sewage sludge. The following eight landfill types (or cor-responding lysimeter types) were selected by all scientists involved in the investigation of landfills without recycling influence (Fig. 2-1):

The Braunschweig-Watenbüttel central facility (total waste)
Landfill type 1 (Lys. 1 MW)
Municipal waste without pre-treatment, no industrial contamination, highly compressed in 2-m-thick layers, installed at about $1-1\frac{1}{2}$ year intervals, intermediate cover of approx. 10 cm soil (silty sand).

Landfill type 2 (Lys. 2 MW)
Municipal waste mixed with the addition of water (200 l per 5 t waste), no ('zero'-stage) to medium high (1st and 2nd contamination stage) industrial contamination, highly compacted installation of about 40-cm-thick lifts at about 4-month intervals, without intermediate cover.

Landfill type 3 (Lys. 3 MW/SSM)
Municipal waste mixed with population equivalent sewage sludge quantities (aerobically stabilised, approx. 75% water content; municipal waste : sludge $= 2 : 1$), no industrial contamination, highly compacted installation concurrent to type 2 with identical municipal waste masses per layer, without intermediate cover.

Landfill type 4 (Lys. 4 MW/SSM)
Sewage sludge waste mix as type 3, highly compacted installation similar to type 1 in about 2-m-thick layers and approx. 10-cm intermediate cover (soil as type 1).

Landfill type 5 (Lys. 5 MW/SSM)
Sewage sludge waste mix as types 3 and 4, but loosely arranged for decomposition on an air-permeable base, sieved after decomposition, highly compacted installation of filter residues, placement of the next decomposition layer on the highly compacted filter residues.

Landfill type 6 (Lys. 6 MW/SSM)
Decomposing sewage sludge waste mix as type 5, however no ('zero'-stage) to medium high (1st and 2nd contamination stage) industrial contamination and highly compacted installation of the total material after decomposition, placement of a new decomposition layer on the highly compacted material, air supply to the additional waste as type 5.

Principal test Braunschweig-Watenbüttel Landfill

From 1976 to 1985 DFG (German Research Foundation): research project 'Water- and Solid Balances in Sanitary Landfills and their Influence on Ground- and Surface Water' (Spillmann, 1986b)

Lys.	1	2	3	4	5	6	7	8	9	10
Municipal waste		MW	MW/SSM	MW/SSM	MW/SSM	MW/SSM	MW/SSL	MW/SSL	MW	MW/SSM
	–	Mixed	Mixed	Mixed	Mixed	Mixed	Placement in lenses	Placement in lenses	–	Mixed
	–	–	–	–	A. decomp., sieving Aerobic decomp.	Aerobic decomp.	–	–	–	Aerobic decomposition
	Highly compacted Soil cover	Thin layer	Thin layer	Highly compacted Soil cover	Thin layer	Thin layer	Thin layer	Highly compacted Soil cover	Highly compacted Soil cover	Thin layer
	No contam.	No/med. contam.	No/med. contam.	No contam.	No/med. contam.	No/med. contam.	No contam.	No contam.	High contam.	High contam.

Building stage 1 1976/1977 1 September–31 October

Building stage 2 1977/1978 1 November–31 October

Building stage 3 1978/1979 1 November–31 October

Building stage 4 1979/1980 1 November–31 October

Demolition of Lys 1, 4 and 8, June 1981

1981/1986 1 November–31 October   Relocation 1981 (Lys. 6/10)

[m] 5 4 3 2 1 0

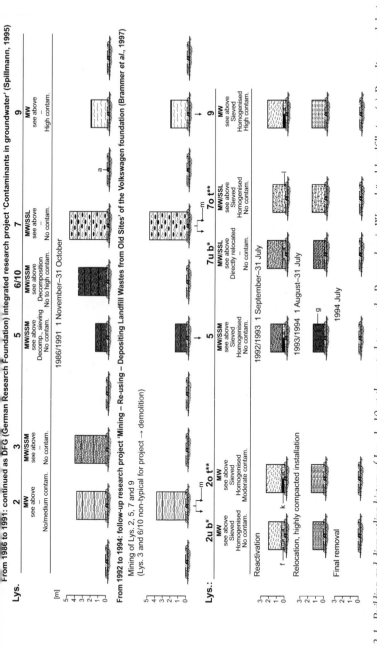

Fig. 2-1. *Building and dismantling history of Lys. 1–10 at the central test on the Braunschweig-Watenbüttel landfill site. (a) Base liner and drainage; (b) compacted installation in about 2-m-high lifts with earth cover; (c) compacted installation in about 0.50-m-high lifts without earth cover; (d) sewage sludge waste mix at the beginning of decomposition; (e) sewage sludge waste mix after about 1 year decomposition; (f) bottom ventilation (bulky waste); (g) sewage sludge waste mix, highly compacted after about 1 year decomposition; (h) sealing cover for determining the retarded leachate run-off; (k) anaerobic/aerobic alternating aeration; (l) highly compacted direct relocation; (m) dividing the lysimeter within reactivation tests; (n) artificial increase of precipitation by irrigation*

Note: b*: bottom; t**: top

11

Landfill type 7 (Lys. 7 MW/SSL)

Municipal waste without pre-treatment, no industrial contamination, installation in about 50-cm-thick layers together with population equivalent, lens-shaped inserted sludge masses of 0.1–0.2 t, time interval about 4 months, without intermediate earth cover.

Landfill type 8 (Lys. 8 MW/SSL)

Municipal waste as type 7, but highly compacted installation in 2-m-thick layers at time intervals of $1-1\frac{1}{2}$ years, intermediate cover with approx. 10 cm soil as type 1.

Two more waste bodies, of the most dissimilar landfill operation types, were constructed (type 1: anaerobic municipal solid waste without sewage sludge, type 6: aerobic sewage sludge waste mix) in order to investigate the effect of extreme industrial contamination.

Landfill type 9 (Lys. 9 MW)

Municipal waste as type 1, but with high industrial contamination (3rd contamination stage).

Landfill type 10 (Lys. 10 MW/SSM)

Municipal waste as type 6, but with high industrial contamination (3rd contamination stage).

Anaerobic deposits were built both as deposits with earth cover in 2-m steps (old type of operation until 1970, or in individual cases until today) and as landfills without earth cover in about 0.50-m steps (new technique, gradually introduced from 1970) in order to obtain information on corresponding contents, but also to gain additional information about the influence of the installation method. Leachate was expected to percolate more evenly through thin deposits without an earth cover while access to air would promote degradation. The aerobic landfill types behaved completely similar in the decomposition phase and exhibited differences only after the installation of the highly compacted decomposed material and/or the sieve residues.

The sources of waste material supply were selected in agreement with the test concept (Chapter 1) to ensure that the organic components remained approximately constant both seasonally and over the expected building period. Therefore municipal waste was acquired from a pure residential housing area in the city of Braunschweig, which is connected to district heating and contains only small

Table 2-1 Separated masses of waste glass and scrap metal from the municipal waste, in % by weight DS waste, used for the central test on the Braunschweig-Watenbüttel landfill

Time	Affected sections of the structure	Collection results in the selected area [t/week]			
		Waste glass	Sheet metal	Municipal waste	Sum
Nov. 1986	Basic structure	Without separation			31.60
July 1987 (Introduction of separation)	3rd lift Lys. 7 4th lift Lys. 2, 3, 4	0.70	0.06	31.6	32.36
March 1988–end of test setup (improved separation)	Addition of waste Lys. 1, 4, 6 5th lift Lys. 7 6th lift Lys. 2, 3	2.50	0.06	29.4	31.96

ornamental gardens. The refuse freighters always used the same routes, which were exclusively in this collection area. The municipal waste was thus almost totally free from ash, commercial waste and large volumes of packaging materials from shops and contained only small amounts of garden wastes. Nevertheless, it was not possible to entirely avoid a change in the inorganic fraction of the municipal waste despite these constant boundary conditions because separation of waste glass and cans was introduced in Braunschweig while the test facility was being built. However, the resulting changes are known for this selected closed collection area (Table 2-1). They have to be considered in the interpretation of the analysis of the initial material.

Tables 2-2 and 2-3 give an overview of the composition of the waste used. Table 2-2 gives an overview of the remaining fractions of the domestic waste as well as a detailed classification to the coarse fraction, while Table 2-3 shows the results of the preliminary waste analyses after pre-sorting. Their average values and extreme values of the complete analyses are illustrated in Table 2-4. The chemical analyses were carried out by Fresenius Institute, Taunusstein.

The analyses indicate that mass distribution of the three fractions $d > 10$, $10 > d > 5$ and $5 > d \, cm$ hardly differ among the three building stages. However, the sorting analyses of the coarse fractions $d > 10 \, cm$ are very different. Thus the metal percentages (which are important to the further consideration of the overall assessment of the heavy metal problem) vary greatly. Also, the fractions of plastic, paper and cardboard vary to a very great extent.

13

Table 2-2 Sieve analysis of the municipal waste used in the central test on the Braunschweig-Watenbüttel landfill and sorting analysis of the fraction d > 10 cm, in % by weight DS waste

Fraction[a] # cm × cm[b]	Building stage 1 Sept. 1986–Oct. 1987			Building stage 2 Nov. 1987–Oct. 1988			Building stage 3 Nov. 1988–Oct. 1989
	Min.	Medium	Max.	Min.	Medium	Max.	Individual value
d > 10	20.0	35.0	43.0	23.8	30.7	35.1	36.6
10 > d > 5	26.6	30.2	37.0	35.4	38.5	43.1	28.6
5 > d	26.6	34.8	51.0	28.0	30.8	33.1	34.8
Coarse material d > 10 cm							
Metal, total	3.9	9.1	18.3	3.0	5.1	7.5	10.8
Ceramics, rock							2.3
Glass	14.6	18.9	30.0	4.7	13.6	18.2	8.5
Plastics	14.9	23.2	30.6	14.5	15.9	18.2	13.3
Wood and leather	3.0	3.6	10.4	6.0	6.6	7.3	5.3
Paper, cardboard	31.0	39.8	49.6	51.6	57.0	65.9	50.2
Kitchen wastes	2.1	5.1	8.0	0.9	1.8	2.3	9.7

(a) all data in % by weight DS waste
(b) # = mesh size

The preliminary waste analyses indicate that on average more than 50% of the pre-dried waste consists of organic material of which approx. 25–30% is total carbon. The easily degraded organic substance was determined with the help of dichromic oxidation and calculated in terms of carbon. This easily degraded fraction was relatively high in all waste samples.

The average heavy metal contamination is comparatively low. Taking the limiting values for lead, cadmium, chromium, copper, nickel, mercury and zinc, specified in the sewage sludge regulation of 1982, as a comparison, the measured average of the heavy metal concentrations are usually markedly below these values. The concentrations of further pollutants such as total cyanide, total phenol, substances extractable by n-hexane or polycyclic aromatic hydrocarbons are in a range usually found in other municipal waste samples.

Sewage sludge was acquired from Bad Harzburg because the included materials are barely affected by commercial influences and no local differences had to be considered. The sludge was simultaneous aerobically stabilised and dewatered in a chamber filter press to approx. 25–30% solid content. It was compacted to a lumpy consistency and it did not cause an odour nuisance. It could thus be deposited into a compacted landfill without difficulty.

Table 2-3 Preliminary domestic waste analyses for the central test on the Braunschweig-Watenbüttel landfill site, initial material (sorted municipal waste, without coarse metal and glass fraction), in % by weight DS waste

Parameter or component	$\bar{x}$ Sample Lys. 1	$\bar{x}$ Sample Lys. 2	$\bar{x}$ Sample Lys. 3	$\bar{x}$ Sample Lys. 4	$\bar{x}$ Sample Lys. 5 + 6	$\bar{x}$ Sample Lys. 7	$\bar{x}$ Sample Lys. 8	$\bar{x}$ Mixed sample waste addition Lys. 1 + 4
					[% by weight DS waste]			
Drying loss, 105°C	2.2	2.5	2.5	2.2	2.8	3.3	2.4	n.d.[a]
Ignition loss, 105–650°C	53.3	57.0	56.8	45.1	53.0	60.7	59.9	57.0
Total carbon	27.4	29.5	28.0	25.5	26.1	32.5	31.5	27.9
Easily degradable org. substance, calculated as carbon (C)	23.0	24.8	23.9	17.5	21.5	31.9	26.5	25.0
Chloride (Cl⁻)	n.d.	n.d.	n.d.	n.d.	n.d.	n.d.	n.d.	0.69
Total phosphorus (P)	0.31	0.23	0.11	0.32	0.19	0.17	0.17	0.18
Total nitrogen (N)	0.83	0.72	0.61	0.81	0.76	0.42	0.74	0.55
Ignition residue, 650°C (about inorg. fraction)	44.5	40.5	40.7	52.7	44.2	36.0	37.7	n.d.

(a) not determined

Table 2-4 Complete waste analysis of the delivered Braunschweig municipal waste, medium and extreme values of the <100 mm fraction, pre-drying of the samples at 50°C (the ignition loss was gradually determined. First the material was heated at 450°C and then at 650°C), all values relate to waste DS

Parameter or component	Average value	Max./min.	Unit
Drying loss, 105°C	2.6	2.8/2.4	% by mass
Ignition loss, 105–450°C	51.4	53.9/49.0	% by mass
Ignition loss, 450–650°C	4.0	4.0/3.9	% by mass
Sodium (Na)	2.50	2.95/2.26	% by mass
Potassium (K)	0.71	0.77/0.65	% by mass
Magnesium (Mg)	0.35	0.44/0.27	% by mass
Calcium (Ca)	2.82	3.16/2.60	% by mass
Manganese (Mn)	0.031	0.037/0.027	% by mass
Iron (Fe)	1.06	1.28/0.62	% by mass
Molybdenum (Mo)	14	Individual value	mg/kg DS waste
Cobalt (Co)	9	Individual value	mg/kg DS waste
Nickel (Ni)	170	265/75	mg/kg DS waste
Selenium (Se)	1.5	Individual value	mg/kg DS waste
Arsenic (As)	3.6	9/1.4	mg/kg DS waste
Lead (Pb)	398	920/95	mg/kg DS waste
Cadmium (Cd)	26	50/4	mg/kg DS waste
Chromium (Cr)	1420	2810/269	mg/kg DS waste
Copper (Cu)	372	532/266	mg/kg DS waste
Mercury (Hg)	2	Individual value	mg/kg DS waste
Zinc (Zn)	672	1018/370	mg/kg DS waste
Phosphate (PO_4^{3-})	0.20	0.21/0.18	% by mass
Total cyanide (CN^-)	32	34/22	mg/kg DS waste
Total boron (B)	48.3	54/37	mg/kg DS waste
Total phenols	169	181/147	mg/kg DS waste
Volatile phenols	7	7/6	mg/kg DS waste
Substances extractable by n-hexane, dried at 105°C	3.70	3.80/3.68	% by mass
Anionactive detergents	40	43/37	mg/kg DS waste
Kationactive detergents	95	98/91	mg/kg DS waste
Non-ionogenic tensides	18	20/15	mg/kg DS waste
Polycyclic aromatic hydrocarbons:			
1,12-benzperylene	0.26	0.42/0.10	mg/kg DS waste
1,2,3-cd indenopyrene	0.75	1.10/0.41	mg/kg DS waste
11,12-benzfluoranthene	0.25	0.38/0.12	mg/kg DS waste
3,4-benzfluoranthene	0.56	1.00/0.12	mg/kg DS waste
3,4-benzpyrene	0.22	0.37/0.06	mg/kg DS waste
Fluoranthene	0.46	0.66/0.26	mg/kg DS waste

Table 2-5 gives an overview of the components of the sewage sludge used. The results of the complete analyses are compiled in Table 2-6. Both tables show the individual measurements combined into average values and, in addition, the extreme values. Based on the analyses, it

Table 2-5 Preliminary sewage sludge analyses of the delivery; average and extreme values

Parameter or component	Average value	Max./min.	Unit
Drying loss, 50°C	71	78/68	% by mass
Ignition loss, 105–650°C	51	59/47	% by mass
Components related to the samples dried at 50°C:			
Total carbon (C)	19.6	22.5/16.7	% by mass
Total nitrogen (N)	2.43	2.92/1.57	% by mass
Easily degradable org. substance, calculated as C	14.4	19.6/9.2	% by mass
Total iron (Fe)	3.03	4.09/2.37	% by mass
Calcium (Ca)	16.3	20.4/10.5	% by mass
Chloride (Cl)	0.84	1.72/0.54	% by mass
Total phosphorus	1.35	1.91/0.58	% by mass

can be said that both water content and the organic fractions of the sludge vary only to a relatively minor extent. Depending upon the kind of sludge pre-treatment, the easily degradable fraction (calculated in terms of C) was rather different. The heavy metal portions were markedly below the limiting values of the German sewage sludge regulation of 1982 and, in addition, were lower than the corresponding contents in the waste. Contamination by other pollutants was in the concentration ranges normal for sewage sludge.

2.3.2 Residual waste after different intensive recycling – Wolfsburg facility

Depending upon the intensity of recycling, current and future landfills will be made up of residual waste of various bio-waste mixes and wastes with the valuable material removed until the German Waste Management Act (Abfallgesetz 2005) (*Federal Gazette – Bundesgesetzblatt,* 2001) is fulfilled. In its initiative called 'Waste 2000', the city of Wolfsburg, a body statutorily responsible for waste disposal, was among the first to introduce waste separation in households as early as 1983, which substantially increased the quality of the secondary raw materials and had achieved a recycling rate of 68% by weight by 1985. It has to be taken into account that composite packing materials made of cardboard/plastic, cardboard/metal and plastic/metal were not accepted at that time as raw materials for recycling. The influence of recycling on volume reduction and leachate contamination was investigated (Chammah *et al.,* 1987; Collins and Spillmann, 1990) sponsored by the Lower Saxony State and the city of Wolfsburg. This programme

Table 2-6 Complete sewage sludge analyses of the delivery; medium and extreme values of the samples pre-dried at 50°C

Parameter or component	Average value	Max./min.	Unit
Drying loss, 105°C	7.0	8.8/4.6	% by mass
Ignition loss, 105–450°C	36.6	37.9/34.6	% by mass
Silicic acid (SiO_2)	6.0	8.1/5.1	% by mass
Titanium (Ti)	0.09	0.10/0.08	% by mass
Aluminium (Al)	2.24	2.77/1.66	% by mass
Sodium (Na)	0.13	0.20/0.10	% by mass
Potassium (K)	0.16	0.23/0.12	% by mass
Magnesium (Mg)	0.96	1.40/0.48	% by mass
Calcium (Ca)	17.0	20.4/14.2	% by mass
Manganese (Mn)	2.77	6.73/0.65	% by mass
Iron (Fe)	3.04	4.06/2.37	% by mass
Nickel (Ni)	36	45/29	mg/kg DS sludge
Arsenic (As)	8.7	14/1.9	mg/kg DS sludge
Lead (Pb)	187	234/106	mg/kg DS sludge
Cadmium (Cd)	12	17/6	mg/kg DS sludge
Chromium (Cr)	70	196/40	mg/kg DS sludge
Copper (Cu)	138	196/87	mg/kg DS sludge
Zinc (Zn)	1519	2104/1080	mg/kg DS sludge
Phosphorus (P)	1.48	1.91/0.53	% by mass
Total cyanide (CN^-)	0.05	0.1/<0.05	mg/kg DS sludge
Easily released cyanide	0.02	0.06/n.d.	mg/kg DS sludge
Free + sulfidic sulfur (S)	1278	1900/480	mg/kg DS sludge
Total boron (B)	59	79/47	mg/kg DS sludge
Total phenols	113	178/75	mg/kg DS sludge
Volatile phenols	23	61/4	mg/kg DS sludge
Substances extractable by n-hexane, dried at 105°C	4.2	12.5/0.51	mg/kg DS sludge
Polycyclic aromatic hydrocarbons			
1,12-benzperylene	0.18	0.20/0.10	mg/kg DS sludge
1,2,3-cd indenopyrene	0.23	0.52/0.06	mg/kg DS sludge
11,12-benzfluoranthene	0.15	0.21/0.07	mg/kg DS sludge
3,4-benzpyrene	0.16	0.22/0.09	mg/kg DS sludge
Fluoranthene	0.40	0.60/0.15	mg/kg DS sludge

n.d. = not detected
Determination limit for easily released cyanide: 0.5 mg/kg

was continued in the integrated project of the DFG using spiking of chemical pollutants and indicator bacteria.

The reductions in mass according to different separation intensities are compiled in Fig. 2-2, the result of sorting the residual waste (RW) according to optimum sorting of dry waste is illustrated in Fig. 2-3. With the exception of the total waste (MW) each of the two recycling stages were investigated both as a common compacted landfill (anaerobic) and as a decomposition landfill (aerobic) using extensive aerobic

Without recycling

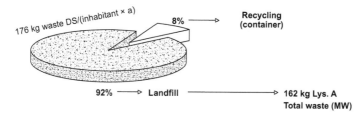

Valuable material recycling

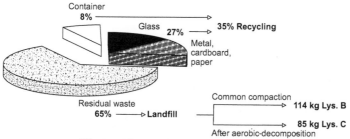

Valuable material recycling and wet waste composting

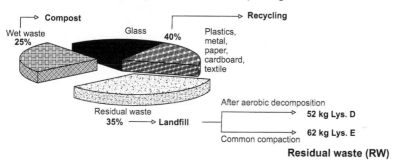

Fig. 2-2 Reduction of the dry mass of the Wolfsburg municipal waste using various recycling measures, in % by weight DS waste

stabilisation, so that the DFG research programme was extended by the following landfill alternatives (relevant lysimeter types):

Wolfsburg facility (residual waste)
Landfill type A (reference) (Lys. A, MW anaerobic)
Municipal waste, city of Wolfsburg without the influence of recycling, industrial contamination, compacted in 2-m layers, 10 cm earth cover of loam.

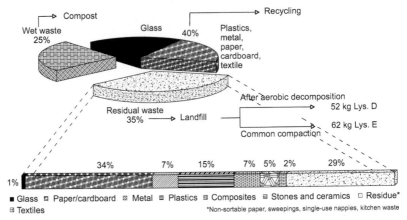

Fig. 2-3 Composition of the residual waste dry mass of the Wolfsburg municipal waste after valuable material recycling and wet waste composting ('biowaste' composting), in % by weight DS waste

Landfill type B (Lys. B, MW no valuable materials, anaerobic)

Municipal waste after manual selection of valuable materials, no industrial contamination, no separation in the household, compacted in 2-m layers, 10 cm earth cover of loam.

Landfill type C (Lys. C, MW no valuable materials, aerobic)

Municipal waste after manual selection of valuable materials, no industrial contamination, no separation in the households, extensive aerobic stabilisation before compaction, no cover.

Landfill type D (Lys. D, RW aerobic)

Residual waste after separation in the households and intensive follow-up sorting, industrial contamination, extensive stabilisation before compaction, no cover.

Landfill type E (Lys. E, RW anaerobic)

Residual waste after separation in the households and intensive follow-up sorting, industrial contamination, compacted in 2-m layers, compacted earth cover of loam.

20

2.4   Constructing the central test facility
Peter Spillmann

2.4.1   Construction conditions to fulfil the physical model laws and boundary conditions

Having a certain composition of the wastes, the physical processes such as compaction during and after installation as well as leachate and gas flow determine the possibility of the biochemical processes. Therefore, the fulfilment of mechanical and fluidmechanical model laws is a prerequisite for the extensive transfer of the results obtained in the test facility. The following laws must be fulfilled for the transfer of fundamental physical processes in the wastes: Cauchy–Riemann's law (waste deformation), Froude's law (movement under the influence of gravitational force), Reynold's law (flow processes under the influence of viscosity) and Weber's law (influence of the surface tension of liquids). Since the material including the liquid used in the model must agree with the actual landfill for biochemical reasons, all these conditions can only be fulfilled at a $1:1$ scale.

It is a specific requirement that boundary conditions at the test facility enable unhindered settlement, but completely prohibit lateral elongation. Furthermore, the influence of air temperature – as far as it affects the degradation processes – must not be greater than in operating landfills, and the potential oxygen supply from the drainage must be reduced or excluded to the same extent as in the large-scale design. It is also a requirement that a hydraulic boundary condition does not increase run-off on the walls of the test facility. The building rate must correspond to that of a landfill and/or allow judgements about different building rates.

2.4.2   Building a lysimeter

Figure 2-4 gives an overview of the principal test facility and Fig. 2-5 illustrates the scheme of a lysimeter. The building principle was already designed and implemented successfully in 1975 in tests that started earlier and then were continued concurrently (Spillmann, 1988). It was based on the idea that the test facility must represent a sectional core from the internal area of a relevant landfill for each landfill type. It should be accessible on all sides and controllable while the influence of the unavoidable boundary should be negligible.

The landfill sectional core with the smallest boundary surface in comparison to its content is a cylindrical lysimeter, whose compressible but tension-resistant cylinder wall prevents the lateral elongation of

21

Fig. 2-4 Test facility on the Braunschweig-Watenbüttel landfill, Building stage 1, eight landfill types as lysimeters including the most important sampling arrangements for solids and leachates (Lys. 9 (type 1) and Lys. 10 (type 6) followed as highly contaminated stages 3 years later)

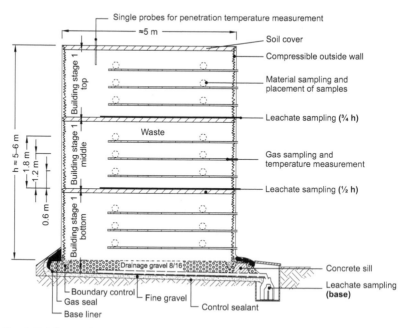

Fig. 2-5 Scheme of a lysimeter (here Lys. 1) on the Braunschweig-Watenbüttel landfill, schematic section alongside drainage

unhindered settlement, and whose thermal insulation limits the lateral heat release to such an extent that heat exchange is also characteristic of the available surface on the landfill. The unavoidable deviation of waste composition at the cylinder wall from the average composition of the landfill (coarse parts must be removed at the boundary) remained below about 10% for non-comminuted municipal waste with a cylinder diameter of 5 m (Spillmann, 1989). The free erection of the cylinder on a double seal entirely fulfils the requirement for accessibility and controllability. The conditions in a lysimeter were designed in such a manner as to be directly transferable to an operating landfill, when composition, dimension, density and building rate of the waste are identical to those of the operating landfill. The mixing effect which, in addition to compaction, can increase density was tested in separated alternatives in order to better determine its effect (e.g. MW, non-mixed, pressing compacted (Lys. 1); MW, mixed, press compacted (Lys. 2)).

The details of lysimeter design and the special casing used for erecting the easily deformed cylinder walls were descibed in an initial detailed publication (Spillmann, 1986b). Those features of design needed to understand the tests are illustrated by the photographs (Fig. 2-6) which show the building process of a lysimeter and its compression during the landfill processes tested.

The first picture shows the substructure (Fig. 2-6(1)). It consists of a sealing film (2 mm ECB) and a control film (1 mm ECB) drained with fine sand which was placed on compacted and smoothed sandy soil (later variants: concrete shielding against rats). The sealing area was hydraulically partitioned by a ridge in the middle, and the boundary run-off separately collected in a chute along the cylinder wall – according to the corresponding sealing areas. The area drainage (8/16 mm particles) also formed the protective layer and a concrete sill (cf. Figure 2-5) prevented the gravel from slipping into the water collecting vessels at the outlet. Figure 2-6(2) shows the building process of a casing designed specifically for this test (double-shell galvanised sheet metal with an expanded rigid plastic core), whose segments were manually joined together to form a 4-m high, bend-resistant cylinder. The tension-proof reinforced, flexible cylinder wall (2 mm ECB with a fine-mesh steel fabric welded-on) was suspended into this casing. The casing ensured the circular form of 5 m diameter during compaction, which equalled the effect of a heavy compactor (Delmag captured hammer) and prevented sharp waste components from penetrating the outer skin. To enable the addition of waste in a lysimeter, the

Fig. 2-6 Building a lysimeter on the Braunschweig-Watenbüttel landfill (Lys. 1–10): (1) substructure; (2) casing, load-bearing walls are being suspended; (3) clamping procedure of the flexible load-bearing wall after the highly compacted installation of the waste and removal of the casing; (4) lysimeter in the final state after the load-bearing wall was equipped with thermal insulation and rain protection

casing was installed on supports. The third photograph illustrates the cylinder with the casing just removed (Fig. 2-6(3)). It is being clamped with (polyester) clamping braids at about 10 cm distances after the removal of the casing. During this clamping procedure (four clamping spots per each clamping ring to remove friction) the outside wall was indented in the area of the clamping braids by blows using a plastic hammer and the inward facing wave formed was kept in its final position by the tightened braid. Thus possible elongation caused by the removal of the casing was reversed and the wavy form was defined which occurs during the settling of the municipal waste. Figure 2-6(4) shows a finished lysimeter including thermal insulation and a rain protection film as well as about 2-m high scaffolding.

The gas seal at the transition of the lysimeter wall to the base liner and the drainpipes at the discharge for anaerobic deposits as well as the fresh air supply for a decomposition landfill are not shown. The gas seals were made of sandwiched aluminium foil and, in addition, a bentonite sill was arranged near the discharges. The air supply to the

24

decomposing sewage sludge waste mix (landfill types 5 and 6) was implemented based on the relevant practice of the 'chimney draught method' (Spillmann and Collins, 1981) (cf. Section 11.4.1).

The supplementary lysimeter battery of the Wolfsburg facility (Lys. A–E) was built and operated to investigate the influence of recycling according to the same principle, but applying smaller dimensions (d = 3.50 m; h = 3.50 m) on the Wolfsburg city landfill site. It was possible to choose smaller dimensions because the large-sized valuable material was removed by recycling.

2.5 Arrangement of the lysimeters in the Braunschweig-Watenbüttel central test facility and schedule of building and dismantling

The arrangement of the lysimeters and the schedule of building and dismantling (cf. Fig. 2-1) has already been reported in detail (Spillmann, 1986b, Section 3.4). Therefore only those processes will be discussed which are needed to understand the long-term effect.

The intended maximum 2 m per year building rate corresponded to the usual building rate of landfills in the German municipal districts. Anaerobic deposits with earth cover (Lys. 1, 4, 8 and 9) and decomposition landfills (Lys. 5 and 6) – as was previously common – were built in 2-m lifts. These 2-m lifts were further divided into four sub-steps for building anaerobic landfills without earth cover (Lys. 2, 3 and 7) at periods of about 3–4 months in order to enable access of oxygen in the starting phase. This operating technology was fairly common in Germany until mid-2005.

Speaking about height increments, it has to be noted that the objective was to deposit approximately the same waste mass per 1 m^2 surface area so as to compare different landfilling techniques. However, different densities of, e.g., a sewage sludge waste mix and the installation of sludge lenses in municipal waste resulted in such large height differences that it was not possible to implement the third addition of waste in full height on some lysimeters because of the limited range of the excavator.

The addition of the two single-step lysimeters 9 (same design as Lys. 1) and Lys. 10 (same design as Lys. 6) to the test facility was necessary because the addition of chemicals to lysimeters 2 and 6 – built in three stages – non-contaminated, low and medium contaminated – initially remained without any recognisable effect and the limit of the contamination had to be determined with the help of the two supplementary lysimeters (Lys. 9 and 10, highly contaminated) (cf. Section 2.6).

For the precise investigation of the medium-term conversion processes in the material of the waste body of representative landfill types, the lysimeters 1, 4, 5 and 8 were dismantled after 5 years of testing and their contents were analysed. After being investigated, the decomposed content of lysimeter 5 was re-installed in a highly compressed state. Thus, lysimeters 2, 3, 5, 6, 7, 9 and 10 remained available for the investigation of the long-term effects.

After completing the building work in 1981, the decomposed material of Lysimeter 10 was relocated in a highly compressed state on top of Lysimeters 6 (denoted as Lys. 6/10 in the following). This was necessary to be able to observe potential transportation and fixation processes of chemicals on the path from the highly contaminated material of Lysimeter 10 into the moderate-contamination top layers of Lysimeter 6 (low to medium contamination) and from there into the non-contaminated bottom layer of Lysimeter 6 (cf. Fig. 2-1 and Fig. 2-7).

All still operating lysimeters were equipped with an irrigation system, starting from September 1982, to increase the rainfall. This device's specialty was that the rainfall volume of a rain event could be increased by about 50% simultaneously and at approximately the same intensity of rain. The evaporation and infiltration conditions of the artificial precipitation are thus identical to those of natural rain.

The supplementary equipment for the investigation of the recycling influence was built in 1986 on the Wolfsburg landfill site in the first stage (Lys. A–E), extended in 1987 and dismantled after 4 years of observation in 1990.

2.6 Experimental chemical and microbial contamination

2.6.1 Contamination with chemicals

Total waste – Braunschweig-Watenbüttel central test

Four typical groups of materials (phenols, cyanides, galvanic sludge and pesticides) were selected from those substances also deposited on municipal waste landfills in the past and were added to both an anaerobic deposit of sludge-free municipal waste (Lys. 2) and a decomposing deposit of population equivalent sewage sludge waste mixture (Lys. 6) in two increased contamination stages (low to medium). Quantities and places of addition are indicated in Fig. 2-7.

Contamination stage of addition	Spiked chemicals, mass [kg]					
	Galvanic sludge	Phenol sludge	Barium cyanide	Simazin	Lindane	Dichrotophos
1 (low contamination)	110	130	100	1.5	0.8	0.5
2 (medium contamination)	600	340	100	3.0	1.6	1.0
3 (high contamination)	1000	500	200	6.5	10.0	1.5

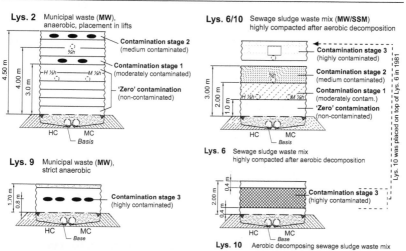

Fig. 2-7 *Place and height of the spiked chemicals for lysimeters 2, 6, 9 and 10. (After decomposition, Lys. 10 was placed in a highly compacted state on top of Lys. 6 in 1981 (nomenclature: Lys. 6/10)); (a) = new sampling point $\frac{1}{1}$h for Lys. 6/10*
HC: *Hygiene Centre = lysimeter half for tests by the Institute for Hygiene*
MC: *Microbiology Centre = lysimeter half for tests by the Institute for Microbiology*

The two stages were selected based on earlier special permissions for the deposition of such materials on municipal waste landfills (cf. older instructions such as 'Disposal of Pesticide Residues' including insecticide, disinfection and wood protective residues by the States' Waste Working Group). The increase of the initially spiked masses (low contamination) in the following spiking steps (medium contamination) depended on the results of the leachate analyses, which indicated only extremely low contamination. Since the medium-contamination stage also failed to exhibit a substantial increase in the contamination in the leachate, another further increased stage (high contamination) was selected which exhibited a multiple of the initial contamination. For this purpose, lysimeters 9 and 10 were re-established. Lysimeter 9 (landfill type 1, municipal waste without pre-treatment, highly compacted and covered with 10 cm of soil) was selected as an anaerobic lysimeter because early seepage pathways can be expected in the non-mixed municipal waste, through which chemicals may be washed out. Unlike Lysimeter 2 (municipal waste, highly compacted installation after homogenisation, without earth cover) into which chemicals had been inserted earlier. This is a modified alternative with coarser channels (non-homogenised waste). Lys. 6 was maintained as an aerobic deposit for the insertion of chemicals. Figure 2-7 displays place and concentration of chemicals in Lysimeters 2, 6, 9 and 10.

Residual wastes – supplementary tests in Wolfsburg

The residual wastes of the supplementary test facility in Wolfsburg (Lys. A–E) were contaminated with the same industrial wastes and/or products as the total wastes of the central facility in Braunschweig in order to determine the influence of recycling. It had to be borne in mind in this respect that the total waste from Wolfsburg was not identical with that of the central facility in Braunschweig (less natural organic substance) and that the location Wolfsburg received higher rainfall. For this reason not only was the residual waste of the extensive recycled material (Lys. D, aerobic; Lys. E, anaerobic) intentionally contaminated, but also the total waste from Wolfsburg (Lys. A, anaerobic). The waste deposits after separation of industrially usable materials but without composting the organic fraction (Lys. B, anaerobic, Lys. C, aerobic) continued to be operated as non-contaminated alternatives.

Experience gained at the Braunschweig facility indicated that waste deposit could also be chemically spiked at a later time without changing the fundamental results, as long as the degradation process had not yet reached the stable methanogenic phase in the anaerobic deposits. The active, unstable phase can be detected from the ratio of biochemical to chemical oxygen demand (BOD_5/COD) and the absolute magnitude of the biochemical oxygen demand of the leachates. These conditions were met by all anaerobic deposits at the time of spiking (cf. Table 2-7: BOD_5/COD > 0.4; $BOD_5 > 105$ mg O_2/l from Lys. A; $BOD_5 > 104$ mg O_2/l from Lys. E).

The non-contaminated bottom layer had a substantial influence on the retention and degradation of the spiked contamination. Spiking was therefore carried out intentionally at about 30 cm depth below the earth cover, so that a filtration path of more than 1.50 m length remained. The contamination in compact form ('lenses') of the anaerobic residual waste corresponded to the spiking of the anaerobic total waste.

The spiking of the residual waste by aerobic pre-treated material before compaction (Lys. D) was performed in 'lenses' after the end of

28

Table 2-7 Contamination parameters of the leachates for the characterisation of biochemical stabilisation of the wastes used in the tests on the Wolfsburg landfill site

Parameter	Unit	Initial phase 1985					Summer 1988				
		Lys. A Anaerobic total waste	Lys. B Anaerobic, no valuable	Lys. C Aerobic, no valuable	Lys. D Aerobic residual waste	Lys. E Aerobic residual waste	Lys. A Aerobic total waste	Lys. B Aerobic, no valuable	Lys. C Aerobic, no valuable	Lys. D Aerobic residual waste	Lys. E Aerobic residual waste
pH value		6.6	7.1	8.6	8.4	7.7	6.9	7.0	8.5	8.0	7.7
Cond.	μS/cm	18 800	19 600	8100	5400	15 600	14 500	16 050	10 000	12 200	13 150
O_2	mg/l	1.7	0.4	9.0	6.7	0.3	0.2	0.2	8.0	8.0	0.2
COD	mg O_2/l	28 800	21 100	750	790	12 000	25 536	21 504	1949	1344	12 768
BOD_5	mg O_2/l	17 900	12 800	8	7	9500	12 600	12 300	4.1	n.d.[a]	5850
Cl	mg/l	1740	1920	1770	1060	2160	1843	2198	3048	4112	2198
Fe	mg/l	320	106	5	2	24	312	56	2	1	21
Ca	mg/l	1520	800	260	140	240	1480	1090	248	1104	262
NH_4-N	mg/l	1000	590	2	1	790	1240	1680	5	7	1060
$\dfrac{BOD_5}{COD}$	–	0.62	0.6	–	–	0.79	0.49	0.57	–	–	0.45

(a) n.d. = not detected

29

Table 2-8 Intentional contamination of the total and residual waste of the lysimeters A to E for the tests on the Wolfsburg landfill, in kg spiked contaminant

Date	Contaminant	Total waste anaerobic degradation (Lys. A)	Residual waste aerobic degradation (Lys. D) [kg]	Residual waste anaerobic degradation (Lys. E)
27.02.1989	Sodium cyanide	33	33	33
01.06.1989	Simazin	2.25	2.25	2.25
01.06.1989	Lindane	1.25	1.25	1.25
07.07.1989	Galvanic sludge	100	100	100
07.07.1989	Phenol sludge	–	42	42

the intensive decomposition phase of the second building stage; it thus represents the most unfavourable case for a decomposition landfill. The mass of the two building stages together corresponded to the mass of an anaerobic building stage, so that the same non-contaminated waste masses were contaminated by the chemicals both in the anaerobic landfill types A and E, and in the aerobic landfill type D. The selected chemicals depend on water as a carrier and cannot migrate on their own.

For the accurate investigation into the movements and degradation processes of the semi-water-soluble simazin and lindane contaminants, they were placed in five sampling points per lysimeter, arranged in a circle, with the contaminated waste in nets and the installation place was marked. Other, similarly administered contaminations were marked but no nets used, as they were easily retrieved due to their large masses. The contamination with phenol sludge had to be limited to the residual wastes (Lys. D and E) because no more identical material was available.

The intentional contamination of the residual waste displayed in Table 2-8 resulted from the conditions explained.

2.6.2 Contamination with indicator germs

The effect of short- and medium-term contamination with indicator bacteria has been described in detail (Jung and Knoll, in Spillmann, 1986b). It was proved that in the case of intensive contamination, e.g. by a sewage vacuum tanker, pathogenic bacteria were removed by the leachate within a short time. Based on these findings, bacteria play no role concerning the long-term effect.

2.7   Measurements and sampling

2.7.1   *Parameters determined*

Comment: the programme of the measurements and sampling has been published in detail dealing with the short- and medium-term processes (Spillmann, 1986b). Therefore only that part is repeated, whose results were used for the assessment of the long-term effects.

The following parameters were determined using physical measurements for the description of landfill types:

- Mass and water content of the delivered municipal waste.
- Mass and water content of the delivered sewage sludge.
- Mass and water content of mined sewage sludge waste mixes.
- Mass of the mined municipal waste, if needed, together with sewage sludge lenses, but separate water contents for municipal wastes and sewage sludge.
- Volume of waste immediately after its compaction.
- Change in volume of waste during storage and at time of removal.
- Rainfall volume on the area of the deposit as well as its spatial and time distribution.
- Discharge quantity and distribution time of leachate with separate determination of boundary discharges and accumulation processes on intermediate covers.
- Waste temperature and its change over time.
- Air flow rate to the decomposing waste at selected dates.

Meteorological data (continuous measurements in the weather station):

- air temperature
- relative humidity.

All further measurements and investigations were performed on samples by the research participants. These parameters are illustrated in the chapters of the individual subjects and the relevant methods are described.

The following samples were taken for detailed investigation:

- municipal waste and sludge at the time of delivery
- municipal waste and sewage sludge waste mixes continuously after installation and at relocation of the decomposed mixes for highly compressed placement
- municipal waste and sewage sludge separately and sewage sludge waste mixes during removal
- landfill gas from different heights within the deposit

- leachate at the base and from different heights within the deposit.

2.7.2 Measurement method

The parameters specified in Section 2.7.1 were measured using the following methods:

Mass determination of incoming and outgoing waste

Weighing by weighbridge in the Braunschweig landfill. Possible measurement tolerances: up to approx. +2% at installation, since the official net weight is automatically deducted from a full tank and, in addition, occasionally when the excavator was feeding the lysimeters, some inert material was included from the road; up to approx. −1% at removal using double weighing of the material attached to containment wall.

Water content of the waste

Drying of large waste samples (cf. Section 2.7.3) at 60°C up to consistent weight; subsequent drying of small sub-samples at 105°C resulted in differences of 1–3% by weight, which was usually caused by changes in the organic substances. Possible measurement tolerances particularly resulted from the inhomogeneity of the waste, in particular the inaccurate determination of coarse waste components, maximum +6% by weight.

Volume of the compacted waste

The cross-section of the waste body was predetermined by the steel casing and steel reinforcement of the side walls while the height measurements were performed by surveying the level at installation and removal.

Measurement tolerances

Height levelling ±1 cm on h = 4.0 m corresponds to 0.3%; casing diameter d = 5.0 m plus 0 to +5 cm = 0–1% assembly tolerance including elongation after casing removal and clamping; volume tolerance at final measurement −0.3% to +2.3%.

Volume change of the layers

Height measurement using a meter rule at the insulated lysimeter and in the relevant profiles at removal. Measurement tolerance: ±10 cm height difference = ±5% per 2 m building stage.

Precipitation
Measurements using Hellmann's rain-gauges, $A = 200\,cm^2$; a rain-gauge properly arranged at 1 m height next to the weather station, two to four rain-gauges on the surface of each lysimeter. Measurement tolerance: approx. 5–10% when properly arranged; proof is based on accurate lysimeter measurements of the Agricultural-Meteorological Station Braunschweig-Völkenrode (personal information to the author).

Discharge
Measurement of the levels in calibrated collecting containers (cf. Fig. 2-5). Measurement tolerance: approx. ±1 litre, measuring interval: one measurement a week corresponds to 0.3% of the annual discharge; possible source of error: leakage in the lysimeter and dripping water at the transition from film to measuring vessel; proof of error-free operation: no leachate on the control film, intact seal after dismantling, parallel and sometimes equal cumulative discharge curves of similar lysimeters over 5 years.

Waste temperature
Penetration measurements using mobile resistance thermometers in stationary driving rods (stationary steel pipes $d = \frac{1}{2}$ in galvanised and painted with bitumen, peaks welded and filled with glycerin as a contact liquid 20 cm high, holes above the glycerin level prevent its rise due to rain water); measurements along a line using stationary resistance thermometers; measurement tolerances under field conditions: resistance thermometer for punctiform measurement approx. ±1°C, continuous resistance thermometers for average measurements approx. ±2°C.

Air temperature and relative humidity
Measurements using a hygrothermograph in a weather station next to the lysimeters. Measurement tolerances: air temperature ±1°C, relative humidity ±2%; regular checks on the devices in the field (temperature) and laboratory (relative humidity, temperature).

2.7.3 Sampling methods
The samples listed in Section 2.7.1 were taken using the following methods:

Material sampling from the delivered wastes
Municipal waste: about 0.5 m^3 per lysimeter was taken by the excavator from the delivered municipal waste in the first year of tests, it was sieved

(100 × 100 mm and 50 × 50 mm mesh sizes) and dried at 60°C in a drying room arranged specifically for this purpose. The fraction >100 mm was sorted by hand. The material groups of coarse fractions determined by sorting were forwarded to the Fresenius Institute for waste analysis. The material groups of the compostable substances >100 mm and all further samples <100 mm were passed on to the Fresenius Institute for chemical analysis. Since the analyses only exhibited small fluctuations in the composition, samples were not taken at each waste addition to the lysimeters in the following years but only at four dates per year during building activity.

Sludge
Due to its greater homogeneity compared to municipal waste, approx. 10 kg sample was taken from each delivery (7 t) and immediately forwarded in a hermetic container to the Fresenius Institute for chemical analysis.

Sampling during the test
During the building process, two 100-mm dia steel pipes were arranged 50 cm apart at one- and two-thirds of the height of each lysimeter half (Fig. 2-8) to enable sampling and insertion of prepared samples. The pipes were removed after completion of the building process through a hole in the cladding and the resulting cavity was filled with the

Fig. 2-8 Installations to take material and gas samples from the lysimeters (Lys. 1–10)

Fig. 2-9 Gauze bag for insertion of bacteria carriers and ampoules to be placed directly in fresh waste

same waste to achieve the same density. This method prevented the waste of the sample insertion from becoming mechanically interlocked with the rest of the lysimeter fill and so could be removed again when required. If doubts arose whether or not the processes in the sample were identical to those in the lysimeter, it was possible to extract material from the rest of the lysimeter fill through this cavity. The openings of the sampling ports were hermetically closed.

While placing the municipal waste, bacteria markers were inserted and removed in that half of the lysimeters intended for tests to be performed by the Institute for Hygiene (the half to the left of the outlet). The marked bacteria carriers (fabric strips and ampoules) – surrounded by fresh waste – were placed into a plastic gauze bag, which was clamped with a high-strength braid (Fig. 2-9). The sample identification was punched into the end of the braid and it was led out from the lysimeter. This enabled systematic sampling and always provided a clear overview of the samples in place. Moreover, additional temperature measuring probes were driven into the lysimeter fill for the assessment of stratification conditions at those sampling points, which were not near to an available general temperature measuring probe.

Sampling during dismantling of the lysimeters

Waste removal from the lysimeters was carried out in 2-m stages, in reverse to installation. Before removing a stage, about a 1-m wide slot was excavated, so that the profiles of the deposits could be assessed and, after a visual observation, characteristic samples taken by hand. In addition, the appearance of the profiles was recorded by colour photographs.

A sample of waste or sewage sludge waste mix contained a minimum of 15 kg and a maximum of 30 kg of material, samples from sludge lenses comprised a minimum of 1 kg. The samples were placed into tear-resistant plastic bags, and the date and sampling point including a sketch of the sampling point were noted in water resistant lettering on the exterior. Eight to ten samples were taken for each lysimeter, they were dried at 60°C up to a constant weight and dispatched to the Fresenius Institute for analysis.

Gas sampling

Two 15-mm dia HDPE tubes were inserted for gas sampling at the same heights as the material and bacteria carrier sampling points, i.e. about 50 cm apart in an arc and were led laterally through the lysimeter wall (Fig. 2-8: a thin PE tube was placed in the middle between the removable steel pipes. The extension of the thin tube – not shown in the picture – runs along a circle). The gas sampling tube was perforated at graduated distances for the flow tests, thus gas could flow evenly into the tube which enabled a representative gas sample to be taken. An exchangeable linear temperature probe within the gas pipe measured the relevant temperature at the same place where the gas samples were taken.

Leachate sampling

Fresh random samples were obtained in such a way that the contents of a sample bottle (2 litres) suspended under the discharge were continuously renewed from the bottom to the top through a funnel, whose end had been extended with a hose down to the bottom of the bottle. Thus the bottle always contained the last 2 litres of the outflow as a fresh sample (cf. Fig. 2-5).

Weekly average samples were collected in large collection containers equipped with measuring scales, which could accommodate the weekly discharge in normal weather. Readings of water volumes were taken at the beginning of the week and a mixed sample was taken. After high rainfall the volumes were additionally measured in the middle of the week and, if necessary, a sample was taken, which was stored cooled until the beginning of the following week and mixed volume equivalently with the sample of the beginning of the week. Overflow due to heavy rain was collected in large-volume measuring barrels. Water collecting vessels were protected by thermally insulated wooden boxes from weather influences and kept ice-free in the winter by a resistance heating (own design).

Two water sampling arrangements were constructed between building stage 1 and 2 at approx. 1.80 m height ($\frac{1}{2}$h, cf. Fig. 2-5) in each lysimeter and one sampling arrangement between building stage 2 and 3 at approx. 3.60 m height ($\frac{3}{4}$h, cf. Fig. 2-5) for the qualitative investigation of leachate contaminants along the filtration path. They were 40-cm wide gravel-filled, sealed chutes which went through the entire cross-section of the lysimeter with the final 50 cm near the lysimeter boundary cladding being covered (see Fig. 2-11(6), chute with opened boundary cover).

2.7.4 Waste removal from the lysimeter for tests

2.7.4.1 Selection criteria for the individual investigation steps

During the measurements on the landfill sectional cores (lysimeters) only indirect assumptions can be drawn about the degradation and stabilisation processes. The direct conclusions are based on the investigation of small samples from a large waste body. Therefore it is necessary to test the material of the entire waste body before a final assessment.

In the first investigation phase the reactions were investigated during construction and placement until the hydraulic equilibrium of precipitation input and leachate discharge for an average location in Germany was reached. At the end of this phase the waste bodies built with an earth cover were removed from the three initially anaerobic, almost concurrent pairs of lysimeters (Lys. 1 and 2, Lys. 3 and 4, Lys. 7 and 8) to perform tests on the materials (cf. Fig. 2-1). The pair of deposits with aerobic pre-treatment (Lys. 5 and 6) was kept as alternatives with chemical contamination (Lys. 6/10) and without (Lys. 5).

In the second investigation phase of the long-term study on the total waste, the latter observations were made more extensive and limited to measurements on the waste bodies. The landfill sectional cores of the residual waste (Lys. A–E in Wolfsburg) were also intensively sampled and then extensively until the hydraulic equilibrium was reached. The landfill sectional cores from residual waste were dismantled after 8 years of observation time with intensive sampling and sent for final storage or disposed of as hazardous waste.

The selected sectional cores from total waste landfills, Lys. 2, 3, 5, 6/10, 7 and 9 in Braunschweig, were further tested in the interdisciplinary research project 'Mining – Re-using – Depositing Landfill Wastes from

Old Sites' sponsored by the Volkswagen foundation. The intensive investigation of the waste bodies before waste removal, reaction measurements in a biochemical reactivation and their progress control after final storage enabled an interpretation of the stabilisation measured over the medium term with regard to the long-term effect. Typical old landfills (Lys. 2, 7, 9 and – as an antipole – a deposit with extensive biochemical stabilisation (Lys. 5)) were only selected for these investigations because of limited finances (cf. Fig. 2-1).

Extensively biochemically stabilised sewage sludge waste mixes with industrial contamination (Lys. 6/10, cf. Fig. 2-1) could only be chemically investigated within the project. Although this landfill type is nontypical for old landfills, it would have been interesting as the final stage of an initially non-stabilised landfill. The non-contaminated, anaerobic sewage sludge waste mix (Lys. 3) is also nontypical for old landfills and was therefore not included in the programme for landfill mining.

2.7.4.2 Execution of waste removal and accompanying measurements

Preparation of the lysimeters for waste removal
When removing the waste from the lysimeters it has to be determined what portion of the contained water mass represents the stored mass. For the separation of this stored mass from the water mass whose discharge is only retarded by the flow resistance of the municipal waste, the surface of the lysimeters 1, 4, 5 and 8 were covered to protect them from rainfall from November 1980 (end of the autumn dry period) until waste removal in May 1981.

2.7.4.3 Removal process and explanation of the cross-sections found

Removal and new installation after the end of the decomposition phase (Lys. 5)
The two pictures in Fig. 2-10 illustrate the removal procedure of a decomposed sewage sludge waste mix, e.g. from Lysimeter 5. The decomposed material – as planned and tested on an operating landfill – is removed from the bottom-ventilated flat windrow and replaced in highly compacted thin lifts. Thereafter, the casing is re-installed (Fig. 2-10(1)) before the material is removed. The material is tested during the material removal (Fig. 2-10(2)). The heavily waved lysimeter wall (reinforced film), compressed from 2 m height to 1.4 m, is replaced

(1)
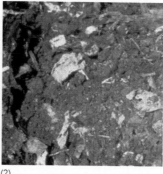
(2)

Fig. 2-10 Reconstruction of Lysimeter 5 (aerobic sewage sludge waste mix) to a highly compacted installation directly after the decomposition phase; (1) removal and placement in the bend-resistant casing; (2) typical section through a decomposed mix after about 12 months decomposition time

by a smooth one. After weighing and placement of the decomposed mix in a thin-layer (approx. 80 cm of the initial 2 m height) a new mix is placed on top of the highly compacted material for decomposition with bottom ventilation (bulky waste). The perfect assembly of the steel casing around the compressed lysimeter proved that the concertina effect must have been formed inward, otherwise it would not have been possible to close the casing cylinder.

Removal of compacted deposits (Lys. 1, 4 and 8)

The photographs in Fig. 2-11 illustrate the sequence of the removal. Before the beginning of the excavator work the rain protection cover of the surface, including gravel, was carefully removed together with the rain lining and thermal insulation of the side walls (Fig. 2-11(1)) and the load-bearing outside skin checked whether elongations had occurred (proof: steel reinforcment intact), whether the compression waves (Fig. 2-11(2)) cause a reduction of the circumference (proof: tension of the braids decreases) and whether leachates had run off later- ally from the wall (sign: rust stains and deposition streaks on the outside wall). Finally, the height of the lysimeter was measured with ±1 cm accuracy using levelling (about 1 point per 1 m^2 surface).

After profile assessment and sampling in a building stage (cf. Fig. 2-11(3) to Fig. 2-11(5) and Section 2.7.3), the degradation continued because only as much material was removed from the extended stage opened by the slot as was needed to recognise deviations in composition and material degradation and to record them by taking samples. Each

Fig. 2-11 Dismantling of selected lysimeters (Lys, 1, 4 and 8). (1) Checking of the load-bearing outside wall (Lys. 4); (2) waving of the outside skin due to sagging (folds curve inward, since the steel reinforcement of the outside skin does not stretch) (Lys. 8); (3) cut through a deposit for sampling (Lys. 8); (4) typical profile of a sludge-free municipal waste after 5 years of predominantly anaerobic degradation (Lys. 1); (5) typical profile of a sewage sludge waste mix after 5 years of anaerobic degradation (Lys. 4); (6) proof of efficacy of leachate sampling arrangement ($\frac{1}{2}$h, Lys. 4) until dismantling (no deposition or incrustation inside the deposit)

stage above the earth cover of the next older stage was cleared away by manual work in such a way that the soil surface, including plant residues, was maintained and the leachate sampling devices that took samples from the inside of the deposit remained intact. Thus the water sampling arrangements were checked for possible depositions and incrustations (Fig. 2-11(6)), the permeability of the cover layers was determined using U100 samplers and soil profiles were described.

40

The description of the cross-sections provided the following assessments

Figure 2-11(3) shows a slot in the municipal waste with sewage sludge lenses (Lys. 8). The sludge lens can be recognised by the black colour on the right in front of the cut. The municipal waste in Lysimeter 8 differed visually from the sludge-free municipal waste of Lysimeter 1 shown in Fig. 2-11(4) only in a much higher water content. Newspapers were still easily readable after 5 years and metal was largely conserved. Brown colouring indicates some aerobic degradation – irrespective of the height within the lysimeter. In the concurrent lysimeters (Lys. 2 and 7) the same observations were made after 15 years of deposition. The main physical difference to the anaerobic sewage sludge waste mix of Lysimeter 4 (Fig. 2-11(5)) was that the clearly recognisable cavities in the sludge-free municipal waste of Lysimeters 1 and 8 had been filled by sludge in the mix.

The deposits also exhibited biochemical differences in as much the mix (Lys. 4) failed to show traces of aerobic degradation and no anaerobic degradation could be visually determined either. Even some metallic cans were still bright after 5 years and plants were not recognisably decomposed either. In the concurrent lysimeter (Lys. 3) extensive degradation processes were only recognisable in the top layers after 15 years of deposition. The state of the anaerobic mix in Lysimeter 4 therefore represents the opposite to the intentional aerobic degradation of the exact same mix in Lysimeter 5 (Fig. 2-10(2)) regarding biochemical degradation. Due to targeted aerobic degradation of the mixes the same material was decomposed to such an extent in only 1 year that only plastics and fragments of glass were still recognisable as wastes.

The base drainage was opened with same care as the intermediate cover. As Fig. 2-11(6) shows, there was no deposition in the crushed stone of the intermediate sampling arrangements inside the lysimeter, while extensive depositions were formed next to the outlet due to air access. (It was only possible to keep the outlet free by constant maintenance.) The drainage crushed stone at the base was completely free of depositions in all four lysimeters and it was only discoloured black in the boundary control of Lysimeter 4. This finding excludes any major air access to the base drainage, or else encrustations would have developed as they did near the outlet of the intermediate sampling arrangements.

The height of the base drainage crushed stone was measured using levelling in the same way as that of the lysimeter surface. The excavated material of Lysimeters 1, 4 and 8 was loaded into containers without an interim storage, weighed on the weighbridge of the Braunschweig landfill and tipped on the landfill.

Waste removal and reconstruction for the testing of reactivation
Waste removal and relevant examinations were carried out similarly to the final removal (e.g. Lys. 1, 4 and 8). The material sieved for the reactivation test (<100 mm) was treated in the same way as the materials for intentional aerobic degradation (Lys. 5 and 6): homogenisation in a batch mixer (system KUKA), separated in the top and bottom lysimeter halves; gas exchange according to the 'chimney draught method' (like Lys. 5 and 6 in the decomposition phase); highly compacted installation similar to the final deposition according to biochemical stabilisation (see details in Brammer *et al.*, 1997).

2.8   Model laws for conversion of the results to different large-scale designs
Peter Spillmann

2.8.1   Conversion model
It was proved when establishing the model design (Section 2.4.1) that hydraulic processes can only be modelled at a 1:1 scale. This scale was adhered to for flat landfills. The results are therefore directly transferable to these landfills.

The influences of different landfill conditions on the physical and biochemical processes were determined by the simultaneous implementation of different operating technologies. This enabled us to derive different model scales for the individual influences. The processes in high landfills can then be predicted from the sum of the individual conversions.

2.8.2   Conversion of water and solid balances

2.8.2.1   *Relationship between climatic water balance and precipitation input into the waste body*
Precipitation and evaporation depend on climate conditions and the evaporation conditions on the landfill surface. Since it was proved that the capillary water rise is negligible, the relationships derived in the investigation can be transferred depending on the surface only

and regardless of the landfill height:

$$V_{i\,model} = V_{i\,nature} \qquad\qquad (2.1)$$

V_i = volume of rainfall input

2.8.2.2 *Storage capacities and their changes by degradation processes*

Initial situation

The storage behaviour of waste differs from that of soil to a large extent because the capillaries of waste materials are separated from each other by coarse minerals and pieces of waste (mulch effect). The storage capacity of a waste body therefore depends on the specific capacities of the individual waste materials per 1 t waste dry substance and the deposited waste mass per $1\,m^2$ landfill surface area, if no local perched water emerges or no barrier layers were installed. The specific storage capacity per 1 t waste dry substance (DS) and its change thus depends on the composition of the initial material and the degradation of the organic substances (1 t org. waste DS can store about 1 t water). When the waste composition is identical and the same operating technology is applied, the initial capacities for each layer are equal. Since the extent of degradation is also equal in similar landfills, the final storage capacities for each layer are also equal. Therefore the initial and final storage capacities for each layer are equal for identical wastes. The initial and final capacities grow with the number of layers and therefore linearly with the height of the landfills:

$$\frac{V_{st_{model}}}{V_{st_{nature}}} = \frac{h_{model}}{H_{nature}} \qquad\qquad (2.2)$$

V_{st} = storage volume = storage capacity
h_{model} = landfill height of the model
H_{nature} = landfill height of the large-scale design

Change by degradation processes – anaerobic degradation

The organic substance contained in the waste – related to the dry substance – primarily consist of paper and stable plant material. Under pressure, paper can also store as much water as its dry weight (Grewe, 1987). As a first approximation, the same capacity can be assumed for stable plant material. Hence it follows that the storage capacities decrease from the maximum at delivery proportionally to the degradation of the organic materials. Mass reduction was determined by weighing. In this way it was found that the degradation is

43

negligibly low during the initial acidic phase (Spillmann, 1986b, Table 3). The mass change important for the water regime occurs through anaerobic degradation during the methanogenic phase. Since the bulk consists of cellulose, the anaerobic degradation provides the following mass change for a complete reaction:

Cellulose + Water → Glucose → Methane + Carbon dioxide

$$(C_6H_{10}O_5)n + H_2O \rightarrow C_6H_{12}O_6 \rightarrow 3CH_4 + 3CO_2 \tag{2.3}$$

The complete anaerobic degradation of 1 t cellulose requires, therefore, about 0.1 t water and approx. 0.3 t methane while approx. 0.8 t of carbon dioxide is generated. The degradation of 1 t cellulose decreases simultaneously with the storage capacity by 1 t water. The process of storage change due to anaerobic degradation processes is thus proportional to the gas generation which can be described for an undisturbed process sufficiently accurately by the following half-life function:

$$G_t = G_e(1-10^{-kt}) \tag{2.4}$$

$G_t[t]$ = generated mass of gas in [t] until time [t]a
$G_e[t]$ = maximum possible mass of gas to be released in [t], dependent on waste composition; the masses G are usually indicated in standard m^3 per 1 t output mass (cf. e.g. Rettenberger, 1992).
$t[a]$ = time in years [a]
k = degradation constant, dependent on deposition conditions

Beginning, extent and end of a rapid anaerobic degradation and extent and cause of delays were measured in this test. Thus, landfill gas production can also be determined for individual waste layers from the mass balances of this test in connection with the biological determination of the degradation path, without being forced to measure the generated gas masses.

The generated methane can be further oxidised microbially if diffusion and convection provide sufficient amounts of oxygen to the surface layer:

Methane + Oxygen → Carbondioxide + Water

$$3CH_4 + 6O_2 \rightarrow 3CO_2 + 6H_2O \tag{2.5}$$

This means that approx. 0.7 t water and further 0.8 t carbon dioxide is generated from 0.3 t methane, which is released from 1 t cellulose, under the provision of 1.2 t oxygen. Deducting 0.1 t of water from methane generation, a maximum 0.6 t water per 1 t degraded cellulose contributes to saturating the storage capacity or to leachate discharge (values rounded to one decimal place).

Change through degradation processes – aerobic degradation
The change of storage capacity due to aerobic degardation of the waste is well-known and, in an ideal case, is identical to anaerobic degradation with secondary oxidation of methane:

Cellulose + Water + Oxygen → Carbondioxide + Water

$$C_6H_{10}O_5 + H_2O + 6O_2 \rightarrow 6CO_2 + 6H_2O \tag{2.6}$$

Approx. 1.6 t carbon dioxide and approx. 0.7 t water are generated from 1 t cellulose under the provision of 0.1 t water and approx. 1.2 t oxygen. The storage capacity simultaneously decreases by 1 t water due to the degradation of 1 t cellulose.

The process of intensive exothermic aerobic degradation can be determined from the process of heat loss. The process of intentional aerobic degradation on a bottom-ventilated flat windrow with rural municipal waste has been described before in the DFG integrated programme (Spillmann and Collins, 1979). For fast and undisturbed conditions the process can be described as a half-life function with a half-life of 6 months for a bottom-ventilated flat windrow (cf. Fig. 2.13):

$$W_t = W_e(1-10^{-0.05t}); W_e = 1.0 \tag{2.7}$$

W_t = sum of the relative heat loss $\left[\frac{\%}{100}\right]$ at time t [months], related to the final value $W_e = 1$ (total decomposition time until heat emission ≈ 0).

There was insufficient knowledge of the aerobic degradation of waste after the methane generation phase due to oxygen diffusion before the DFG integrated programme. The determination of these processes was therefore an objective of the DFG research programme.

2.8.3 Aerobic stabilisation of the waste body due to atmospheric oxygen diffusion

2.8.3.1 Initial conditions
The small masses of the 5–6 m high landfill sectional cores did not produce any measurable gas flow after the end of the intensive methanogenic phase. They correspond to a landfill, which is at least still passively degassed at the end of the usable gas generation phase and the diffusing oxygen is therefore not displaced or bound by residual gases. The progress of aerobic stabilisation by diffusing oxygen was

measured in the integrated programme on different landfill types with different amounts of chemical waste contamination. The aerobic stabilisation has a substantial influence on long-term emissions. A model law must be established in order to be able to transfer these measurements onto high operating landfills.

2.8.3.2 Model law for aerobic stabilisation by oxygen diffusion

Area of application

Based on the results of preceding tests (Spillmann and Collins, 1979), the process was divided into an aerobic primary stabilisation and an aerobic secondary stabilisation:

- The primary aerobic stabilisation is defined as the aerobic biochemical degradation of organic material with clearly proved oxygen consumption. It begins from the end of anaerobic degradation in the methanogenic phase and lasts until the end of the intensive aerobic degradation. It can be chemically characterised among other things by the fact that iron is reduced to iron sulfide under air exclusion at a laboratory scale. This phase is subject of the model law.

- The secondary aerobic stabilisation is defined as a slow change of humic-like substances into more stable, water-insoluble compounds. This phase is characterised chemically by the fact that methane generation in the laboratory is near the detection limit and iron oxides are not reduced to iron sulfides after air exclusion under laboratory conditions. This material can also be kept aerobic under landfill conditions after extensive compaction. This phase is not the subject of the model law.

Hypothesis

The process of oxygen consumption during the primary aerobic stabilisation can be linearised with sufficient accuracy. Under this assumption there is a mathematical analogy between Terzaghi's consolidation theory (primary consolidation) and the primary aerobic stabilisation. The model laws are therefore identical:

$$\frac{t_{model}}{t_{nature}} = \left[\frac{h_{model}}{H_{nature}} \right]^2 \tag{2.8}$$

t = time [h]
h, H = flow path [s in Darcy's law in Terzaghi's model]

h, H = diffusion path [z in Fick's law, diffusion distance in the model of 'aerobic stabilisation']

Proof

(1) Proof of the mathematical analogy:

Terzaghi's consolidation theory describes the compression of a water-saturated cohesive soil under surcharge by squeezing out the pore water until the elimination of the excess pore water pressure (primary settlement, Fig. 2-12, top). The subsequent compression without excess pore water pressure (secondary settlement), which usually drags on for a long time, has a moderate extent and which has no negative influence on shear strength, is not the subject of the theory.

In Terzaghi's consolidation model the flow resistance of the particles of cohesive soils is modelled as a rigid plate with capillary channels. The

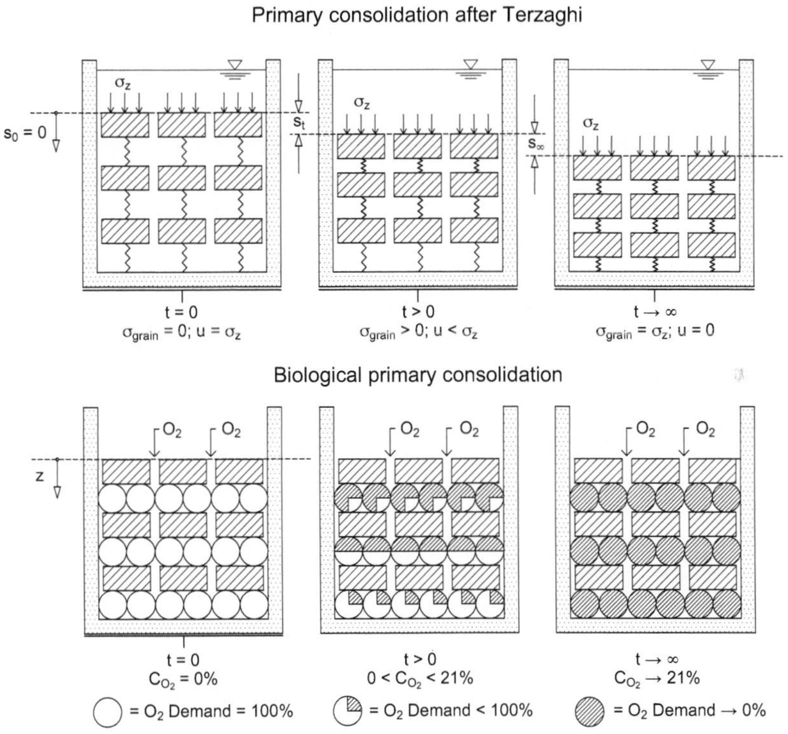

Primary consolidation after Terzaghi

Biological primary consolidation

Fig. 2-12 Derivation of the model law for the biological stabilisation from the analogy to Terzaghi's consolidation model. (Top) Primary consolidation according to Terzaghi: t = time; s = flow path; σ = stress; u = water pressure; σ_{grain} = grain stress. (Bottom) Primary biological stabilisation: O_2 = oxygen; C = concentration; t = time; z = diffusion path

elasticity of the particle structure is modelled by springs with linear characteristics, which keep the rigid plates parallel at constant distances, as long as the system is not loaded. These gaps correspond to the cavities in the soil saturated with water and are included in the model as water-filled cavities. If the system is loaded, first a constant hydrostatic pressure develops in the whole system. A pressure difference only develops along the capillaries in the first plate, so that water flows off from the first cavity. The flow process is described by Darcy's law. Proportionally to the water discharge from the first chamber the plate distance is reduced. Thus the springs are compressed, they exert a pressure on the rigid plate of the second chamber, produce a pressure difference along the capillaries of the second plate and induce a flow from the second chamber into the first. Since water is incompressible, water from both the first chamber and the second chamber must flow through the capillaries of the first plate. Since the number of capillaries available to water discharge is equal in all plates and exhibit equal cross-sections, the pressure difference must also increase linearly proportionally to the flow rate along the capillaries. This process continues gradually, until water flows off from the bottommost chamber. It ends if the entire surcharge is carried by the springs and the hydrostatic water pressure is nearly zero. This process is called primary settlement and describes the compression of cohesive soils until the end of the reduction of excess pore water pressure. The mathematically accurate solution of the problem is well understood and standard soil mechanics knowledge and is discussed in great detail in the relevant text books (cf. e.g. Schmidt, 1996).

It yields the model law:

$$\frac{t_{model}}{t_{nature}} = \left[\frac{h_{model}}{H_{nature}}\right]^2 \tag{2.9}$$

t = time, h = special flow path, as a rule, 'height' of the layer

Flux $q\left[\frac{g}{cm^2 \cdot s}\right]$ is described by Darcy's law in Terzaghi's model:

$$q = v \cdot \rho \left[\frac{g}{cm^2 \cdot s}\right] \tag{2.10}$$

$$v = k_f \cdot I = k_f \cdot \frac{dh}{ds} \left[\frac{cm}{s}\right] \tag{2.11}$$

with $q\left[\frac{g}{cm^2 \cdot s}\right]$ = flux, $\rho\left[\frac{g}{cm^3}\right]$ = density, $v\left[\frac{cm}{s}\right]$ = seeping velocity for laminar flow, $k_f\left[\frac{cm}{s}\right]$ = resistance constant for laminar flow and $I = \frac{dh}{ds}$ = pressure gradient = change of pressure head h along the flow path s.

The analogy to the aerobic biological stabilisation of a waste body consists of the circumstance where the water in Terzaghi's model is replaced by gas and the primary aerobic biological degradation capacity takes the place of the springs (Fig. 2-12 bottom).

Analogous to the pressure difference along the capillaries of the upper plate (Terzaghi) the concentration differences of the gases work at both ends of the diffusion channels ('biological consolidation').

The diffusion flux is described by Fick's law:

$$J_g = -D_p \frac{dC_g}{dz} \left[\frac{g}{cm^2 \cdot s} \right] \qquad (2.12)$$

(e.g. from Jury et al., 1991, pages 203–204)

$J_g \left[\dfrac{g}{cm^2 \cdot s} \right]$ = diffusion flux

$D_p \left[\dfrac{cm^2}{s} \right]$ = diffusion coefficient in a porous material

$C_g \left[\dfrac{g}{cm^3} \right]$ = gas concentration

$z \, [cm]$ = local coordinate

Both laws relate to mass transport (flux) calculatorily to the entire cross-section considered including all solids (filtration velocities). Mathematically both equations are identical. The process reproduces itself in Terzaghi's model in deeper layers, if springs are loaded by the discharge of pore water. In biological consolidation the process spreads into deeper layers, if oxygen consumption decreases with increasing stabilisation of the organic substances and the diffusive gas exchange can also reach the second and then other layers. As the discharge volumes of the individual layers are added up in Terzaghi's model, so the diffusion flow rates are added up in the model of biological consolidation. If the change of oxygen consumption can be linearised sufficiently precisely, a complete mathematical agreement exists between Terzaghi's model and that of biological consolidation.

(2) Proof of time-related linearisation of aerobic degradation processes:

The aerobic degradation proceeds proportionally to energy release. In the preparation phase of the integrated programme, energy release was investigated for municipal waste in a bottom-ventilated flat windrow (analogous to Lys. 5 and 6 of the integrated programme) up

49

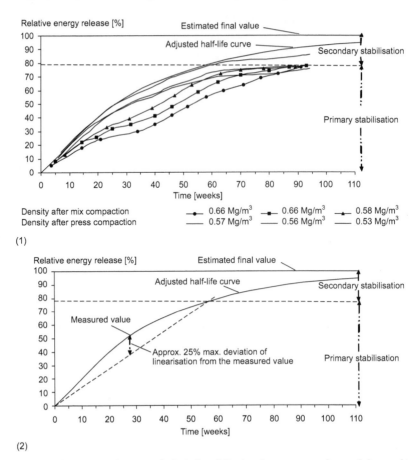

Fig. 2-13 Derivation of primary biological stabilisation from energy release of the aerobic degradation process. (1) Relationship between energy release and biological stabilisation of bottom-ventilated flat windrows of identical contents but different waste density: the sum of energy release is the same for each stabilisation stage; time requirement depends on oxygen supply, the half-life is 0.5 year at optimum supply of the static windrow; (2) linearisation of the half-life curve up to primary stabilisation

to the end of the primary aerobic stabilisation with and without distur-bances (Spillmann and Collins, 1979). It follows from the cumulative curves of relative energy release (Fig. 2-13(1)) that the disturbance-free aerobic degradation runs with about a half-life of 0.5 years for this landfill type.

The cumulative curves of the relative energy release of all six alterna-tives approach the same final value regardless of delays (the increase at the end of the test is steeper after a delay). Since the materials of all alternatives, also those with substantially retarded degradation, reached the primary aerobic stabilisation at the end of the test, the final value of the bottom curve indicates the minimum value, which must be achieved for the primary aerobic stabilisation. The sum of the energy release of the optimum aerobic degradation process during the test period exceeded the minimum sum for primary stabilisation. If one linearises the relative cumulative energy curve of the optimum degradation process up to the minimum value of primary aerobic stabilisation, the straight line only deviates from the measured cumulative curve in the most unfavorable point by about 25% (Fig. 2-13(2)). The average deviation is considerably smaller. The linearisation is therefore sufficiently precise for an estimation.

In the real, anaerobic degraded landfill material the energy content of the easily degradable materials is released chemically bound as methane. Heat energy available for the subsequent aerobic degradation is therefore considerably smaller than for the intentional aerobic degradation according to Fig. 2-13(2).

2.8.3.3 *Approximate comparison of long-term oxygen input into old landfills – by filtrating precipitation – with oxygen supply due to diffusion*

The filtrating precipitation can be assumed to be oxygen-saturated as a first approximation. The groundwater recharge in soils is calculated by the difference of precipitation and evaporation. The key factors for the evaporation are the climatic conditions, the species of plant cover, the storage capacity of the soil penetrated by roots and the capillary water rise in the range penetrated by roots. The calculations of the Lower Saxony State Office for Ecology (NLÖ) indicate that the average groundwater recharge is at least 140 mm under a grassland on loess and a maximum of 330 mm under a pine forest on a highly permeable sandy soil to the north of the low-rain area near Braunschweig with an average annual precipitation of approx. 700 mm (300 mm in summer and 400 mm in winter) (Ohlrogge, 1998).

When transferring the results to old waste deposits without sealing covers, it can be assumed that they are affected by their plant cover over the long term. Neumann (1978, 1981) showed in his extensive investigations 30 years ago that a root penetration on waste deposits, comparable to that of natural soil, can be achieved when seeds are planted directly into the waste or the cover layer is permeable (sand). Since mineral (sand) covers and compacted waste surfaces with no cap are more or less equivalent with regard to their evaporation behaviour (Spillmann, 1986b), a maximum of 330 mm long-term rain water input can be assumed for an average north German location with plant cover. The saturated oxygen content of the rain is approx. 11 mg O_2/l at 10°C (DIN 38 408, Part 22). This yields a maximum annual oxygen input of

$$O_{2\,\text{rain input}} = 330 \frac{dm^3}{m^2 \times a} \times 11 \frac{mg\,O_2}{dm^3} \approx 3600 \frac{mg\,O_2}{m^2 \times a} = 3.6 \frac{g\,O_2}{m^2 \times a}$$

for an average landfill location in the north German plain.

Air input in the form of seepage fronts does not occur in the waste, because preferred paths are produced within the waste (Spillmann and Collins, 1978; Spillmann, 1986b).

Oxygen input by diffusion has been described in deriving the model in Section 2.8.3.2 (Equation 2.12):

$$J_g = -D_p \frac{dC_g}{dz}$$

$$D_p \left[\frac{cm^2}{s}\right] = \text{effective diffusion coefficient for porous materials}$$

$$C_g \left[\frac{g}{cm^3}\right] = \text{mass concentration of gas in soil air}$$

$z\,[cm] = $ local coordinate

Figure 2-14 describes the effective diffusion coefficient D_{s,CO_2} (index s indicates soil as a porous material) for carbon dioxide as a function of air pore volume ε in soils. This coefficient can be used for O_2 as the minimum value with sufficient accuracy (Richter and Großgebauer, 1987).

The air pore volume ε refers to the open pore space available for the diffusion flux. Since the coarse pore space of compacted waste is approx. 30% by volume and sand behaves like compacted waste concerning the diffusion of water vapour (evaporation), the maximum effective diffusion coefficient for sand of

$$\text{Sand } D_{s,O_2} > D_{s,CO_2} = 15 \times 10^{-3} \frac{cm^2}{s} \approx 50 \frac{m^2}{a} \text{ for } \varepsilon = 0.3$$

can be used as a first approximation for the diffusion of oxygen in compacted waste, if no sealing horizons, e.g. intermediate cover of cohesive soil, are present. In the simplest case the active aerobic degradation zone, in which the oxygen concentration C_{O_2} [g/m²] can

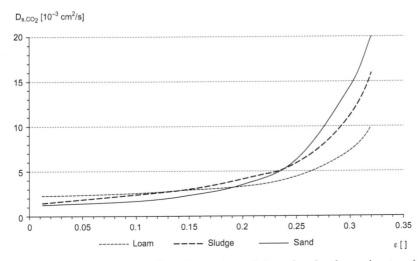

Fig. 2-14 *Effective diffusion coefficient D_{s,CO_2} (s = soil) for carbon dioxide as a function of the air pore volume ε for different soils*

be assumed to be zero, moves slowly toward depth z [m]. In this simplest case of progress of aerobic stabilisation assumed for estimation, oxygen consumption on the diffusion path z is equal to zero. This assumption yields the smallest possible concentration gradient and thus the smallest diffusion flux of outside air into the waste body. The diffusion flux of oxygen J_{g,O_2} is under these conditions:

$$J_{g,O_2} = D_{s,O_2} \frac{\Delta C_{O_2}}{\Delta z} \left[\frac{g}{m^2 \times a} \right] \tag{2.13}$$

$$D_{s,O_2} \left[\frac{m^2}{a} \right] = 50 \frac{m^2}{a} = \text{effective diffusion constant for compacted waste}$$

$$\Delta C_{O_2} \left[\frac{g}{m^3} \right] = C_{O_2\,begin} - C_{O_2\,end}$$

concentration difference of oxygen between beginning and end of diffusion path Δz [m]

$$C_{O_2\,begin} = C_{O_2\,air} \approx 0.21 \times 1300 \frac{g}{m^3} = 270 \left[\frac{g}{m^3} \right]$$

$$C_{O_2\,end} = C_{O_2\,active\,zone} = 0$$

$$\Delta C_{O_2} = 270 \left[\frac{g}{m^3} \right]$$

$\Delta z \, [m] = $ diffusion path

$z_{begin} \, [m] = z_{landfill\ surface} \, [m] = 0$

$z_{end} \, [m] = z_{active\ zone} \, [m] = $ distance of the top consumption zone from the landfill surface

$$J_{g,O_2} = 50 \times \frac{270}{z_{end}} \left[\frac{m^2 \times g}{a \times m^3 \times m} \right]$$

Thus the diffusion flux is

$$J_{g,O_2} = 50 \times \frac{270}{100} = 135 \frac{g}{m^2 a}$$

for the consumption zone at a depth of 100 m in the landfill.

The oxygen supply from precipitation is with $3.6 \frac{g O_2}{m^2 a}$ negligibly small compared to the diffusion flux in cavity-rich waste bodies in deep landfills.

3

Characterisation of long-term effects using physical measurements on water and solids balance

Peter Spillmann

3.1 Changes in mass

3.1.1 Changes in mass of the total waste with and without population equivalent sewage sludge

Table 3-1 displays the mass of the deposited solids, their deposition conditions and mass reductions together with the change in water content. It has to be taken into account, when evaluating the results, that the alternatives with intermediate earth layers (Lys. 1, 4 and 8) had already been removed after 4 years of storage, while similar alternatives without any earth cover only degraded during the 16 years of observation. In addition, it must be realised that the water content exhibited a considerable range of variation in individual material samples so that only the order of magnitude of the water mass and thus the changes in solid material could be determined. The estimation of the mass of the cover layers is also accompanied with tolerances. Despite these restrictions, the trends and the order of magnitude of the changes can be clearly recognised.

After 4 years of storage, the substantial differences in the initial moisture content between sludge-free municipal waste (Lys. 1, 2 and 9) and population equivalent sewage sludge waste mixes (Lys. 3, 4) reduced (initial values: 26–30% by weight without sewage sludge and 42–44% by weight with sewage sludge addition; final value regardless of the initial moisture content: approx. 45% by weight). Sewage sludge addition in 'lenses' caused local leachate accumulation to exceed 50% by weight and increased the water content of the municipal waste (Lys. 7 and 8). This water content of the initially anaerobic deposited wastes did not change during another 12 years of deposition.

Table 3-1 Change of density and mass in the deposited wastes in Lysimeters 1–10 on the Braunschweig-Watenbüttel landfill, water balance (R = Storage fraction caused by rainfall)

Lysimeter	Placement				Removal				Change of mass			Recharge		Storage-content ΔW	R.-storage fraction
	$m_f^{[a]}$ [t]	$WC^{[b]}$ [%]	$DS^{[c]}$ [t]	$W^{[d]}$ [t]	m_f [t]	WC [%]	DS [t]	W [t]	ΔDS [t]	ΔDS [%]	ΔW [t]	[t]	[mm]	[mm]	[mm]
1[c] Anaerobic, no sludge	58	26	43	15	55	45	30	25	−13	30	+10	6.5	330	510	+180
2[f] Anaerobic, no sludge	58	26	43	15	58	43	33	25	−10	23	+10	5	250	510	+340
3 Anaerobic, sludge	84	44	47	37	No balance, simultaneous to 4										
4[e] Anaerobic, sludge	94	43	54	40	93	45	51	42	−3	6	+2	1.5	80	100	+20
5[f,g] Aerobic, sludge	43	44	24.3	19	30	30	21	9	−3	13	−10	1.5	80	−510	−590
6 Aerobic, sludge	77	42	45	32	No balance, simultaneous to 5										
7[f] Anaerobic, 'lenses'	74	43	42	32	64	50	32	32	−10	24	0	5	250	0	−250
8[e] Anaerobic, 'lenses'	66	42	38	28	70	58	29	41	−9	24	+13	4.5	230	660	+430
9[f] Anaerobic, no sludge	27	30	19	8	31	45	17	14	−2	≈10	+6	1	50	300	+250
10 Aerobic, sludge	25	44	14	11	No balance, simultaneous to 6										

(a) Wet mass; (b) Water content, related to wet mass; (c) Dry substance; (d) Water; (e) Removal after 4 years; (f) Removal after 16 years; (g) Compacted mass after decomposition, loss of decomposition about 20% DS, Σ decomposition about 30% DS

The water content of the decomposed mixes (Lys. 5, 6 and 10) decreased after compaction from 42% by weight to values between 30 and 35% by weight. This trend was also observed on the decomposed sewage sludge waste mixes in the supplementary simultaneous tests (Spillmann, 1989, Table 38).

The dry substance of the anaerobic non-mixed waste was reduced by more than 20% by weight under the test conditions of a flat landfill without perched water, a short acidic phase and a transition to partial aerobic degradation within 3 to 4 years. Weight measurements failed to find any extensive degradation over the following 10 years. No degradation was measurable in mixes with a long-lasting acidic phase (preservation) within 4 years. The analysis of these mixes (Lys. 3) after another 10 years of storage indicated that in the most favourable case the state of degradation of the top 1–1.5 m thick layer corresponded to the condition of a decomposed mix of the same composition after about 2 years.

In the intermediate range (approx. 1.5–2.5 m under the surface) degradation processes were clearly recognisable. The base range remained conserved in principle even after 14 years of storage. Measurements were not performed because this material was not used in the subsequent programme.

Since extensive aerobic stabilisation of the top layer is made possible by oxygen diffusing from the surface, based on the model law on aerobic stabilisation (Section 2.8.3), the result can be applied to similar compacted, initially anaerobic deposits of greater depth using the formulae

$$t_{landfill} = t_{model} \times \frac{H^2_{landfill}}{h^2_{model}}$$

$$t_{model} = 14 \text{ years}$$

$$h_{model} = \text{approx. } 1\text{-}1.5 \text{ m}$$

$$\max t_{landfill} = \frac{14a}{1.0^2 \, m^2} \times H^2_{landfill}[m^2] = 14 \times H^2_{landfill}[a]$$

$$\min t_{landfill} = \frac{14a}{1.5^2 \, m^2} \times H^2_{landfill}[m^2] = 9.4 \times H^2_{landfill}[a]$$

which means that the base of a 100 m deep landfill with equivalent materials will be stabilised to the same extent as the top layer of this waste in about 90 000–140 000 years.

The permanently aerobic deposit (Lys. 5) not only reduced the moisture content of the mix, but also the dry substance by more than 25% by weight. The result is reached in $1\frac{1}{2}$–2 years.

It should be noted for the degradation rate assessment measured here (always related to DS), that the coarse fraction of the municipal waste (d > 100 mm) contained up to 60% by weight of degradable material (mainly paper and cardboard) and the ignition loss of the fine fraction (d < 100 mm) was about 50% by weight. Degradation rates above 25% by weight of the total initial dry substance are possible over the long term after nutrient compensation under favourable conditions (e.g. Lys. 5: 33% by weight DS waste) by the addition of sewage sludge, because the degradation of 25% by weight DS waste corresponds to only about 50% by weight of the dry substance deemed degradable. This degradation rate is well known from composting technology. After a thorough processing, this degradation was exceeded (approx. 60% by weight of degradable substance) in the supplementary simultaneous tests (Spillmann, 1989, Table 38). Therefore, when transferring the degradation measured values to other waste compositions, first the mass ratio of the total waste dry substance to the degradable material has to be determined and then the degradation rate must be related to the degradable substance:

Attainable absolute mass reduction under landfill conditions:

$$\Delta DS = DS_{total} - (DS_{non\text{-}org.} + DS_{resid.\,org.}) \approx 0.5 \times DS_{org.}$$

Relative degradation rate:

$$\Delta DS/DS_{total} = (DS_{non\text{-}org.} + DS_{resid.\,org.})/DS_{total} \approx 0.5 \times DS_{org.}/DS_{total}$$

DS = Mass of the waste dry substance
Δ = Mass difference
DS_{total} = Dry mass of all wastes
$DS_{non\text{-}org.}$ = Dry mass of minerals and plastics
$DS_{resid.\,org.}$ = Dry mass of organic substances not degraded over the long term
$DS_{org.}$ = Mass of the wastes, which are classified as 'biodegradable' after sorting or determined as 'ignition loss' at max. 550°C without plastics plus ash content of the organic substance (paper).

Time needed for degradation depends largely on the deposition conditions (see comparison between Lys. 2 and Lys. 6 in this chapter and further details in Chapter 11).

3.1.2 Change in mass in residual wastes with different recycling influence

Table 3-2 shows the placed and removed masses. It should be noted for the interpretation of the measured values that the anaerobic landfill

Table 3-2 Density and mass change of the residual waste without sewage sludge of the lysimeters A–E on the Wolfsburg landfill, water balance

Lysimeter	Placement				Removal							Changes				
	$m_f^{(a)}$ [t]	$WC^{(b)}$ [%]	$DS^{(c)}$ [t]	$W^{(d)}$ [t]	m_f [t]	WC [%] min.	WC [%] max.	DS [t] min.	DS [t] max.	W [t] min.	W [t] max.	ΔDS [t]	ΔDS [%]	Recharge/storage [t]	Recharge/storage [t]	Storage [mm]
A Anaerobic total waste	17.6	28	12.7	4.9	17.0	32	39	10.4	11.6	5.4	6.6	1.1 ÷ 2.3	9 ÷ 18	0.55	1.1	+115
B Anaerobic, no valuable material	18.2	29	13.0	5.2	≈15	32	39	9.2	10.2	4.8	5.8	2.8 ÷ 3.8	22 ÷ 29	1.4	−0.8	−84
C Aerobic, no valuable material	10.5 / 6.9 / Σ17.4	30 / 26 / ≈29	7.3 / 5.1 / Σ12.4	3.2 / 1.8 / Σ5.0	15.0	32	39	9.2	10.2	4.8	5.8	2.2 ÷ 3.0	18 ÷ 24	1.1	−0.3	−31
D Aerobic residual waste	10.9 / 5.0 / Σ15.9	25 / 26 / 25	8.2 / 3.7 / Σ11.9	2.7 / 1.3 / Σ4.0	15.6	32	39	9.5	10.6	5.0	6.1	1.3 ÷ 2.4	16 ÷ 20	0.65	1.3	+135
E Anaerobic residual waste	14.3	19	11.6	2.7	21	32	39	12.8	14.2	5.8	7.2	0	0	0	4.5	+470

(a) Wet mass; (b) Water content, related to wet mass; (c) Dry substance; (d) Water

59

types (A, B and E) were covered with a compacted soil layer of sandy loam (also after decomposition, Lys. D). Mixing the soil with the waste cannot be completely avoided during removal. Measurement variations of water content determination are also rather high. Therefore masses were calculated both for the lowest and the highest measured water content, so that at least the mass change trend can be clearly recognised from a comparison of the ranges. This finding is important in that similar high mass losses were measured after 7 years of degradation in a waste with a compostable fraction as in the total waste of the Braunschweig facility after the same time ('zero' alternative, Lys. A, degradation 9–18% by weight DS waste; anaerobic without valuable material, Lys. B, degradation 22–29% by weight DS waste). Even an extensively sorted residual waste (Lys. D) still contained a considerable amount of degradable substance so that between 16 and 20% by weight of the total residual waste dry substance was degraded under favourable conditions of a targeted aerobic degradation.

It was not possible to calculate a water reservoir balance because of the uncertainties in the mass determination by weighing. However, it can be assessed from rainfall and discharge balances (Section 3.2.2).

3.2 Water balance

3.2.1 Calculation of water balance from climatic water balance

It has already been determined within the first 4 years of observation that only about 20 mm of a rainfall evaporates from a compacted waste surface with or without a sand cover but without vegetation. This is because the capillaries between the waste components are interrupted and therefore capillary attraction is missing which is the prerequisite of a higher evaporation rate (mulch effect).

This observation was confirmed in all subsequent years. This measurement result makes it possible to calculate the rainfall input into the landfill surface from the climatic water balance, although the climatic balance is calculated from the sum of the rainfall less potential evapotranspiration (= maximum possible total evaporation of the vegetated surface). Figure 3-1 shows this calculation procedure by an example for Braunschweig.

For two consecutive hydrologic years (= 1 November of the first year to 31 October of the second year) the climatic water balance is plotted over the time abscissa as a sum curve in the 1st calculation step (balance after Haude, 1955): infiltration = rainfall less maximum possible

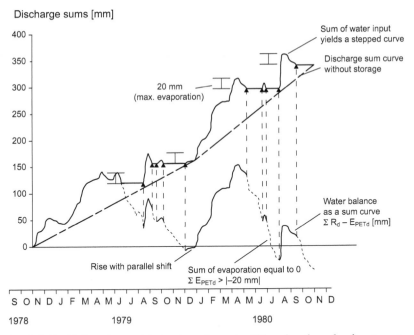

Discharge sums [mm]

Fig. 3-1 Establishing the precipitation input into a landfill surface from the climatic water balance after Haude (1955) to establish the landfill leachate recharge; maximum evaporation term applied = |−20 mm|. PETd = Potential daily evapotranspiration

climatic evaporation of an optimally watered grass stock in full growth. In the 2nd calculation step, only those evaporation subtotals were considered as subtractions from the negative sections of the sum curve which did not exceed the sum of the preceding rainfall and was limited to a maximum $|E| = |−20\,\text{mm}|$. One obtains a sum curve as a result which consists of the parallel shifted positive sections of the climatic balance whose peaks were reduced by maximum $−20\,\text{mm}$. Rainfall periods with $R \leq 20\,\text{mm}$ appear as horizontal lines when an evaporation period follows with $|\text{pot}.\,E| \geq |−20\,\text{mm}|$. Connecting the minima of this stepped curve, one obtains the sum of the water input into the landfill surface. At the same time, it is the discharge sum curve of a waste body which is water-saturated, when it neither releases water by water production from biochemical degradation, nor by consolidation and it compensates for the seasonal fluctuations by retention. If the retention is low, the discharge sum curve resembles the entry curve (= stepped curve). The total sum curve remains the same (Fig. 3-1). The discharge measured on landfill sectional cores from total waste at the Braunschweig facility (low to average rainfall) corresponded to

61

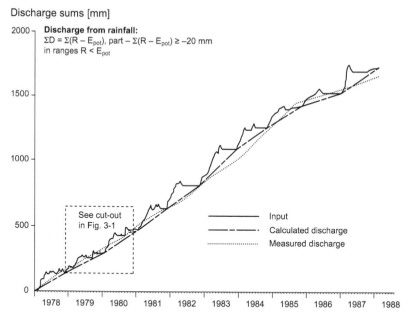

Fig. 3-2 *Comparison of calculated rainfall input with the measured leachate discharge after storage saturation over 10 years*

the straight lines sufficiently well in all parallels with and without sewage sludge, so that it was possible to determine the discharge sum curve of many years from the climatic water balance (dashed line in Fig. 3-2) based on this calculation procedure (solid line). At the Wolfsburg facility (more intensive rainfall than in Braunschweig) the discharge followed the stepped curve of the input with a delay of about 1 to 2 months.

Storage, remobilisation and water recharge also have an influence on the water balance and the discharge process in the initial phase of the deposition due to degradation processes. They have been described in Section 2.8.2.2: the same water mass is released from a water-saturated waste with or without addition of sewage sludge since the water-saturated dry mass has been degraded anaerobically.

Water recharge is a consequence of aerobic degradation processes which may be direct or indirect (indirect aerobic degradation = anaerobic degradation to methane and carbon dioxide, afterwards aerobic degradation of methane to carbon dioxide and water in the top waste layers). The maximum possible water recharge per 1 t of degraded waste DS, derived from the complete aerobic degradation of cellulose, equals 0.6 t.

It follows from this estimation that approx. 1 t of stored water and 0.6 t of water from water recharge, i.e. a total of 1.6 t of water is released from water-saturated organic substances by primary or secondary aerobic degradation of 1 t of water-saturated dry substance. This yields a water recharge of approx. 0.1 t of water per 1 t of waste DS and 0.3 t of released water from saturated waste related to the measured degradation of approx. 20% by weight of the entire waste mass in deposits with a short acidic phase and clearly aerobic degradation in the top layers.

Degradation is inhibited in dry waste. Therefore, a storage phase must first take place after the deposition of the fresh waste. After sufficient moisture penetration, mobilisation of water and maybe water recharge sets in with the degradation of organic substances. Discharge only corresponds to the rainfall input over the long term after the end of the degradation and consolidation processes. Therefore, water balance has to be divided into three development phases:

- storage phase
- mobilisation and consolidation phase
- equilibrium phase.

According to the measurements on operating landfills (e.g. Ehrig, 1986) and the discharges determined on landfill sectional cores, about 10–15% of the rainfall flows off through short cuts in coarse channels in the storage phase simultaneously to the storage processes. However, up to 40% of some heavy rain events can flow off within a short period (see details in Chapter 4). The capillary storage capacity is reduced when an intensive degradation begins after overcoming the acidic phase. Recharged water from direct or indirect aerobic degradation saturates the remaining available storage capacity and the discharge increases to values of the rainfall input in this phase.

The mobilisation and consolidation phase begins when already saturated storage material is degraded after the saturation of storage capacity. Then more water flows off than corresponds to the rainfall input. The extent of the excess and its process depend on local landfill conditions and can very substantially affect the transported masses by leachate.

Degradation and consolidation (= decrease of pore volume due to surcharge) are completed in the equilibrium phase and the content of the waste body is biologically no longer or quantitatively insignificantly degradable. The leachate discharge depends on the rainfall input into the waste body alone in this final state.

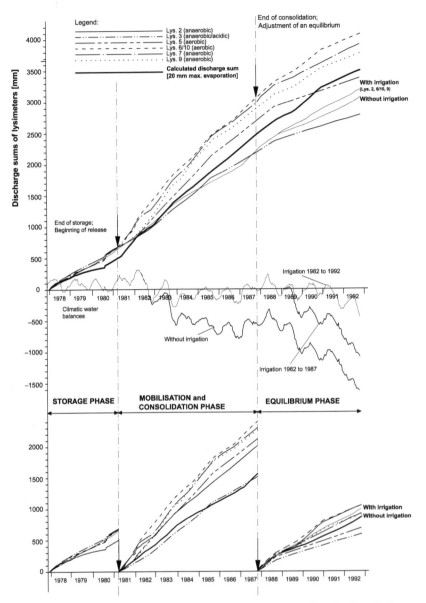

Fig. 3-3 Climatic water balance after Haude and sum of the resulting leachate discharge from various waste bodies (Lys. 2, 3, 5, 6/10, 7 and 9, Braunschweig-Watenbüttel landfill)

The time-profile of these three phases has been verified quantitatively at the Braunschweig facility for completely different waste compositions. The storage phase has already been reported on in detail (Spillmann and Collins, in Spillmann, 1986b).

Figure 3-3 shows the sums of all discharges which were liberated above the climatic water balance of the location during the entire test. The discharge sum is indicated in bold which has been calculated as a discharge according to Fig. 3-1 from the climatic water balance after the end of storage. It has to be taken into account in the interpretation that rainfall was temporarily increased by spray irrigation. It is clearly recognisable that the calculated discharge sums fully agree with the actual measured values in the first and last third, while the calculated discharge sum curve stays far behind nearly all measured discharge sums in the medium third. The discharge sum curve was therefore illustrated in three sections separately under the overall display.

Section 1 of Fig. 3-3 (storage phase) shows the sums of the discharges which flowed out from water-saturated waste bodies without reactivation of the storage. They agree with the rainfall input in all alternatives, calculated according to Fig. 3-1 (bold line). However, a test of the water content after 4 years indicated that only sewage sludge waste mixes with a long acidic phase were placed in a water-saturated state and they neither stored nor produced water (Lys. 3). All other anaerobic deposits had been placed in an unsaturated state, but they generated water by secondary aerobic degradation thus saturating the storage capacity. Water produced in the initial phase in an operation using targeted aerobic degradation evaporates due to the high decomposition temperature of about 80°C (see details in Spillmann, 1986b, Sections 4.2 and 4.3).

Any further degradation going beyond the saturation phase (mobilisation and consolidation phase Fig. 3-3, Section 2) and the consolidation, produced a leachate discharge which exceeded the rainfall-related discharge in all variants with the exception of the anaerobic, long-term acidic sewage sludge waste mix (Lys. 3, no degradation) over a period of 6 years.

In the equilibrium phase following the mobilisation and consolidation phase (Fig. 3-3, Section 3) the discharge was identical to or lower than the rainfall input over 6 years up to the beginning of the removal of the landfill sectional cores. (The bypass of the climatic water balance shown in the diagram results from the artificial irrigation of the deposits spiked with industrial wastes. The irrigation was necessary to compensate for a dry period untypical for Germany.)

It should be taken into account in the interpretation and translation of these results to large-scale facilities that only short preservation processes took place under the test conditions, and coarse channels from bulky and commercial waste drained water in a non-directly channelled way. Therefore the processes described here proceed within the old type waste bodies at a different speed, different intensity and usually more slowly than in the test specimens. However, a forecast of the water balance of old landfills is possible from the local climatic water balance, the composition of the wastes and the degree of degradation and storage saturation of the wastes determined by boreholes. For new landfills with controlled degradation and non-perched water drainage, the water balance can be predicted from these findings when the compositions of the wastes, the climatic water balance and the properties of the final cover are known with sufficient accuracy. (The specification '30% fine garbage' occurring frequently in sorting analyses is not sufficient for the calculation of the water balance, see, e.g., Franke (2003) and Schneider (2005) for application.)

3.2.2 Influence of recycling on the rainfall-discharge behaviour

The sums of the rainfall inputs were calculated using the same method as in Braunschweig (Section 3.2.1 and Spillmann, 1986b) and compared with the discharge sums over 7 years (Fig. 3-4). Taking the first 3 years as a storage phase and starting with the comparison of the 4th year of observation, the discharge sum curves of three landfill sectional cores covered with sandy loam, yield almost an identical curve up to the end of the test over a time interval of 4 years. The difference between decomposing waste with valuable material recycling (type C) and decomposing residual waste after separate biowaste collection and valuable material selection (type D) can be traced back to the smaller fraction of degradable water-adsorbing substance and the low water requirement resulting after the addition of waste. Taking an evaporation of approx. 20 mm for each rain event in the calculation of the rainfall input (derived from the measurements in Braunschweig, open waste surfaces or sand cover without vegetation), the sum of the discharges of the open decomposition of the waste including biowaste (type C) over $5\frac{1}{2}$ years is almost in full agreement with the calculation. Putting a higher evaporation of approx. 25 mm according to the compacted loam cover (as indicated in Fig. 3-4), the calculation agrees with the sums of anaerobic waste bodies. Recycling has the influence that a markedly higher storage capacity is available in the residual waste after separate collection of biowaste, as long as the plastic/aluminium-paperboard composite packages, which were not yet accepted for recycling at the

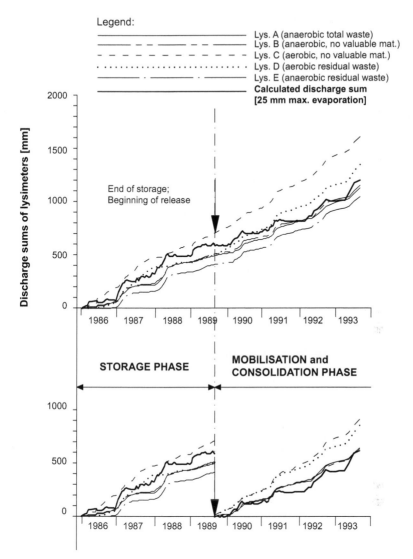

Fig. 3-4 Comparison of the influence of different recycling measures and different operating techniques (anaerobic/aerobic) on the discharge sums (Lys. A–E, Wolfsburg landfill); maximum evaporation term = |25 mm|

time of the investigation, are not consistently separated together with the valuable material.

The discharge sum curves could be represented by a straight line with sufficient accuracy over more than 15 years of measurement on total waste in Braunschweig (rain-poor area, $\sum R < 600$ mm/a, low rainfall intensity). However, the discharge sum curves in Wolfsburg showed a parallel profile to the sum curve of the rainfall input in the winter term, but with a delay of about 1–2 months.

Since this effect was not only measured in the processes in the residual waste, but also in the process of the total waste in Wolfsburg, this observation can rather be traced back to another location with higher rain intensity than to the influence of recycling. This observation suggests for the design of leachate purification plants that the discharges at an average German location or a location with increased rainfall will follow the seasonal rainfall inputs, as soon as the storage capacity is saturated. The discharge is then increased by the amount of release and water recharge because of the reduction of storage capacity.

3.2.3 Example: calculation of the water balance of a waste body

Figure 3-5 shows a comprehensive example of how to calculate leachate output. The calculation refers to a modern anaerobic landfill on a sealed and drained base with leachate treatment and features the following operational aspects:

1. The layers are placed so slowly that the initial acidic phase is limited to about half a year. Thus the contamination in the leachate is greatly reduced (see details in Chapter 5) and the highly problematic precipitation of lime in the drainage system and sewage treatment plant is avoided.

2. Waste placement is carried out in such a way that waste is directed to a new area after a layer depth of 10 m is achieved and the next 10 m is placed about 15 years later. When the storage capacity is saturated, the amount of leachate increases causing an increase in the amount of soluble pollutants transported to the sewage treatment plant (see details in Chapter 6). Therefore there is an opportunity to eliminate the pollutants by leachate treatment. In addition, subsidence also provides additional landfill space.

3. The waste body is actively degassed. Thus water regeneration due to methane oxidation in the upper waste layers is negligible.

The example calculation is based on a landfill operation modelled by Lys. 2, all quantities are calculated for a $1 \, m^2$ waste layer:

- 20 cm lifts without an earth cover, $1.5 \, t/m^2 = $ approx. $2 \, m/m^2$ layer depth of municipal waste in a year with a free storage capacity of $500 \, l/m^2$ per year;

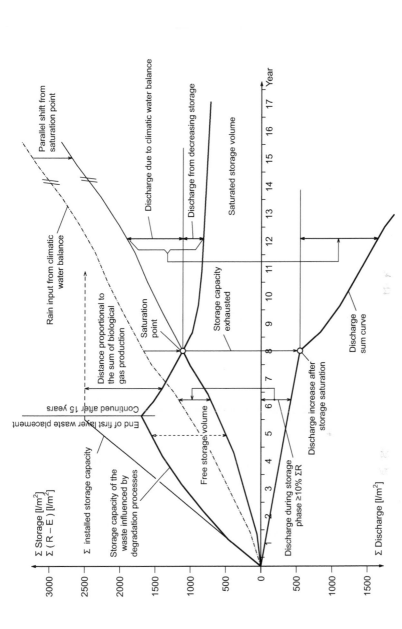

Fig. 3-5 Diagram illustrating the water balance calculation of a municipal waste landfill using sum curves (simplified calculation example) (R: rain; E: evaporation)

- Reduction of the storage capacity to $160\,l/m^2$ with a half-life of 2.5 years;
- biological methane production is directly proportional to the reduction of the storage capacity;
- $1\,l/m^2$ storage reduction $= 1\,kg/m^2$ degradation of organic dry substance corresponds to $0.3\,kg/m^2$ methane production (see Section 2.8.2.2), i.e. $(500 - 160) \times 0.3 = $ approx. $100\,kg/m^2$ total methane production per each layer, half-life 2.5 years.

The following steps are required for the calculation of leachate discharge (Fig. 3-5):

1. Determination of storage capacity as a function of time:
   The broken line illustrates the sum curve of the installed storage capacity based on the above set of operational data. In the case of an acidic phase lasting more than 5 years, it is equal to the available initial capacity up to the beginning of the methanogenic phase. If, as assumed in this example, the acidic phase can be restricted to a maximum of $\frac{1}{2}$ year, the reduction of the storage capacity must be calculated according to the degradation rate for each layer (in this instance: 2.5 years half-life).
   Calculation steps:
   1st year: complete storage 1st layer
   2nd year: complete. storage 2nd layer + reduced storage 1st layer after 1 year biological reduction
   3rd year: complete. storage 3rd year + reduced storage 1st layer after 2 years biological reduction + reduced storage 2nd layer after 1 year biological reduction
   This sequence yields the thick curve illustrating the storage capacity influenced by degradation processes as a function of time.
2. Rainfall input
   The sum curve of rainfall input (dash-dotted line) can be calculated from the climatic water balance and the amount of rainfall available for evaporation (limited to about $20\,mm = 20\,l/m^2$). The leachate discharge being about 10% to 15% of the rainfall under average German conditions will then be subtracted from this sum during the storage phase. However, up to 30% of heavy rain can flow off directly through coarse channels before the storage is saturated (see details in Chapter 4). If heavy rain occurs frequently, this discharge must be additionally considered. The sum curve of these differences yields the sum curve of storage saturation (continuous line) which intersects the sum curve of

storage capacity in the selected example after about 7.5 years after the start of the methanogenic phase and indicates saturation of the waste body (short acidic phase of $\frac{1}{2}$ year omitted). From then on not only the total rainfall input flows off as leachate but water stored in the degradable material as well. Thus the supply to the sewage treatment plant increases drastically which may be combined with surges of soluble pollutants (see Chapter 6). A temporary abatement of this process can be achieved by covering the waste with a grassed over permeable earth layer during the resting phase, thus increasing evapotranspiration. The rainfall input can then be calculated as a 'groundwater recharge' for vegetated soil of limited thickness, e.g. using the HELP model. An impervious cover is counterproductive because the then inevitable water shortage may inhibit degradation or even stop it all together. It is still more important that in this case water-soluble pollutants are not mobilised during the operation of the sewage treatment plant but at an unpredictable time much later.

If the initial acidic phase cannot be avoided and it lasts for, say, 5 years, then the first available storage volume is shown by the broken line. If the landfill is filled continuously up to the final height of 50 m using the old method, the storage capacity does not become saturated during operation, even after methane production starts under the assumed German weather conditions. A large, unmanageable storage volume develops whose uncontrollable degradation can release large quantities of leachate with large amounts of pollutants at an unpredictable time later on. Furthermore it must be noted that leachate discharge is low in the storage phase, but, in contrast to the methanogenic phase, it is characterised by extremely high amounts of organic and mineral contaminants during the acidic phase (cf. Chapter 5). Lime precipitation in the drainage system can lead to leachate accumulation and pose a hazard to the stability of the landfill. Lime precipitation can also cause leachate accumulation in landfills with no base liner. Their sealing effect however is not sufficient to protect the groundwater. Landfills without a base liner and with a high rate of fill will only contaminate the groundwater with pollutants after such a long time that it can hardly be predicted. However, monitoring of landfills has usually stopped by then since stabilisation is currently estimated to be about 30 years.

The crucial influence of buoyancy on the stability of a waste deposit results from the low particle density of about 1.5 t/m^3 for municipal solid waste (Spillmann, 1988, Table 29) which means that only 0.5 t/m^3 is available to produce friction. About 0.5 t–0.6 t DS are contained in 1 m^3 landfill volume (Spillmann, 1988, Fig. 37) so that only 0.25 t–0.30 t mass per 1 m^3 landfill volume is available to cause friction taking into account buoyancy. With an angle friction for wet waste of about 20° (Spillmann, 1988, Fig. 19) the waste only yields a maximum friction force of 0.3 t/m^3 × tan 20° × g = 0.3 × 0.37 × 10 = 1.1 kN/m^3 which is only a fraction of the laterally acting force of 1 m^3 of perched water.

To sum up, in combination with degradation processes, storage processes in the organic substance of the waste exert a major influence on the waste's water balance. Water balance in turn has a major effect on degradation processes, mobilisation of pollutants and stability of the landfills. The correct calculation of these processes is therefore an important objective of landfill design and operation. Calculation models, which have been derived for soils and thus fail to take account of these processes, are certainly not suitable for calculating the water balance of waste deposits containing a substantial organic fraction. The HELP model frequently used for municipal waste landfills also belongs to this group! This model has been developed for soil layers and is therefore only applicable to this medium.

If waste disposal is preceded by an extensive biological stabilisation on a separate area so that a permanently aerobic waste body (Lys. 5) is produced, leachate discharge depends on the rainfall input alone at the time of landfilling due to the climatic water balance (dash-dotted curve). It is slightly increased when the stabilised waste contains more than 30% by weight water before compaction and water content is reduced to about 30% by weight by the surcharge (consolidation).

4

Detection of water movements, evaporation processes and water regeneration using environmental isotopes ^{2}H and ^{18}O

Piotr Maloszewski, Heribert Moser, Willibald Stichler and Peter Trimborn

4.1   Introduction to the investigation method

In natural waters, among 10^6 water molecules with the isotope composition of ^{1}H$_2$ ^{16}O there are about 2000 molecules with ^{1}H$_2$ ^{18}O containing the heavy oxygen isotope ^{18}O and about 320 molecules ^{2}H^{1}H^{16}O, in which one of the two hydrogen atoms ^{1}H is replaced with the heavy hydrogen isotope ^{2}H (deuterium). The ^{2}H and ^{18}O contents are subjected to fluctuations, which occur through temperature-dependent isotope fractionation in all phase transitions (e.g. evaporation, condensation) because the heavier molecules, particularly at low temperatures, condense more easily and find it more difficult to evaporate due to their greater inertia. The consequence of this is that any precipitation (rainfall, snow, hail, etc.) is characteristically marked chronologically and locally by its ^{2}H and ^{18}O contents. The temperature effect of the isotope fractionation results in annual changes of isotope contents: low ^{2}H and ^{18}O contents are observed in the winter and high ones in summer. However, due to the different origin and genesis of rainfall, substantial fluctuations in isotope contents may occur in consecutive rainfall events. Figure 4-1 shows the chronological profile of ^{18}O contents in the rainwater of the Braunschweig-Watenbüttel and Wolfsburg measuring stations over the observation period. The ^{18}O content is, as usual, characterised by the so-called δ value which indicates the ‰ deviation in isotope content related to the standard V-SMOW (Vienna Standard Mean Ocean Water). The measuring accuracy is

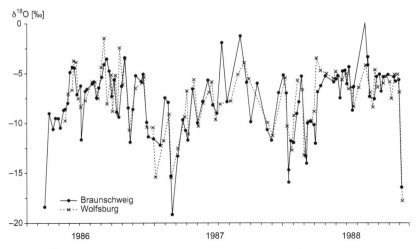

Fig. 4-1 ^{18}O *contents in precipitation as a function of time at the measuring stations on the Braunschweig-Watenbüttel and Wolfsburg landfill (non-weighted values)*

1‰ for δ^2H and 0.15‰ for δ^{18}O (2σ criterion). More detailed data on measurement technique and the fundamentals of isotope hydrological methods can be found in the relevant literature (e.g. Moser and Rauert, 1980; IAEA, 1983).

The measurements of isotope contents in precipitation and leachate at landfill sectional cores (lysimeters, construction cf. Chapter 2) pursued two goals:

- Determination of hydraulic parameters for certain lysimeters (landfill sectional cores)
- Investigation of isotope fractionation processes due to chemical and biological processes.

To determine the hydraulic parameters, the precipitation hydrograph curves marked by isotope tracers are compared as input functions with the hydrograph curves at the sampling points of the lysimeters as output functions. This comparison provides data for infiltration rate, transit time, flow velocity, hydrodynamic dispersion and ratio of total water volume to the mobile volume.

Isotope fractionation effects were also to be expected in the water due to chemical and biological degradation processes of organic wastes because the isotope composition of water generated by degradation differs from that of precipitation. These effects had to be investigated since they may provide tracers in the leachates enabling their identification in groundwater.

74

4.2 Performing of the investigations

4.2.1 Test programme

In the period from 1986–1988, rainwater samples from the Braunschweig-Watenbüttel and Wolfsburg measuring stations and water samples from the following lysimeter sampling points were tested for their ^{18}O and ^{2}H contents:

Total waste without and with population equivalent sewage sludge (Braunschweig-Watenbüttel landfill test site)

Landfill type	Sampling procedure (cf. Chapter 2, Fig. 2-5)		
Lys. 2: Mixed domestic waste, in 50-cm layers without biochemical pre-treatment, compacted industrial contamination in 'lenses'	$\frac{3}{4}h$	$\frac{1}{2}h$	base
Lys. 3: Waste-sewage sludge mix in 50-cm layers without biochemical pre-treatment, high compaction, long acidic phase with preservation effect	$\frac{3}{4}h$	$\frac{1}{2}h$	base
Lys. 5: Waste-sewage sludge mix, biochemically degraded to a very large extent, then very extensively compacted	$\frac{3}{4}h$	$\frac{1}{2}h$	base
Lys. 6/10: Waste-sewage sludge mixed with very high industrial contamination, respiration activity periodically $\rightarrow$ 0 and then very highly compacted	$\frac{3}{4}h$	$\frac{1}{2}h$	base
Lys. 7: Domestic waste, 'sewage sludge lenses', without biochemical pre-treatment, compacted	$\frac{3}{4}h$	$\frac{1}{2}h$	base
Lys. 9: Domestic waste, high industrial contamination in 'lenses', without biochemical pre-treatment, compacted, approx. 40 cm soil capping			base

The sampling points $\frac{3}{4}h$ and $\frac{1}{2}h$ are at the depths under the lysimeter surface indicated in Table 4-1 (Section 4.3.1); the sample from the base does not contain the discharge from the boundary regions.

Domestic residual waste without sewage sludge (Wolfsburg landfill test site)

Landfill type		Measuring point
Lys. A:	Total waste, anaerobic	Base
Lys. B:	Recycling of valuable materials, anaerobic	Base
Lys. C:	Recycling of valuable materials, aerobic stabilisation before compaction	Base
Lys. D:	Recycling of valuable materials and biowaste, aerobic stabilisation before compaction	Base
Lys. E:	Recycling of valuable materials and biowaste, anaerobic	Base

Samples from the base are free from boundary discharges.

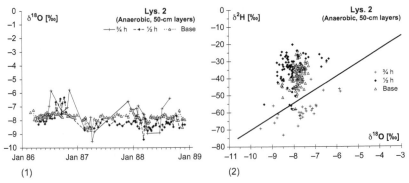

Fig. 4-2 Lysimeter 2, domestic waste, high compaction, medium contaminated, high-compaction placement in 40-cm layers, Braunschweig-Watenbüttel landfill. (1) ^{18}O content in leachate at sampling points base, $\frac{1}{2}$h and $\frac{3}{4}$h (cf. Fig. 2-5); (2) δ^2H–δ^{18}O relationship in leachate at sampling points base, $\frac{1}{2}$h and $\frac{3}{4}$h. Solid line is the 'precipitation line': δ^2H = 8 × δ^{18}O + 10

For comparison, further informative measurements of ^{2}H and ^{18}O contents were performed on samples of circulating water from laboratory waste columns in the Braunschweig Technical University, Institute for Municipal Water Management during controlled biodegradation tests. These samples were taken weekly from September 1987 to November 1988.

It was found in preliminary tests that pre-treating the lysimeter samples with activated carbon, which was necessary to eliminate the high organic content for sample preparation, did not change the amount of ^{2}H and ^{18}O. The measurement results obtained are illustrated in Figs 4-2–4-12. The ^{18}O contents were only used for the hydraulic assessment

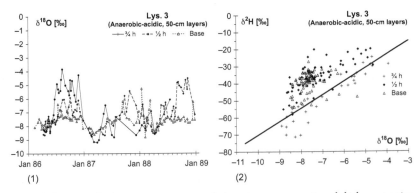

Fig. 4-3 Lysimeter 3, domestic waste sewage sludge mix, non-contaminated, high-compaction placement in 40-cm layers, Braunschweig-Watenbüttel landfill. (1) ^{18}O content in leachate at sampling points base, $\frac{1}{2}$h and $\frac{3}{4}$h (cf. Fig. 2-5); (2) δ^2H–δ^{18}O relationship in leachate at sampling points base, $\frac{1}{2}$h and $\frac{3}{4}$h. Solid line is the 'precipitation line': δ^2H = 8 × δ^{18}O + 10

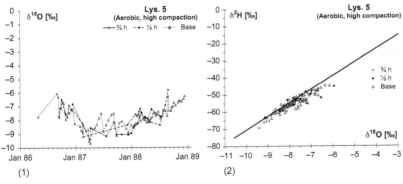

Fig. 4-4 Lysimeter 5, rotted domestic waste sewage sludge mix, non-contaminated, high-compaction placement of sieved residues, Braunschweig-Watenbüttel landfill. (1) ^{18}O content in leachate at sampling points base, $\frac{1}{2}h$ and $\frac{3}{4}h$ (cf. Fig. 2-5); (2) $\delta^2H–\delta^{18}O$ relationship in leachate at sampling points base, $\frac{1}{2}h$ and $\frac{3}{4}h$. Solid line is the 'precipitation line': $\delta^2H = 8 \times \delta^{18}O + 10$

described in the next section of this chapter. Altogether about 4000 samples were analysed for the investigation.

4.2.2   Assessment methods

Procedure
Assessment of the test results took place methodically in two steps:

In the first step, based on monthly average values, curves of the isotope contents determined over the entire 3-year observation period

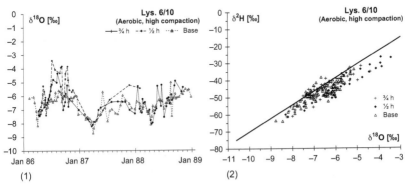

Fig. 4-5 Lysimeter 6/10, rotted domestic waste sewage sludge mix, high contamination, high-compaction placement of all wastes, Braunschweig-Watenbüttel landfill. (1) ^{18}O content in leachate at sampling points base, $\frac{1}{2}h$ and $\frac{3}{4}h$ (cf. Fig. 2-5); (2) $\delta^2H–\delta^{18}O$ relationship in leachate at sampling points base, $\frac{1}{2}h$ and $\frac{3}{4}h$. Solid line is the 'precipitation line': $\delta^2H = 8 \times \delta^{18}O + 10$

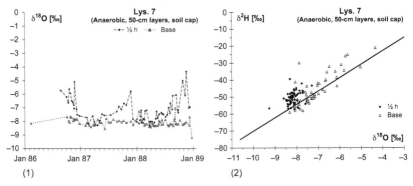

Fig. 4-6 *Lysimeter 7, domestic waste with sewage sludge in 'lenses', no industrial contamination, placement in 50-cm layers, Braunschweig-Watenbüttel landfill (no discharge at $\frac{3}{4}h$). (1) ^{18}O content in leachate at sampling points base, $\frac{1}{2}h$ and $\frac{3}{4}h$ (cf. Fig. 2-5); (2) δ^2H–$\delta^{18}O$ relationship in leachate at sampling points base, $\frac{1}{2}h$ and $\frac{3}{4}h$. Solid line is the 'precipitation line': $\delta^2H = 8 \times \delta^{18}O + 10$*

were assessed and so the long-term hydraulic parameters of water flow through lysimeter were determined. In the second step a lysimeter simulation was undertaken for certain rainfall events with relatively large rainfall, where a value is measured to determine the transit time of this amount of water which deviates strongly from the normal annual variation of isotope contents.

Mathematical modelling of long-term processes

To determine the hydraulic parameters of a groundwater system such as mean transit time, hydrodynamic dispersion and the volume of

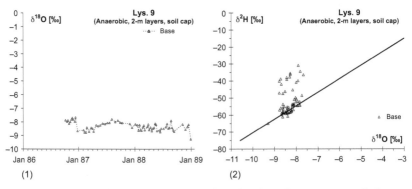

Fig. 4-7 *Lysimeter 9, domestic waste, extremely high industrial contamination, high compaction, approx. 2-m layer with a cap of sand containing silt, Braunschweig-Watenbüttel landfill. (1) ^{18}O content in leachate at sampling point 'base' (cf. Fig. 2-5); (2) δ^2H–$\delta^{18}O$ relationship in leachate at sampling point 'base'. Solid line is the 'precipitation line': $\delta^2H = 8 \times \delta^{18}O + 10$*

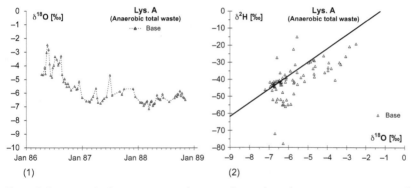

Fig. 4-8 *Lysimeter A, domestic waste without recycling, industrial contamination, compacted in 2-m layers with a loam cap, Wolfsburg landfill. (1) ^{18}O content in leachate; (2) δ^2H–$\delta^{18}O$ relationship in leachate. Solid line is the 'precipitation line': $\delta^2H = 8 \times \delta^{18}O + 10$*

mobile and total water from measured tracer concentration curves, appropriate mathematical flow models have to be used. In most cases so-called black box models (e.g. Maloszewski and Zuber, 1982; Maloszewski *et al.*, 1983) are used for this purpose. These models are based on the assumption of a steady state (i.e. flow rate is constant) distribution of transit times of the relevant tracer in a water saturated environment. In a one-dimensional flow field the tracer transport is generally described using the dispersion model whose transport equation reads:

$$D_L \frac{\partial^2 C}{\partial x^2} - v \frac{\partial C}{\partial x} = \frac{\partial C}{\partial t} \qquad (4.1)$$

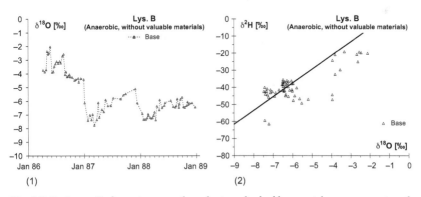

Fig. 4-9 *Lysimeter B, domestic waste after selection of valuable materials, non-contaminated, compacted in 2-m layers with a loam cap, Wolfsburg landfill. (1) ^{18}O content in leachate; (2) δ^2H–$\delta^{18}O$ relationship in leachate. Solid line is the 'precipitation line': $\delta^2H = 8 \times \delta^{18}O + 10$*

79

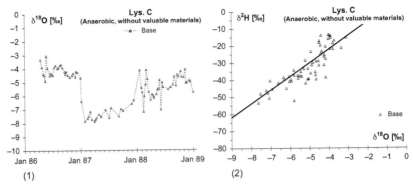

Fig. 4-10 Lysimeter C, rotted domestic waste after selection of valuable materials, non-contaminated, no cap, Wolfsburg landfill. (1) ^{18}O content in leachate; (2) δ^2H–$\delta^{18}O$ relationship in leachate. Solid line is the 'precipitation line': $\delta^2H = 8 \times \delta^{18}O + 10$

where $C(x, t)$ is the tracer concentration in groundwater, D_L the longitudinal dispersion coefficient, v the water flow velocity and x the length of the flow distance. Having a variable input tracer concentration, $C_{inp}(t)$, the solution of Equation 4.1 becomes:

$$C_{out}(t) = \int_0^t C_{inp}(t')g(t - t')\, dt' \qquad (4.2)$$

where $C_{out}(t)$ is the tracer concentration in water at a distance of x from the point of input and

$$g(t') = \left[4\pi P_D t'/T\right]^{-1/2} \exp\left[-\left(1 - t'/T\right)^2 / \left(4P_D t'/T\right)\right]/t' \qquad (4.3)$$

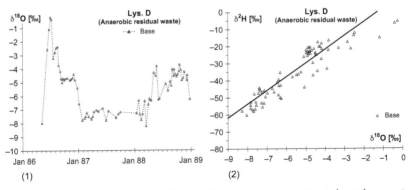

Fig. 4-11 Lysimeter D, rotted residual waste after intensive post-sorting, industrial contamination, no cap, Wolfsburg landfill. (1) ^{18}O content in leachate; (2) δ^2H–$\delta^{18}O$ relationship in leachate. Solid line is the 'precipitation line': $\delta^2H = 8 \times \delta^{18}O + 10$

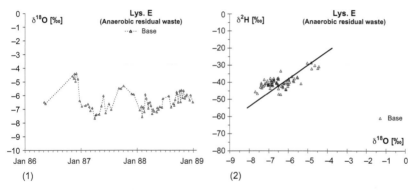

Fig. 4-12 Lysimeter E, residual waste after intensive post-sorting, industrial contamination, compacted in 2-m layers with a loam cap, Wolfsburg landfill. (1) ^{18}O *content in leachate; (2)* $\delta^2H–\delta^{18}O$ *relationship in leachate. Solid line is the 'precipitation line':* $\delta^2H = 8 \times \delta^{18}O + 10$

is the weighting function (solution of Equation 4.1 for Dirac impulse $\delta(t)$) with the parameters

$$T = x/v = T_0 \text{ (average flow time of water)} \tag{4.4}$$

and

$$P_D = D/vx \text{ (dispersion parameter)}$$

$$vx/D = Pe \text{ (Pecklet number)} \tag{4.5}$$

Since unsaturated conditions prevail in a lysimeter, the dispersion model must be modified accordingly. The basic assumption is that an unsaturated groundwater system contains two water components, i.e. mobile and stagnant (immobile) water in narrow pores. Between the two water components a diffuse exchange of tracer takes place. Thus parts of previously immobile pore water are replaced with mobile water including tracers (in this case, isotopes) in the pores. The portion of immobile water extends the average flow time of the tracer and increases the longitudinal dispersion. After a sufficiently long flow time the whole stagnant water is fully labelled by tracer transported within mobile water. Maloszewski and Zuber (1985) and De Smedt et al. (1986) have included this effect into the dispersion model. This is, however, accompanied with a change of the meaning of the parameters in Equations 4.2–4.5.

In a saturated environment the mean transit time of water $T = t_0$ and the dispersion parameter $P_D = D/vx$ can be calculated directly when adjusted to the measured output function $C_{out}(t)$ by solving the

81

convolution integral (Equation 4.2). In an unsaturated environment the mean transit time of the tracer $T = t_t$ and an apparent dispersion parameter $P_D = (D/vx)^*$ can only be calculated in this way (Maloszewski *et al.*, 1990; Herrmann *et al.*, 1987).

The mean transit time of the tracer is combined with the mean transit time of water by

$$t_t = nt_0/n_m \qquad (4.6)$$

where n denotes the portion of pore water volume of the total (mobile + stagnant) and n_m that of mobile water. The apparent dispersion parameter $(D/vx)^*$ is purely a mathematical parameter and describes the variance of flow time distribution of the tracer in the system as a consequence of dispersion and diffusive exchange between mobile and stagnating water. When the average discharge Q is known, Q and parameter t_t enable the average total volume V of the water in unsaturated system to be determined:

$$Qt_t = Qt_0n/n_m = V_mV/V_m = V \qquad (4.7)$$

where V_m is the average volume of mobile water in the system. Thus:

$$t_t/t_0 = n/n_m = V/V_m = R_p \qquad (4.8)$$

To determine both mean transit times (and volumes) separately, both portions n and n_m must be known. However, from the measured values of lysimeter tests only the overall portion n can be determined as a result of applying the dispersion model. The portion of mobile water stays, however, unknown. From the mean transit time of the tracer t_t, together with the corresponding flow distance x from the lysimeter surface (depth), the average flow velocity of the tracer becomes:

$$v_t = x/t_t \qquad (4.9)$$

which is R_p times smaller as real water flow velocity (v) and the tracer dispersivity from the apparent dispersion parameter $(D/vx)^*$ is:

$$(\alpha_L)_t = x(D/vx)^* \qquad (4.10)$$

4.2.3 Assessment of test results to describe long-term procedures

The required hydraulic parameters of the individual lysimeters were determined using calculation programmes based on the equations derived in Section and with the help of the least squares method. ^{18}O contents in the various precipitations illustrated in Fig. 4-1 (monthly

Table 4-1 *Values of hydraulic parameters of lysimeters 2, 3, 5, 6/10, 7 and 9 on the Braunschweig-Watenbüttel landfill, derived from measurements of ^{18}O contents in water samples over the period from 1986–1988*

Lysimeter sample		Depth [m]	T [month]	P_D [-]	v_t [m/month]	$(\alpha_L)_t$ [cm]	Q_{out} [mm/year]	n [%]
2	$\frac{3}{4}$h	0.50	25	0.0020	0.020	0.10	–	
Anaerobic,	$\frac{1}{2}$h	1.95	40	0.0025	0.049	0.49	–	
50 cm	Base	3.25	68	0.0022	0.048	0.72	182	31.7
3	$\frac{3}{4}$h	0.65	24	0.0010	0.027	0.07	–	
Anaerobic,	$\frac{1}{2}$h	2.05	37	0.0010	0.055	0.20	–	
50 cm	Base	3.45	66	0.0015	0.052	0.52	129	20.6
5	$\frac{3}{4}$h	–	18	0.0015	–	–	–	
Aerobic,	$\frac{1}{2}$h	0.75	26	0.0010	0.029	0.08	–	
high comp.	Base	1.35	38	0.0010	0.036	0.14	164	38.5
6/10	$\frac{3}{4}$h	2.45	25	0.0010	0.098	0.25	–	
Aerobic,	$\frac{1}{2}$h	2.25	26	0.0010	0.086	0.23	–	
high comp.	Base	3.10	38	0.0010	0.082	0.30	207	21.1
7	$\frac{1}{2}$h	2.25	38	0.0010	0.059	0.23	–	
Anaerobic,	Base	3.85	72	0.0025	0.053	0.96	191	29.8
50 cm + soil								
9	Base	1.80	48	0.0020	0.038	0.36	182	40.4
Anaerobic, 2 m								

T = average retention time of the tracer between lysimeter surface and sampling point
P_D = dispersion parameter
v_t = average flow velocity of the tracers ^{18}O
$(\alpha_L)_t$ = average longitudinal dispersivity of the tracer ^{18}O
Q_{out} = average discharge of the lysimeter
n = total water content in the lysimeter
Base = measured in the centre
h = height of lysimeter (cf. Fig. 2-7)

average values weighted with the volume of rainfall) were used as an input function. The weighted monthly average values of the ^{18}O contents in water at the respective sampling points of the lysimeters determined from Figs 4-2–4-12 (diagram (1)) served as output functions. The results for total waste (Lys. 2, 3, 5, 6/10, 7 and 9, landfill sectional cores in Braunschweig-Watenbüttel) are displayed in Table 4-1 and for residual waste (Lys. A–E, landfill sectional cores in Wolfsburg) in Table 4-2. Figures 4-13 and 4-14 show examples of the calibration of the calculated curves to the measured $\delta^{18}O$ contents.

For the investigation of residual waste (Wolfsburg landfill) special assumptions had to be made to adapt the model, since the operation of these landfill sectional cores was so delayed that during the

Table 4-2 Values of hydraulic parameters of lysimeters A to E on the Wolfsburg landfill, derived from measurements of ^{18}O contents in water samples over the period of 1986–1988 (explanation of abbreviations in Table 4-1)

Lysimeter	Depth [m]	T [month]	P_D [-]	v_t [m/month]	$(\alpha_L)_t$ [cm]
A Anaerobic total waste	2.75	50	0.001	0.055	0.28
B Anaerobic, no valuable mat.	2.40	49	0.001	0.049	0.24
C Aerobic, no valuable mat.	1.60 (3.00)	36	0.001	0.044	0.16
D Aerobic residual waste	2.03	38	0.001	0.053	0.20
E Anaerobic residual waste	2.80	50	0.001	0.056	0.28

observation period from 1986 to 1988 no steady conditions of water and tracer transport had been reached. This had the consequence that, at the beginning of the measurement period, fractions of old water, stored before the observation period started, flowed out from the residual waste. This old water had markedly different ^{18}O contents, about approx. 4‰ higher, in comparison to rain water infiltrating during the observation period. In order to describe these circumstances, a simple mixture model has been used:

$$Q_{out}C_{out} = Q_{inf}C_{inf} + Q_{old}C_{old} \tag{4.11}$$

$$Q_{out} = Q_{inf} + Q_{old} \tag{4.12}$$

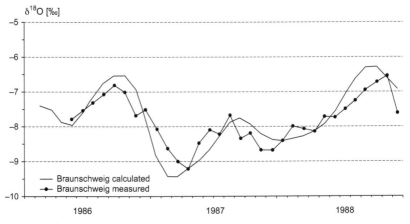

Fig. 4-13 ^{18}O content at the discharge of Lysimeter 5: measured and calculated using the dispersion model (rotted domestic waste sewage sludge, non-contaminated, high-compaction placement of sieved residues, Braunschweig-Watenbüttel), sampling point: 'base'

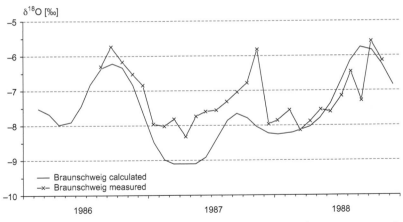

Fig. 4-14 ^{18}O content at the discharge of Lysimeter 7: measured and calculated using the dispersion model (domestic waste with sewage sludge in 'lenses', no industrial contamination, placement in 50-cm layers, Braunschweig-Watenbüttel), sampling point: $\frac{1}{2}$h

where C_{inf} is the average ^{18}O content of precipitation infiltrating within the observation period, C_{old} is the average ^{18}O content of 'old' water (calculated from ^{18}O contents of samples tested at the output of the system and taken at the beginning of the observation period) and C_{out} is the average ^{18}O content of discharged water (average ^{18}O contents without measurement data used for the calculation of the 'old' ^{18}O contents). Q_{inf}, Q_{old} and Q_{out} stand for the corresponding discharge values. The solution of Equations 4.11 and 4.12 reads:

$$P_{inf} = Q_{inf}/Q_{out} = (C_{old} - C_{out})/(C_{old} - C_{inf}), \qquad (4.13)$$

where P_{inf} is the ratio of rain water infiltrating within the observation period to the entire discharge Q_{out}. The data for the calculation of P_{inf} are compiled in Table 4-3. Flow parameters for residual waste (Wolfsburg plant) were then determined accordingly, taking into account the additional component due to presence of 'old' water (Equation 4.2):

$$C_{out}(t) = P_{inf} \int_0^t C_{inp}(t')g(t - t')\,dt + (1 - P_{inf})C_{old} \qquad (4.14)$$

4.2.4 Measurement of short-term hydraulic events

Input function
A special input function $C_{in}(t)$ was calculated to select individual rainfall events which involve a flow pattern deviating from the long-term

85

Table 4-3 Data used to calculate the ratio P_{inf} of rain water infiltrating within the observation period to the total discharge Q_{out} of lysimeters A–E on the Wolfsburg landfill

Lysimeter	C_{old} [%]	C_{out} [%]	C_{inf} [%]	P_{inf} [%]	Q_{out} [mm/year]	Q_{old} [mm/year]	Q_{inf} [mm/year]
A Anaerobic total waste	−4.04	−6.23	−8.30	51	69.3	34.0	35.3
B Anaerobic, no valuable mat.	−3.75	−6.44	−8.30	59	71.7	29.4	42.3
C Aerobic, no valuable mat.	−4.22	−6.29	−8.30	51	103.0	50.5	52.5
D Aerobic residual waste	−3.73	−6.49	−8.30	60	78.1	31.2	46.9
E Anaerobic residual waste	−4.00	−6.40	−8.30	56	58.8	25.9	32.9

C_{old} = average ^{18}O content of 'old' water available in the lysimeter (see text)
C_{out} = average ^{18}O content of discharged water within the observation period (without the values used to calculate C_{old})
C_{inf} = average ^{18}O content of rainwater infiltrating within the observation period
P_{inf} = amount of infiltrating rainwater
Q_{out} = discharge of water discharged within the observation period
Q_{old} = discharge of 'old' water
Q_{inf} = discharge of rainwater infiltrating within the observation period

hydraulic behaviour of waste deposits. This takes into account, in addition to the amount of rainfall, the deviation of ^{18}O contents in the corresponding individual rainwater samples from the long-term average value during the entire observation period:

$$C_{in}(t) = (C_n - \overline{C_{in}}) / \sum P_n + \overline{C_{in}} \qquad (4.15)$$

where:

P_n is the volume of the rainfall event n,
C_n is the ^{18}O content of the sample of the rainfall event n,
$\overline{C_{in}}$ indicates the average value of the isotope content of the rainfall events over the entire observation time of approx. 3 years weighted with the volume of the precipitation,
$\sum P_n$ describes the average monthly rainfall during the observation period of approx. 3 years.

Figure 4-15 shows the amount of rainfall and the input function of ^{18}O contents calculated from the values of the Braunschweig-Watenbüttel rainfall measuring station using Equation 4.15. Significant input signals can be observed due to the greater rainfall around the turn

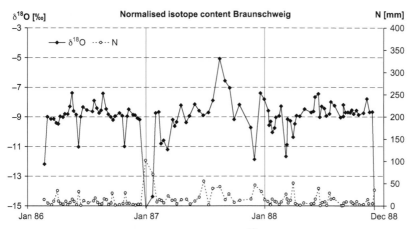

Fig. 4-15 *Volume of rainfall and associated weighted* [18]*O content (see Equation 4.15) of precipitation samples of the Braunschweig-Watenbüttel measuring station*

of the year 1986/1987 and in August, November and December 1987, as well as in March 1988.

Output functions
For the comparison of the input function with the [18]O contents in the discharge of the lysimeters their concentrations were converted in a similar way to those of [18]O contents in the rainwater. The weighting took place using the current discharge values measured.

4.3   Results

4.3.1   Long-term processes
The following general statements can be made from the values of the hydraulic parameters for the total waste (Lys. 2, 3, 5, 6/10, 7 and 9, Braunschweig-Watenbüttel plant) indicated in Table 4-1:

- The high mean transit times of the rainwater in the lysimeters of 3 to 6 years and the average flow velocities resulting from them correspond to the values measured in the water-unsaturated zone of sandy aquifers.
- Relatively small average flow velocities occur in rotted, very highly compacted domestic waste sewage sludge mixes with no industrial contamination (Lys. 5), and in domestic waste with extremely high industrial contamination (Lys. 9). A direct explanation is only available for highly contaminated domestic waste. Its water content

significantly increased within the measurement period in both lysimeters by approximately 40% as opposed to other landfill types (21–31%). An increased water accumulation effect seems to have occurred here.

- Regardless of mud application (none, mix, lenses), waste bodies built in thin layers (40–50 cm) (Lys. 2, 3 and 7) exhibited relatively high values of average longitudinal dispersivity (0.5–1.0 m) as opposed to the other landfill types (0.2–0.4 m). This might be a consequence of the layered construction of these lysimeters and the resulting character due to texture of the lysimeter body.

- Rotted, domestic waste sewage sludge mixed with high industrial contamination (Lys. 6/10) exhibited the relatively highest flow velocities at relatively low water content. That the mean transit time values at sampling points $\frac{3}{4}$ h and $\frac{1}{2}$ h are almost identical is due to the small depth differences of these points. The results are opposite to those of a rotted mix without industrial contamination (Lys. 5). The contradiction can be derived from the consolidation pressure that occurred through the relocation and compaction of Lysimeter 10 (3rd contamination grade) to Lysimeter 6 (1st and 2nd contamination grade) (see Fig. 2-7).

Residual waste values shown in Table 4-2 (Wolfsburg landfill) indicate identical hydraulic conditions to those for total waste (Braunschweig-Watenbüttel Landfill).

Rotted domestic waste without valuable materials (high water content, Lys. C) was extensively compressed and supplemented with a new rotting layer from 1.6 to 3.0 m in October 1987. The relatively small mean transit time is probably due to the consolidation processes. It was left out of the calculation of an average flow velocity because of the different heights of the lysimeters during the observation period. It is noticeable that the fraction of 'old' water with approximately 50% is relatively high in this material. This substantiates the logical assumption that 'old' pore water was pressed out by consolidation pressure.

4.3.2 Short-term water movements in the total waste (Lys. 2, 3, 5, 6/10, 7 and 9, Braunschweig-Watenbüttel)

Figure 4-16 illustrates the weighted output functions of ^{18}O contents, described above, in the discharge from the total waste (Lys. 2, 3, 5, 6/10 and 7) with exception of the extremely contaminated domestic waste (Lys. 9) and their discharge curves. The average values of ^{18}O

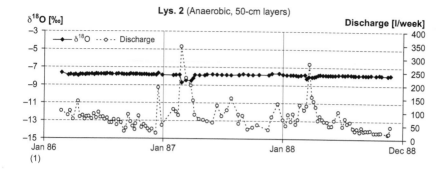

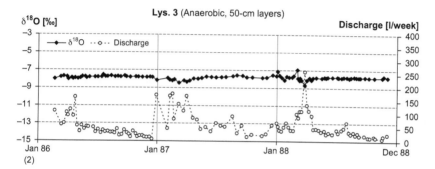

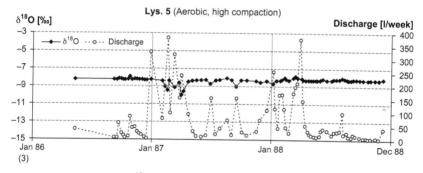

Fig. 4-16 Discharge and ^{18}O contents weighted with discharge in accordance with Equation 4.15 in discharge from lysimeters 2, 3, 5, 6 and 7, without any recycling influence, Braunschweig-Watenbüttel landfill (sampling point: 'base'). (1) lysimeter 2 (highly compacted domestic waste, medium contaminated, high-compaction placement in 40-cm layers); (2) lysimeter 3 (domestic waste sewage sludge mix, no contamination, high-compaction placement in 40-cm layers); (3) lysimeter 5 (rotted domestic waste sewage sludge mix, no contamination, high-compaction placement of sieved residues); (4) lysimeter 6 (rotted domestic waste sewage sludge mix, medium contaminated, high-compaction placement of total waste); (5) lysimeter 7 (domestic waste with sewage sludge in 'lenses', no industrial contamination, placement in 50-cm layers)

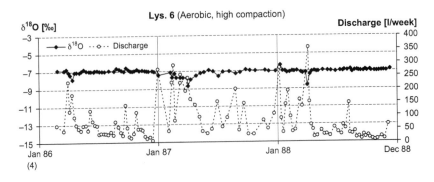

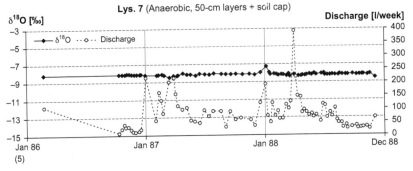

Fig. 4-16 Continued

contents evident from these are different for the individual landfill types. A typical landfill system cannot be proved. The range of the weighted ^{18}O contents and thus the extent of the influence of lysimeter discharge by fractions of rapid flow-through rain water increases with increasing density of the deposits (Lys. 7, 2, 3, 5 and 6/10).

Comparing the significant deviation of the ^{18}o contents of the average values in the output functions with those of the input function (Fig. 4-15), the transit time can be estimated from the time difference of the respective peaks of that rainfall event, which flows away through large cavities by way of a shortcut. The discharge hydrographs of anaerobic and aerobic waste sewage sludge mix of lysimeters 3 and 6/10, shown in Fig. 4-16(2) and Fig. 4-16(4), indicate that the heavy rainfall around the turn of the year 1986/1987 caused an immediate increase in discharge. The $\delta^{18}O$ content meanwhile shows that this rain water arrives only after some weeks and then not entirely directly at the discharge.

The fractions can be obtained from a simple calculation of the mixture from the amplitudes of the peaks of the corresponding ^{18}O

Table 4-4 Estimation of flow times and the fractions of the rain water percolating rapidly through the lysimeters in heavy rain events

Lysimeter	Flow time [months]	Fraction [%]	No. of events
Braunschweig			
Lys. 2 (anaerobic)	0.8 ± 0.3	12 ± 8	3
Lys. 3 (anaerobic-acidic)	1.1 ± 0.5	11 ± 6	4
Lys. 5 (aerobic)	1.0 ± 0.2	24 ± 12	2
Lys. 6 (aerobic)	0.9 ± 0.2	21 ± 6	2
Lys. 7 (anaerobic)	1.6	3	1
Wolfsburg			
Lys. A (anaerobic total waste)	0.7 ± 0.2	13 ± 9	2
Lys. B (anaerobic no valuable mat.)	0.6 ± 0.2	25 ± 3	2
Lys. C (aerobic no valuable mat.)	0.6 ± 0.2	38 ± 12	2
Lys. D (aerobic residual waste)	0.5 ± 0.2	37 ± 10	2
Lys. E (anaerobic residual waste)	0.9 ± 0.2	15 ± 3	2

contents in the input and output function. The relevant values of these estimations for the individual landfill sectional cores are displayed in Table 4-4.

4.3.3 Short-term water movements in residual waste (Lys. A–E, Wolfsburg)

Hydrographs of the discharges and the weighted ^{18}O contents in the discharge of residual waste shown in Fig. 4-17 provide averages between −6‰ and −7‰ for ^{18}O contents. These values – increased in relation to the average of the rainfall during the observation period – can be explained by an admixture of older water stored in the waste body (see Section 4.3.1). Significant differences also occur with these landfill types in the range of ^{18}O contents: total waste, anaerobic (Lys. A), residual waste (Lys. E), both with an overall fill height of approximately 280 cm, exhibit the smallest fluctuations and concomitantly the smallest influence of the discharge by rapid percolation of rainwater. This also corresponds to the high average retention time indicated in Table 4-4. The fluctuation of ^{18}O contents then increases within the residual waste after valuable material recycling (anaerobic (Lys. B) and aerobic (Lys. C)) and residual waste (aerobic (Lys. D)). However, compaction of rotted layers and an additional amount of deposited waste in October 1987 might possibly have caused the ^{18}O hydrographs to form in the two latter lysimeters. They have exhibited no fluctuation since this time (with one exception). One can come to the conclusion

Long-term hazard to drinking water resources from landfills

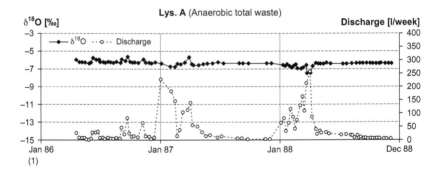

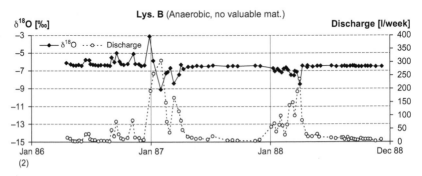

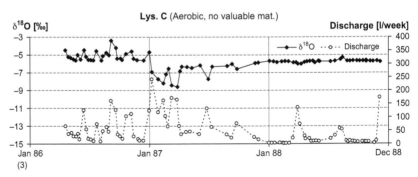

Fig. 4-17 Discharge and ^{18}O contents weighted with discharge in accordance with Equation 4.15 in lysimeter discharge of lysimeters A–E with recycling influence. Wolfsburg landfill. (1) lysimeter A (domestic waste without recycling, industrial contamination, compacted in 2-m layers with loam cap); (2) lysimeter B (domestic waste after selection of valuable materials, no contamination, compacted in 2-m layers with a loam cap); (3) lysimeter C (rotted domestic waste after selection of valuable materials, no contamination, no cap); (4) lysimeter D (rotted residual waste after intensive post-sorting, industrial contamination); (5) lysimeter E (residual waste after intensive post-sorting, industrial contamination, compacted in 2-m layers with a loam cap)

92

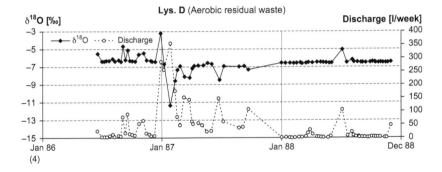

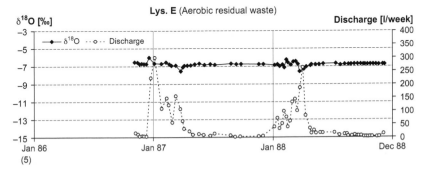

Fig. 4-17 Continued

that during this observation period only a very small discharge fraction from direct rain water run-off was present.

The rainfall events around the turn of the year 1986/1987 and in the winter and spring of 1988 could only be evaluated to determine flow time and fractions of run-off rain water. The results are illustrated in Table 4-4. In lysimeters B and D a sudden increase of the values of the ^{18}O contents occurred in December 1986. This confirms the conclusion already drawn in Section 4.3.1 that the 'old' water enriched in ^{18}O content is first squeezed out from the waste body before infiltrating rain water reaches the discharge in the course of subsequent months during the observation period.

4.4 Interpretation of the results

4.4.1 Recapitulatory assessment of the hydraulic tests

The hydraulic statements resulting from the measurements of $\delta^{18}O$ contents show that the largest part of rain water that fell on the lysimeters remained in the lysimeter for a number of years. Transit times largely depend on the lysimeter filling, boundary layers and changes in the structure during the observation period (addition of waste, compaction). The observation time of approx. 3 years was short relative to transit times of up to 5 years. Thus, in particular, the lysimeters on the Wolfsburg Landfill, which were only commissioned shortly before the test, did not enable steady-state conditions to develop.

However, in the case of heavy rainfall events, approximately 40% of the rain water washes off directly into the discharge, i.e. within a few weeks, thus it only stays in the waste for a short period and may only have little chemical or biochemical influence on the deposits.

4.4.2 Proof of water generation due to degradation processes

Proof method

Changes of the ^{2}H and ^{18}O contents of precipitation can occur when it percolates through the waste, by evaporation, or by chemical and biochemical reactions of the rain water with the waste. It has to be taken into account that the water remains in the waste for several years (see Section 4.3) and thus correspondingly long reaction times are available.

Evaporation, which is essentially effective in the upper layers of the waste, enriches the ^{2}H and ^{18}O contents. This normally takes place at thermodynamic equilibrium in the waste layers near the surface. However, when this does not happen, then ^{18}O becomes more strongly enriched than ^{2}H. This has the consequence that the measuring points in the δ^2H/δ^{18}O diagram (Figs 4-2–4-12; diagram (2)) deviate downward from the 'precipitation line'.

Biochemical reactions in the wastes, in which water is involved and/ or where water develops, in general lead to the enrichment of ^{2}H contents in the total water content of the waste body. The measurement points in the δ^2H/δ^{18}O diagram thus move into the range above the precipitation lines.

Results

Isotope effects read from the δ^2H/δ^{18}O diagrams in Figs 4-2–4-12 (diagram (2)) can only be interpreted in general terms and in such a way as to obtain qualitative conclusions. Detailed findings have to be determined together with the biochemical processes in the deposits (Chapter 7). Marked differences were expected between total waste and residual waste.

94

Total waste (Braunschweig-Watenbüttel Landfill)

The $\delta^2 H/\delta^{18} O$ diagrams of the lysimeters (Figs 4-2–4-7, diagram (2)) show significant differences. Thus the measured values of isotope contents in water from all sampling points of rotted waste sewage sludge mixes (Lys. 5 and 6/10) are below the 'precipitation lines', which indicates a considerable influence from evaporation. The measured values of the samples from the anaerobic domestic waste mix (Lys. 2) and the long-term anaerobic waste sewage sludge mix (Lys. 3) can be differentiated between the individual sampling points. Measured values were mainly observed below the 'precipitation lines' at the sampling points $\frac{3}{4}$h near the surface as in rotted mixes (Lys. 5 and 6/10), which suggests evaporation. The measured values of the sampling points at a greater depth ($\frac{1}{2}$h of deposit) and the base are to a large extent above the 'precipitation line' and indicate a biochemically induced increase in the $^2 H$ contents which only becomes effective after flowing through the upper range of the waste deposit. The measured values in sampling points at the base are also much higher than the 'precipitation lines' for domestic waste with sewage sludge lenses (Lys. 7) and extremely contaminated domestic waste (Lys. 9), also indicating biochemical influence.

The marked differentiation of anaerobic deposits (Lys. 2 and 3), as opposed to stable mixes (Lys. 5 and 6/10), must be attributed to the different stage of progress in biological waste stabilisation. The different waste composition (domestic waste (Lys. 2) as opposed to the waste–sewage sludge mix (Lys. 3)) provides no significant differences in isotope contents of the leachates.

Residual waste (Wolfsburg landfill)

$\delta^2 H/\delta^{18} O$ diagrams of the measured values in the leachate discharge of residual waste deposits (Wolfsburg plant (Figs 4-8-4-12, diagram (2)) hardly exhibit any measured values above the 'rainfall lines', in contrast to the total waste (Braunschweig-Watenbüttel landfill). This suggests that $^2 H$ content of these leachates has only been affected to a minor extent – if at all – by biochemical reactions.

Comparative measurements in laboratory biodegradation tests

Figure 4-18 shows the $\delta^2 H/\delta^{18} O$ ratio in the water of samples taken for informative investigations from the waste columns of the Institute for Municipal Water Management. Throughout the above, the measured values from the 'precipitation lines'; the $^2 H$ contents of domestic waste leachate in columns 1, 3 and 5, which contain a high fraction

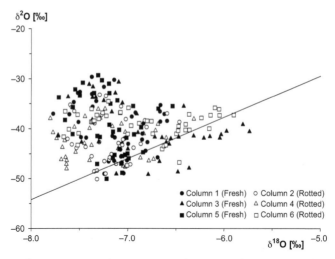

Fig. 4-18 Laboratory waste columns, Institute for Municipal Water Management of the Braunschweig Technical University. $\delta^2H/\delta^{18}O$ ratio in samples from circulating water of the waste columns. The indicated line is the 'precipitation line' $\delta^2H = 8 \times \delta^{18}O + 10$

of fresh waste in relation to the compost fraction, are significantly higher than those of columns 2, 4 and 6 with a relatively high compost fraction. This might be due to the increased methane generation in fresh waste.

4.5 Summary of the results from the isotope investigation

Results obtained so far show that, depending upon the material and structure of lysimeters, ^2H and ^{18}O contents in the water percolating through the lysimeter may be influenced typically by biochemical reactions in deeper layers and by evaporation in upper layers. Thus the leachate leaving the wastes obtains a label which may provide hints about physicochemical and biochemical processes within the deposits. It can also be used as a tracer around municipal waste landfills to check the possible spread of leachate in aquifers.

In order to further clarify the isotope separation effects in waste deposits affected by biochemical processes, detailed investigations are needed which were not possible to perform within this programme: isotope content measurements on the solid substance of the lysimeter filling and on the gases generated in the lysimeter, e.g. methane. Results of preliminary tests on the lysimeters in Wolfsburg are displayed in Table 4-5. They show strongly depleted values of ^2H content together

Table 4-5 ^{13}C *and* ^2H *contents in methane and carbon dioxide from solid samples taken from lysimeters B and D, Wolfsburg landfill*

Sampling point	Methane		CO$_2$
	δ^2H [‰]	$\delta^{13}C$ [‰]	$\delta^{13}C$ [‰]
Lys. B (anaerobic, no valuable mat.) top	−232.7	−48.3	−1.9
Lys. D (aerobic residual waste) middle	−234.9	−50.6	−13.2
Lys. D (aerobic residual waste) bottom	−235.4	−50.6	−14.0

with the corresponding ^{13}C contents in three methane samples and the ^{13}C contents of CO_2 in the same gas sample.

Further investigations on waste columns with circulating water should also be performed to quantitatively determine isotope separation effects for a better interpretation of measurements on landfills.

5

Characterisation of flow path emissions using waste-water parameters

Klaus Kruse and Peter Spillmann

5.1 Objectives and methods of the investigations

5.1.1 Objectives of waste-water analysis

The conventional parameters of waste-water analysis (Section 5.1.3) will characterise the naturally occurring contamination that has to be degraded and/or eliminated in the water. The objective is to achieve a Class 2 quality for the receiving stream. Landfill leachates contain such contamination because even natural materials cannot be discharged into nature if they harm the environment at an unnaturally high concentration.

The type and extent of emissions from natural materials in landfills are considerably affected by the composition of wastes and the biological processes in the waste body so that, when designing the sewage treatment method, these effects must be taken into account. The same applies to the groundwater contamination forecast. The key differences are characterised by three different biochemical degradation phases:

1. Acidic degradation phase: the primary degradation of fresh organic substances (hydrolysis) produces so much organic acid that methanogenic bacteria requiring an alkaline environment cannot degrade the acids. The sewage plant must biologically degrade a large amount of organic materials and control carbonate incrustations in this phase. The same substances cause similar activity in contaminated groundwater.
2. Methanogenic phase: acid production and degradation of acids are in equilibrium. Nitrification of ammonium and a subsequent denitrification of nitrate are crucial for the waste-water technique. In activated groundwater, the distance increases along which

98

mineral nitrogen compounds and difficult-to-degrade organic substances can be detected, therefore 'self-cleaning' is no longer measurable, only dilution takes place.

3. Air phase: after methane production has ceased, oxygen diffuses into the landfill. If no stagnant leachate is there, a permanent aerobic state develops. Substances that are extremely difficult to biodegrade will determine the waste-water technique which has to be used. Humic substances identical to naturally occurring materials are removed by precipitation onto particle surfaces, in the activated groundwater, while new organic compounds are produced that can travel large distances without being degraded (see Spillmann *et al.*, 1995). Self-cleaning is below a measurable level, only dilution reduces the various concentrations.

Due to relationships between the biochemical stabilisation state of the deposited waste and the contamination characteristic of the leachates, waste-water analysis incorporates the following objectives:

1. To what extent can an improvement in landfill technology reduce the transported mass of contaminants to be degraded by technical treatment and/or self-cleaning of the soil? The aim is to skip the first and possibly second design stage of the waste-water treatment plant, and to be able to control constant and low-level contamination in groundwater using preventive measures.
2. To what extent do industrial wastes disturb the biological stabilisation of natural wastes and thus increase the mainly long-term emissions?
3. Will commercial recycling of valuable materials (glass, sheet metal, plastic, composite materials, paper) increase the emissions by enrichment of organic wastes and, conversely, does a consistent separation of organic wastes make biological stabilisation of the residual waste before deposition redundant?

5.1.2   Selection of the tested landfill types

Landfill types were selected for the investigation in such a way that a landfill sectional core was tested for each important placement technology within the current landfill technologies (see Fig. 2-1). They represent the bandwidth of the landfill degradation phases: acidically preserved, degraded in the methanogenic phase and extensively stabilised. The waste composition was the same for all three landfill types, thus differences in leachate contamination alone are due to the different landfill technologies.

5.1.3 Selection of the analytical parameters

Analysis of the leachate allows the contamination in the water to be determined and the processes necessary for complete degradation and stabilisation selected. The following parameters have been proved in practice:

Alkalinity (pH > 7) – acidity (pH < 7)
indicate under which conditions the degradation processes take place and whether preservation methods or mobilization processes occur due to extreme conditions.

Conductivity
is the most important cumulative parameter for the sum of the salts and organic acids.

Chloride content
Indicator for the degradation of organic substances because chlorides make a major contribution to developing osmotic pressure in the living cells. Degradation of organic substance releases chlorides which are then preferably discharged due to their high solubility.

Sum of the organic contamination
Natural organic substances pollute the leachate with difficult-to-degrade organic compounds, which can be seen by the high chemical oxygen demand while the biochemical oxygen demand remains insignificant: $BOD_5/COD < 0.01$.

Mineral and organic nitrogen compounds
Characteristic values that provide conditions for organic substance mineralisation are the mineral nitrogen compounds (ammonium nitrogen NH_4-N and nitrate nitrogen NO_3-N) and organically bound nitrogen, calculated from the difference of total Kjeldahl nitrogen (TKN) less ammonium nitrogen: $N_{org} = TKN - (NH_4\text{-}N)$. Organic nitrogen being an important component of protein indicates biochemical activity in the leachate. Ammonium nitrogen is the mineral degradation stage of protein nitrogen under anaerobic conditions, which is oxidised to nitrate nitrogen under aerobic conditions, if the oxygen surplus is sufficient. The mineralisation stages therefore are not only products of protein degradation, but also reliable indicators for the influence of oxygen on the degradation process.

Nitrogen compounds and chlorides are not only stability indicators, but are also major natural contaminants of water.

100

Sulphates and carbonates

Other natural water contaminants from wastes are sulphates (e.g. aggressive to concrete), hydrogen carbonates and carbonates (a cause for hardening and precipitation).

When interpreting these test results it has to be noted that atmospheric oxygen cannot be completely excluded from the crushed stone in the base drainage of the landfill sectional core despite careful sealing measures. Oxidisable substances detected at the base, such as NH_4-N or the sum of dissolved organic carbon (DOC), represent therefore minimum contamination, to which the initial contamination in the leachate of an anaerobic old landfill of the same type is only reduced in the receiving stream or the sewage pond or in an approx. 50–100 m-long flow distance in a porous aquifer near the surface (see Spillmann *et al.*, 1995). In order to indicate the order of magnitude of the initial contamination before oxidation in the crushed stone of the base drainage, the contamination of the leachates from inside of the waste body were also analysed in random samples and shown together with the contamination at the base.

5.1.4 Scheme of illustration

The following form of illustration has been selected for the results in Fig. 5-1 and Fig. 5-2 when considering the assessment of long-term contamination:

The measured values of the individual parameters, which were measured in waste bodies of the same waste composition, are comparably represented together in a diagram using clearly different symbols.

The sequence of the parameters was selected based on the comparison criteria and the expressiveness of the stability criteria:

- The curves of the discharges are the basis of emission comparison.
- The pH values characterise the environment of the degradation processes.
- Conductivity and chloride provide information on the leaching behaviour and the grade of mineralisation of the cumulative transported mass as a function of the leachate curve.
- The cumulative parameters of the organic substances provide clear information on the extent and degradability of water-soluble organic substances in the deposited waste in connection with the concentrations of nitrogen compounds and the change of these measured values along the filtration path.

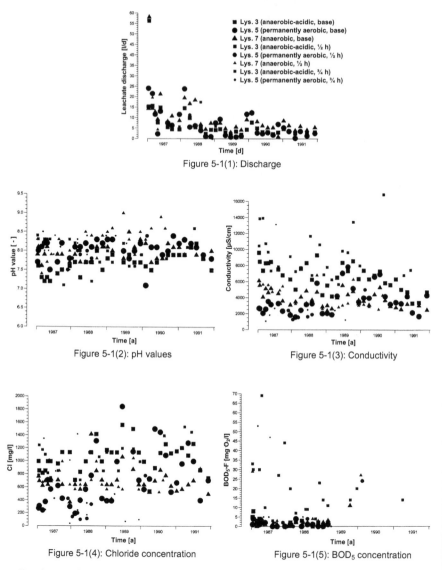

Fig. 5-1 Discharge and concentration curves of selected leachate parameters of industrially non-contaminated landfill sectional cores (Lys. 3, 5 and 7) on the Braunschweig-Watenbüttel landfill

- Eluate investigations (without demolition wastes) indicate that sulphates are only dissolved after extensive mineralisation in an aerobic environment (Spillmann, 1993a). They show that heavy metals can be dissolved by the oxidation of insoluble sulphides.

102

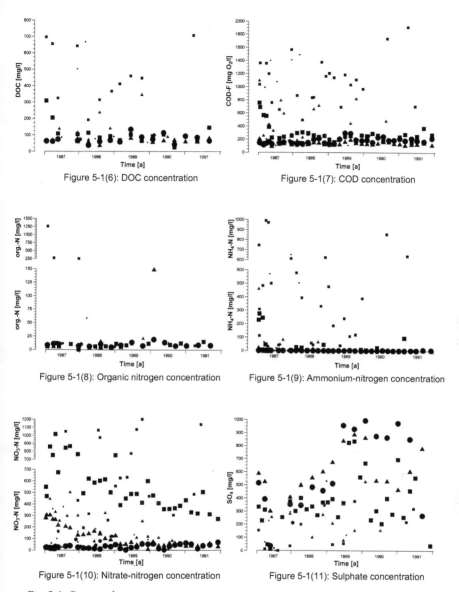

Figure 5-1(6): DOC concentration

Figure 5-1(7): COD concentration

Figure 5-1(8): Organic nitrogen concentration

Figure 5-1(9): Ammonium-nitrogen concentration

Figure 5-1(10): Nitrate-nitrogen concentration

Figure 5-1(11): Sulphate concentration

Fig. 5-1 Continued

- Hydrogen carbonates and carbonates, even without demolition waste, are important mineral components of the waste. They are extensively mobilised not only in the acidic initial phase, but can also be dissolved by carbonic acid in the advanced stage of degradation (pH value decreases from 8 to <7). Carbonate-bound heavy metals can thus be mobilised.

103

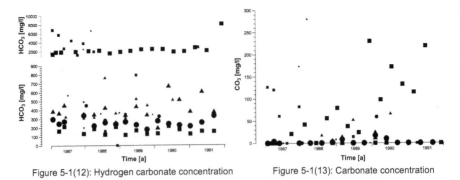

Figure 5-1(12): Hydrogen carbonate concentration   Figure 5-1(13): Carbonate concentration

Fig. 5-1 Continued

5.2 Leachate contamination of natural origin from municipal solid waste without recycling influence and without industrial contamination

5.2.1 Selection of non-contaminated landfill types

To investigate industrially non-contaminated wastes, the following landfill types were spiked with low-concentrations of industrial background contamination:

- Lys. 3: high-compaction, anaerobic sewage sludge waste mix with a long acidic initial phase.
- Lys. 7: compacted domestic waste with sewage sludge 'lenses' (short acidic phase, local stagnant leachate on sewage sludge).
- Lys. 5: biochemically extensively stabilised sewage sludge waste mix, permanently aerobic waste body even after an extremely high compaction.

5.2.2 Illustration and explanation of the results

The results are illustrated in Fig. 5-1 as explained in Section 5.1.3.

To a large extent, the leachate discharges agree regardless of landfill technology. The ratio of the contaminant concentrations in this case is therefore also equal to the ratio of the transported masses characterising the environmental impact. An illustration of the transported masses is therefore not needed to compare the deposits. The sums of the transported masses are taken into account when necessary for interpretation of the results.

104

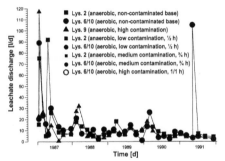

Figure 5-2(1): Discharge

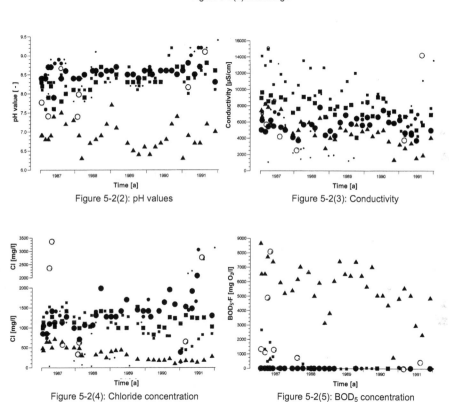

Figure 5-2(2): pH values

Figure 5-2(3): Conductivity

Figure 5-2(4): Chloride concentration

Figure 5-2(5): BOD₅ concentration

Fig. 5-2 Discharge of landfill sectional cores with industrial contamination (Lys. 2, 6/10 and 9) on the Braunschweig-Watenbüttel landfill

Alkalinity (pH > 7) – acidity (pH < 7)

The pH values (Fig. 5-1(2)), without exception, indicate that the degradation processes take place in the weakly alkaline range (7 < pH < 8.5). Apart from a few exceptions, the higher values 8 < pH < 8.5 can be

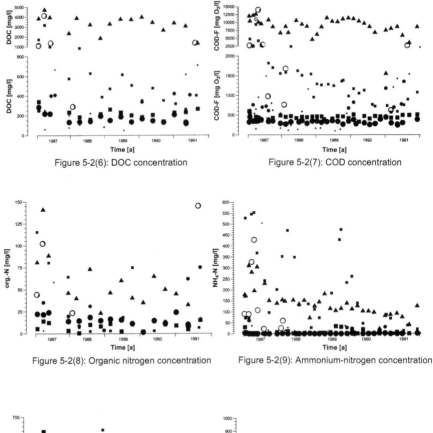

Figure 5-2(6): DOC concentration

Figure 5-2(7): COD concentration

Figure 5-2(8): Organic nitrogen concentration

Figure 5-2(9): Ammonium-nitrogen concentration

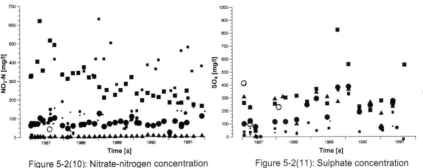

Figure 5-2(10): Nitrate-nitrogen concentration

Figure 5-2(11): Sulphate concentration

Fig. 5-2 Continued

measured in the permanently aerobic, extensively mineralised sewage sludge waste mix (Lys. 5) leachate within this bandwidth. The lower values 7 < pH < 7.5 appear in the long-term anaerobic sewage sludge waste mix of the same composition (Lys. 3). During the observation periods the values moved closer to each other.

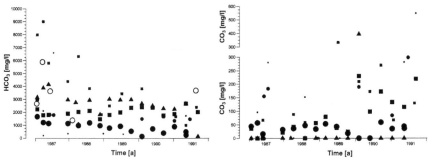

Figure 5-2(12): Hydrogen carbonate concentration   Figure 5-2(13): Carbonate concentration

Fig. 5-2 Continued

Conductivity

At the beginning of the observation period, conductivity (Fig. 5-1(3)) of the leachate from the stabilised, permanently aerobic waste mix (Lys. 5) was only approximately 25% of the conductivity of the leachate from the long-term anaerobic waste mix (Lys. 3). At the end of the observation period, the conductivity values of the base discharges moved close to each other at a low level (approximately 4000 μS/cm). However, one can conclude from the size of the conductivity of the leachate samples from inside the waste body that the soluble, conductive substances from long-term anaerobic wastes had not reached the stability level that the same waste reached immediately after biochemical pre-treatment even after 15 years. The high values inside the waste body are therefore decisive for the forecast of a non-aerated discharge.

Chloride content

The level of chloride contamination (Fig. 5-1(4)) behaves similarly, but not as clearly as that of conductivity. However, it permits the same conclusions. With 500 to 1500 mg/l it is also very high in the 'stabilisation phase' in comparison to the non-contaminated surface or groundwater.

Sum of the organic contamination

The easy-to-degrade organic substance (Fig. 5-1(5)), measured as the 5-day biochemical oxygen demand (BOD$_5$), fell below the reliable measurement range in nearly all samples of the base discharge and was still not detected in the base discharge after approximately 2 years of observation.

The DOC (Fig. 5-1(6)) reached approximately DOC < 70 mg/l within a range of 25 mg/l < DOC < 150 mg/l (corresponding to approximately 200 mg/l COD), which is the legal limiting value for discharge into the receiving stream (Waste Water Provision 2001). The contamination of the base discharges is the same from all three landfill types. The contamination of the samples from the waste body were in some cases, however, about ten times higher in the non-permanent aerobic deposits (Lys. 3 and 7). This high contamination was probably reduced by contact with oxygen in the base. The higher contamination from the inside of the waste body is characteristic of the discharge forecast from the non-aerated base.

With the exception of some compounds, e.g. ammonium, the sum of the oxidisable substances, characterised by the COD (Fig. 5-1(7)) polluted the base discharges with 100–300 mg O_2/l, i.e. about an order of magnitude less than old landfills in the stable methanogenic phase. This corresponds to the legal limiting COD value of 200 mg/l for discharge into the receiving stream (Waste Water Provision 2001). The effect of atmospheric oxygen must also be considered here. The permanently aerobic mix exhibited the lowest contamination (Lys. 5). The moderately compacted municipal waste deposit with sewage sludge lenses (Lys. 7) took a central position in the early years of this measurement series. The main difference between the deposits was that the chemical oxygen demand of 500–2000 mg O_2/l inside the long-term anaerobic deposited mix (Lys. 3) was within the range of old landfills. Also, the contamination peaks inside the deposits of domestic waste with sewage sludge lenses (Lys. 7) repeatedly reached the long-term levels of 800 mg O_2/l in the spring (large amounts of leachate, development of anaerobic zones). In the permanently aerobic sewage sludge waste mix (Lys. 5), the peak values of COD concentrations were so low, even inside the deposit (maximum of 300 mg O_2/l), that the measurements were terminated after 3 years to economise on the expenses. The discharge concentrations usually stayed below the limiting value of COD < 200 mg/l for discharge into the receiving stream.

It has to be noted for the interpretation of the degradation performance at the base that leachates with extremely high organic contamination with BOD_5 < 10 mg/l from the inside of the waste body also contained organic substances which were extremely difficult to degrade. Nevertheless, the low oxygen supply must have produced an aerobic trickling filter in the base of the lower part of the waste body, which is strongly isolated from the air. This was then able to reduce the high residual COD contamination from anaerobic degradation by about half an order of magnitude – an achievement that cannot be realised by aeration alone in a biological purification plant. Low organic contamination in the leachates from anaerobic landfills can therefore only be achieved using an advanced clarification technique (e.g. ozonation to enhance biochemical degradability) or sometimes under old landfills in a porous aquifer with a permeable overburden near the surface after about a flow distance of 100 m.

Mineral and organic nitrogen compounds

The sums of the organic nitrogen loads called 'org. N' in the base discharges are insignificant in all three variants except individual load peaks in anaerobic waste bodies (Fig. 5-1(8)).

The contamination with ammonium nitrogen (NH_4-N) (Fig. 5-1(9)) shows a clearly differentiated picture. Values between 50 and 600, maximum up to 1000 mg $NH_4 - N/l$, were measured inside the long-term anaerobic mix (Lys. 3). The maxima again show oxygen deficiency in the spring (large leachate discharge). Approximately 10% of the concentration peaks reached the base. Inside the domestic waste deposit with mud lenses (Lys. 7), peaks up to 100 mg $NH_4 - N/l$ occurred in the first 3 years of observation, while concentrations exceeding 1 mg $NH_4 - N/l$ were rarely measured in the permanent aerobic mix (Lys. 5). These values are characteristic of the discharge from non-aerated drainage systems and groundwater contamination under old landfills.

Base discharges affected by oxygen also differ markedly. The concentrations from the permanently aerobic mix (Lys. 5) and the municipal waste deposit with sewage sludge lenses (Lys. 7), where degradation mainly became aerobic, were generally below 1 mg $NH_4 - N/l$. 10 mg/l was not reached in peaks either. In the long-term anaerobic mix leachate (Lys. 3), approx. 50 mg $NH_4 - N/l$ occurred simultaneously. Nitrification of ammonium is achieved in modern purification plants and also occurs in aquifers near the surface.

It has to be noted when assessing the inorganic bound nitrogen in the leachate that the nitrogen mass remains constant during the process of oxidising the ammonium to nitrate in an aqueous solution with sufficient surplus oxygen. The intermediate stage of nitrite is usually insignificant concerning the mass balance. However, further reactions take place in the waste body. Ammonium can be adsorbed to organic materials under aerobic degradation conditions when the wastes are sufficiently biochemically stable and sufficient oxygen is provided (parameter: $BOD_5 < 100$ mg/l). If it is oxidised to nitrate, this easily soluble compound is discharged within a short time and at a high concentration

109

in the leachate (Spillmann and Collins, 1978, 1979). Anaerobic micro-ranges cannot be excluded in highly compacted, permanently aerobic waste bodies – they are even considered necessary where nitrate can be denitrified to gaseous nitrogen in them. The effect of this process was proved in industrially non-contaminated landfill sectional cores and investigated in more detail in a simultaneous test: leachates from a 4-m high, permanently aerobic sewage sludge waste mix contained less than 60 mg NO_3/l nitrate over many years. At the same time, their ammonium content was below the detection limit of 4 mg/l. Ammonium could not be released in suitable concentrations $(NH_4 - N < 4$ mg/l) in the eluate of the removed material (Spillmann, 1993a). Nitrogen was thus neither adsorbed as ammonium, nor discharged as a nitrate.

If the discharge of the mineral nitrogen was the same regardless of nitrification, the sum of nitrate-nitrogen and ammonium-nitrogen should be the same for all leachates. This is true for the contamination inside the long-term anaerobic sewage sludge waste mix (Lys. 3) but not for the base discharges and the other waste bodies. The base discharges delivered less nitrogen than measured in the flow inside the waste bodies (Fig. 5-1(10)). This is not a contradiction, since denitrification of the nitrate is possible due to the brief change of aerobic and anaerobic conditions. The minimum was measured both inside the waste bodies and in the base discharge of the permanently aerobic mix (Lys. 5). The other base discharges approached that of the extensively stabilised material of increasing age. It can be concluded from this for the long-term forecast that organic substances can also be degraded anaerobically in certain parts of the permanently aerobic deposit.

Hence it follows that this deposit has not yet reached the geological final phase, i.e. the maximum oxidation and/or humification stage possible in top soil free of stagnant water. However, the anaerobic–aerobic interaction enabled denitrification to take place inside the biochemically stabilised waste body, which can only just be achieved in aquifers near the surface with a permeable overburden (see Spillmann *et al.*, 1995) and requires the supply of easily available carbon sources in biological purification plants.

Sulphates and carbonates
Sulphate contamination (Fig. 5-1(11)) increased with an increasing aerobic stabilisation within the test period. The amount of contamination with 1000 mg SO_4/l is rather high (Drinking Water Provision 2001: <240 mg/l, geogen < 500 mg/l). It has not been caused by demolition

waste – as in normal industrial landfills. A decreasing tendency of this contamination could not be recognised within the investigation period. It can be concluded from this that sulphides are already oxidised to sulphates during the operational period in the biochemically extensively stabilised permanently aerobic waste mix (Lys. 5). If the waste contains insoluble sulphidic metal compounds which can be mobilised, these are discharged in permanently aerobic landfills during the normal operational period of the purification plant and eliminated using waste-water methods.

Hydrogen carbonates (Fig. 5-1(12)) behaved in exactly the opposite way to sulphates. It has to be considered when assessing the absolute contamination that sewage sludge deposited at a population equivalent mass ratio had been conditioned with large amounts of hydrated lime (approx. 100 g conditioning substance for 100 g sludge dry substance). The sludge fraction in the waste, and thus the lime content, were equally high in all waste bodies. The contamination of the base discharges is similarly high in all three alternatives but can be clearly distinguished. Contrary to the opinion that less hydrogen carbonates go into solution from compact sludge lenses (Lys. 7) than from a sewage sludge waste mix (Lys. 3 and 5), this type of landfill provided a significant contamination of approx. 300–700 mg/l of hydrogen carbonate over the entire observation period. The leachates from the permanent aerobic mix (Lys. 5) contained approx. 200–300 mg/l hydrogen carbonate, those from the less stabilised mix (Lys. 3) no more than 250 mg/l. This statement only applies to the oxygen-affected base discharge.

Conditions inside the waste bodies changed to such an extent that contamination with hydrogen carbonates in the less stable mix (Lys. 3) cannot be plotted using the same linear scale as for all other contaminants. Only two out of 11 analyses resulted in under 500 mg/l, seven out of 11 yielded more than 1000 mg/l, and the maximum value amounted to nearly 7000 mg/l. These leachate contaminants in the less stabilised mix reached orders of magnitude of the initial acidic phase of a landfill, although all pH values of the leachate samples were within the alkaline range. Since the mass flow of water is constant, it follows from the change in concentration that approx. 80–90% of hydrogen carbonates must be precipitated in the bottom range of the waste body. Thus, impermeable horizons in the waste body or the drainage system becoming incrusted may develop despite the characteristic values of the base discharge indicate a stable condition with low risk of incrustation.

The concentrations of carbonates (Fig. 5-1(13)) in the leachates of the long-term acidic sewage sludge waste mix (Lys. 3) reached the same concentrations as those of hydrogen carbonates in the base discharge. The concentrations of carbonates were insignificant in other leachates. The sum of the carbonates was at the lowest in the base discharge of the extensively stabilised mix (Lys. 5), and at its highest in the poorly stabilised mix of the same composition (Lys. 3). Thus, it can be concluded that the carbonate discharge is determined by the operational technology. Only biochemically stabilised waste enables the fixation of carbonates over the short and long term in the waste body to such an extent that the contaminants in the leachate do not cause problematic incrustations and the concentrations do not exceed those of carbonate-containing groundwater (usually $<300\,mg/l$) either inside the waste body or in the discharge of leachate.

5.3   Leachate contamination from natural substances under the influence of typical industrial residues

5.3.1   *Objective of the analysis and selection of landfill types*

Ehrig (1986) suggested in the first years of the investigation that the deposits of industrial residues tested in this test may lead to a considerably increased residual COD in the initial phase. In the following comparison, the extent to which these deposits affect the long-term emissions of waste products of natural substances in the leachate should be tested.

First, landfill sectional cores with various contaminants and different operational technologies were compared with one another and then the difference between those with no industrial contamination was determined. It must be taken into account that waste compositions are different in this comparison – unlike in the landfill types with no industrial contamination. Sludge-free municipal waste deposits were investigated under anaerobic conditions and the population equivalent sewage sludge waste mix was tested under aerobic conditions. The three stages of industrial contamination were first distributed to two waste bodies: non-contaminated, low and medium contamination to the first body (Lys. 2 and Lys. 6); the high-contamination stage separately to the second body (Lys. 9 and Lys. 10). In the further part of the investigation, after decomposition ceased, the initially separated high-contamination stage of the aerobic mix (Lys. 10) was extensively compacted and placed on the rotted and highly compacted first body

(Lys. 6) (see Fig. 2-1). The relocation of the fully rotted mix and its compaction and placement on older layers, which had been treated using the same method, comes from the operational technology. The already fully rotted older body (Lys. 6) was selected as a sub-layer in order to be able to measure the filtration effect and degradation performance of the low-contamination stage (corresponding to an analysis value at $\frac{1}{2}$h, see Fig. 2-7) and the medium-contamination stages (corresponding to an analysis value at $\frac{3}{4}$h) and the non-contaminated stage (corresponding to an analysis value at the base). In the following, Lys. 6, with Lys. 10 on the top denoted as Lys. 6/10, will be dealt with: the analysis values of the high-contamination deposit correspond to the measured values at $\frac{1}{1}$h (former base of Lys. 10). A relocation of the anaerobic deposit (Lys. 9) was not included in the operational technology of this landfill type and was not justifiable from health and safety aspects either.

5.3.2 Illustration and explanation of the results

The results are illustrated according to the corresponding explanations in Section 5.1.3 and Fig. 5-2 in the same sequence and form as for the waste deposits with no industrial contamination. Discharges from the base are labelled by large black symbols, corresponding random samples from the low and medium contaminated stage ($\frac{1}{2}$h and $\frac{3}{4}$h) from the waste body with similar but smaller symbols. Only the base discharges were tested from the deposits with a high contamination. Since the base discharge of the highly contaminated aerobic deposit (Lys. 6/10, $\frac{1}{1}$h) also indicates the contamination of the entire aerobic waste body (Lys. 6/10) including the consecutive degradation process, the form of the symbol has been chosen according to the actual position of the base discharge, but labelled with a white circle in the centre.

The leachate discharges agree to such an extent that the various concentrations of the contamination also determine their transported masses. It is therefore sufficient to compare the contaminant concentrations.

Alkalinity (pH > 7) – acidity (pH < 7)

The environment characterised by the pH value (Fig. 5-2(2)) is not uniform. The stages with low and medium contamination (Lys. 2 (anaerobic), $\frac{1}{2}$h and $\frac{3}{4}$h; Lys. 6 (aerobic), $\frac{1}{2}$h and $\frac{3}{4}$h) approached a clearly alkaline level (pH $\approx$ 8.5) with increasing age. There was an insignificant deviation of the highly contaminated aerobic stage

(Lys. 6/10, $\frac{1}{1}$h) towards the neutral side, as far as values were available, because of the low discharges. In contrast, the environment of the highly contaminated anaerobic deposit (Lys. 9, base) fluctuated within the neutral range (6.3 < pH < 7.8). Most values (21 out of 32) were measured in the weakly acidic range. It can be concluded from this that the extremely highly contaminated deposit (Lys. 9, base) remained in the acidic phase until the end of the investigation.

Conductivity

The conductivity (Fig. 5-2(3)) of the leachates of the four variants reached roughly the same orders of magnitude but could clearly be distinguished. The conductivity of the base discharge of the permanently aerobic deposit (Lys. 6/10, base) remained within the range of about 4000 to 8000 µS/cm for the entire period of observation, the samples inside this waste body exhibited only negligibly stronger contamination. The high-contamination maximum stage of the aerobic landfill (Lys. 6/10, $\frac{1}{1}$h) reached peak values of up to 20 000 µS/cm, but was reduced while passing through the low-contamination waste body (Lys. 6/10, $\frac{3}{4}$h and $\frac{1}{2}$h) (typical for the degradation of organic, conductive substances). The conductivity of the anaerobic sludge-free municipal waste deposit with zero to medium contamination (Lys. 2, base, $\frac{1}{2}$h and $\frac{3}{4}$h) decreased in the base during the observation period from an initial range of 8000–10 000 µS/cm to a range of 5000–7000 µS/cm. The contamination decreased inside the landfill from a range of 13 000–15 000 µS/cm to a range of 4000–8000 µS/cm.

Differences in the base discharge indicate the presence of degradation processes. Conductivity of the highly contaminated anaerobic deposit (Lys. 9, base) amounted to a maximum 7000–9000 µS/cm and decreased during the observation period to 1000–3000 µS/cm. This trend contradicts the pH values. Accordingly, they show that organic acids and salts of the highly contaminated anaerobic phase (Lys. 9, base) may have caused the highest conductivity. This result is consistent when the extent of degradation has been restrained by the chemical contamination and the degradation chain leading to methane was disrupted.

Chloride content

The chloride concentrations (Fig. 5-2(4)) behave only slightly akin to the conductivity of base discharges. The contamination of the leachates from the sludge-free, highly contaminated anaerobic deposit (Lys. 9, base) decreased during the observation period from approximately 500–800 mg Cl^-/l to a range of 100–300 mg Cl^-/l. The moderate chloride release also indicates a restrained degradation. The leachates of the aerobic deposit with identical chemical contamination of the sewage sludge waste mix (Lys. 6/10, $\frac{1}{1}$h) contained up to 3500 mg Cl^-/l in the same observation period. Even if their discharge was tested to a lesser extent, the far higher contamination level can clearly be recognised. The sewage sludge cannot be the cause of this difference, since the chloride concentrations of the leachates from the first two comparable contamination grades were equally high regardless of the sludge admixture. The increasing tendency in the moderately contaminated aerobic landfill sectional core (Lys. 6/10, base, $\frac{1}{2}$h and $\frac{3}{4}$h) is due to the leachate from the highly contaminated stage.

Sum of the organic contamination
In this long-term test, the easily degradable organic substances from the highly contaminated anaerobic deposit (Lys. 9, base), characterised by the BOD_5 (Fig. 5-2(5)), also reached a contamination level that normally only flows from freshly deposited wastes with no industrial contamination (9000 mg O_2/l, decreasing to 2000–5000 mg/l). The leachate contamination from the maximum aerobic highly contaminated stage (Lys. 6/10, $\frac{1}{1}$h) reached (after a short, roughly equal peak of 8000 mg O_2/l) a maximum of 1000 mg O_2/l, which is similarly a very high value for a permanently aerobic waste body. These high organic contaminations from the highly contaminated aerobic deposit (Lys. 6/10, $\frac{1}{1}$h) were reduced to 20 mg O_2/l in the passage through the low-contamination waste body of the low- to medium-contamination stages (Lys. 6/10, $\frac{1}{2}$h and $\frac{3}{4}$h). This value is identical to the BOD_5 of the leachate from the transition between the moderately contaminated anaerobic municipal waste (Lys. 2, low to medium contamination, $\frac{1}{2}$h and $\frac{3}{4}$h) and the non-contaminated anaerobic municipal waste (Lys. 2). Hence, it follows that the high-contamination input of the maximum stage (Lys. 6/10, $\frac{1}{1}$h) was reduced in the lower two contamination stages of the permanently aerobic waste mix (Lys. 6/10, $\frac{1}{2}$h and $\frac{3}{4}$h) to the same contamination level, which flowed out from the first two contamination stages of the municipal waste (Lys. 2, $\frac{1}{2}$h and $\frac{3}{4}$h), without having to reduce an additional high contamination input.

In contrast to the municipal waste indirectly affected by industrial contamination, the biochemical oxygen demand in all intentionally contaminated landfill sectional cores at the base did not decrease below the detection limit.

The DOC (Fig. 5-2(6)) contaminated the leachate from the high-contamination anaerobic deposit (Lys. 9, base) within a range of 1000–5000 mg/l. The measurements provided values between 2000 and 4000 mg/l in 13 out of 16 analyses. An obvious decreasing tendency was not recognised. The contamination level in the leachate from the highly contaminated aerobic stage (Lys. 6/10, $\frac{1}{1}$h) were – apart from an initial peak of 4300 mg/l – within a range of approx. 500–1500 mg/l. In the underlying aerobic body of the low- and medium-contamination stages (Lys. 6/10, $\frac{1}{2}$h and $\frac{3}{4}$h) this contamination is reduced to 250–420 mg/l and up to the non-contaminated base by nearly half an order of magnitude to 100–200 mg/l (initial peak: 290 mg/l). The base discharge from the moderately contaminated anaerobic waste body (Lys. 2) with 120–260 mg/l (initial peak 340 mg/l) exhibits a higher contamination level than the leachate from the base of the aerobic deposit. The contamination of the leachate between the non-contaminated base and the moderately contaminated stage ($\frac{1}{2}$h) of the permanent aerobic waste is also substantially higher at 400–1000 mg/l (peak: 3200 mg/l) than that at the same sampling point of the aerobic comparison body (Lys. 6/10, $\frac{1}{2}$h). It has to be emphasised in connection with this comparison that the maximum-contamination stage of aerobic wastes, i.e. Lys. 10, was placed on a moderately contaminated aerobic waste, Lys. 6, to form Lys. 6/10. Thus, the maximum-contamination leachates of Lys. 10 also had to be cleaned by the non-contaminated layers of Lys. 6. In contrast, the extremely highly contaminated leachates of the maximum-contamination anaerobic deposit, i.e. Lys. 9, flowed off separately.

The COD (Fig. 5-2(7)) of the leachate from the highly contaminated anaerobic deposit (Lys. 9, base) was 10 000–14 000 mg O_2/l, slowly decreasing to 3000–9000 mg O_2/l, and corresponded to fresh compacted waste. The leachate contamination of the aerobic comparison body (Lys. 6/10, 1/1 h) also reached a contamination peak of 14 000 mg O_2/l at the beginning of the measurement. The level was outside this peak contamination in the range of about 1000–3000 mg O_2/l and corresponded to that of a stable anaerobic landfill. This contamination – extremely high for an aerobic waste body – was degraded on its way through layers of moderate industrial contamination (Lys. 6/10, $\frac{3}{4}$h and $\frac{1}{2}$h) to about 800 mg O_2/l (peaks up to 1400 mg O_2/l). On the way through the industrially non-contaminated lower level (base) the contamination was further decreased to 240–350 mg O_2/l (peaks: 450 mg O_2/l). In the moderately contaminated anaerobic waste body (Lys. 2) the organic substances from the first two industrial contamination stages ($\frac{3}{4}$h and $\frac{1}{2}$h) caused a chemical oxygen demand of 500–3000 mg O_2/l in the leachate (peak: 9500 mg O_2/l with a decreasing trend). Thus, the leachate contamination was clearly higher inside the waste body of this anaerobic type of landfill (Lys. 2), even without the high contamination from the maximum stage (Lys. 9), than in the aerobic waste body (Lys. 6/10) with maximum contamination. In the non-contaminated anaerobic lower level of Lys. 2 (base), the chemical oxygen demand of the leachate was reduced to 240–350 mg O_2/l (peaks: 450 mg O_2/l) in this type of landfill, too, under oxygen influence and it reached the same value as the discharge from the aerobic waste body (Lys. 6/10, base). In the interpretation of these degradation processes it has to be taken into account that the easy-to-degrade part of the organic substance, characterised by the BOD_5, is very small. Therefore, such an extensive degradation using biological degradation alone in purification plants cannot be achieved according to the state of the art.

Mineral and organic nitrogen compounds

The organically bound nitrogen (total Kjeldahl nitrogen, TKN) (Fig. 5-2(8)) less ammonium-nitrogen (NH_4-N)) was approximately 20–90 mg N/l in the leachate of the highly contaminated anaerobic deposit (Lys. 9, base) (peaks around 140 mg/l) and the trend was slightly falling. Contamination in other base discharges dropped below 25 mg org. N/l without exception. Higher values in the aerobic mix may have been caused by the influx of the maximum contamination grade (Lys. 6/10, $\frac{1}{1}$h).

When assessing the inorganically bound nitrogen, the explanations in Section 5.1.4 have to be taken into account. The inorganically bound nitrogen was exclusively discharged as ammonium (approx. 200–300 mg NH_4-N, decreasing to 50–100 mg NH_4-N) in the leachate of the high-contamination grade of the anaerobic deposit (Lys. 9, base). Due to the moderate biochemical stabilisation of the wastes ($BOD_5 \gg 100$ mg/l) nitrate formation is only possible in the discharge. The concentration clearly fell below that of a normal landfill in the methanogenous phase (to 1000 mg NH_4-N/l), from which it can be concluded that the degradation was inhibited.

The leachate from the low- and medium-contamination sludge-free anaerobic deposit (Lys. 2, $\frac{1}{2}$h and $\frac{3}{4}$h) contained 0–700 mg NH_4-N (Fig. 5-2(9)) and 0–650 mg NO_3-N (Fig. 5-2(10)) on the seeping path towards the non-contaminated base section. The concentrations behaved in the opposite sense, i.e. the sum of the discharged nitrogen always remained in the same order of magnitude and decreased within 5 years from only 500–550 mg N/l to approx. 300–400 mg N/l. It is clearly higher than that in the highly contaminated deposit. This confirms the assumption that degradation was inhibited in the highly contaminated deposit (Lys. 9). On the way to the base discharge, the ammonium was oxidised to nitrate to a large extent (NH_4-N 0–35 mg/l, NO_3-N 150–650 mg/l). The mass of the nitrogen discharged from the base decreased from approx. 500 mg/l (range: 300–650 mg/l) to about 200 mg/l (range: approx. 150–250 mg/l) somewhat faster than inside the waste body.

The leachate of the high-contamination stage of the aerobic sewage sludge waste mix (Lys. 6/10, $\frac{1}{1}$h) still contained roughly 20–100 mg NH_4-N/l after a peak of 450 mg NH_4-N/l, over an observation period of more than 14 years. The nitrate content was insignificant even after 4 years, i.e. the aerobic degradation, was not sufficient in the aerobic stage with maximum industrial contamination for nitrification to occur. On the seeping path through the stages with medium and low industrial contamination (Lys. 6/10, $\frac{3}{4}$h and $\frac{1}{2}$h) the ammonium-nitrogen was nitrified to insignificant concentrations <40 mg N/l NH_4-N. The nitrate-nitrogen increased from 10–150 mg N/l to 10–250 mg N/l in the opposite direction to the ammonium-nitrogen trend. The range of the nitrogen discharge increased from 100–150 mg N/l to 30–250 mg N/l. However, the medium overall contamination remained with approximately 130 mg N/l constant. On the way through the lower level, with no industrial contamination, the ammonium was nitrified to trace levels so that only traces of ammonium-nitrogen <4 mg NH_4-N/l were measured in the base discharge. The entire nitrogen contamination in the base discharge was within a range of 30–130 mg NO_3-N/l as nitrate-nitrogen. For comparison, the values are characteristic within the limits of 50–80 mg N/l (approx. 75% of the values). The discharge of the nitrogen from the base of the aerobic deposits (Lys. 6/10, base) is therefore less than the transported mass, which seeps from the highly contaminated aerobic deposit (Lys. 6/10, $\frac{1}{1}$h), and also less than that in the passages of the medium and low contamination in Lysimeter 6/10 ($\frac{1}{2}$h and $\frac{3}{4}$h).

Sulphates and carbonates

The sulphate contents (Fig. 5-2(11)) exceeded 400 mg/l only in three samples of the deposits tested within 5 years and are insignificant as direct water contamination. As indicators for the degradation processes and future sulphate emissions, they provide clear information in connection with the hydrogen carbonates about the degree of degradation. The parameters of all base discharges are characterised by an increase of the range from 0–310 mg/l to 40–800 mg/l. The clear distinction between base discharges from anaerobic deposits (200–310 mg/l) and those from aerobic deposits (70–125 mg/l) cannot be seen later. That the sulphate transport was very low inside both the anaerobic and aerobic deposits is nevertheless clearly recognisable. Hence, it follows that the degradation had not progressed even in permanently aerobic deposits (Lys. 6/10) up to sulphate mobilisation.

Contrary to sulphates, the leachates were very strongly contaminated with hydrogen carbonates (approx. 500–3000 mg/l in about 80% of the samples) (Fig. 5-2(12)) and to some extent with carbonates (Fig. 5-2(13)) and created a high incrustation risk. It is remarkable that equal amounts of carbonate (1000–4000 mg HCO_3/l) were dissolved in the leachate of both aerobic and anaerobic industrially contaminated waste bodies, although only the aerobic deposit contained sewage sludge conditioned with lime. It is noticeable that peak loads up to 9000 mg HCO_3/l were measured inside the sludge-free, mainly anaerobic deposit (Lys. 2, medium-contaminated stage, $\frac{3}{4}$h) and that, alone, the contamination level of the base discharge from the sewage–sludge waste mix conditioned with a high amount of lime (Lys. 6/10, base) dropped below 1000 mg HCO_3/l over the long term. Hence, it follows that the degree of biological stabilisation determines the incrustation potential of the leachate in wastes with industrial contamination, too, regardless of the lime content in the waste. If one compares the hydrogen carbonate contamination with the ammonium contamination, a coupling of both quantities can be observed. Taking into account that 1 mol of HCO_3^- develops from the degradation of organic substances when 1 mol NH_4^+ is created, i.e. the generation of 1 mg NH_4-N is coupled with the formation of 4 mg HCO_3^-, the contamination of the leachate is linked with hydrogen carbonates to an extent that the mineral components are sufficient to buffer the degradation process. The high mobility of the hydrogen carbonates and the carbonates is therefore proof that all deposits tested were not yet biochemically stabilised and the highly contaminated anaerobic deposit (Lys. 9, third highly contaminated anaerobic deposit, base) still corresponded to a fresh deposit.

A comparison with sulphate contents shows that hydrogen carbonates and sulphate behave in the opposite sense in this phase of degradation and therefore sulphate contamination was still to be expected particularly from the anaerobic deposit with high industrial contamination (Lys. 9, 3rd highly contaminated anaerobic deposit, base).

5.4 Comparison of leachate contamination from waste bodies with and without industrial contamination to determine the industrial influence on the degradation process

Comparing the leachate contamination from industrially non-contaminated landfill sectional cores with those from industrially contaminated deposits clearly prove that an increasing industrial contamination slows down the biochemical stabilisation. If, as is common in the current landfill technology, fresh waste is immediately compacted and air access is sustainably reduced – e.g. by an earth cover layer – the delay effects add up. The delays were quantitatively proved in nearly all parameters (see Fig. 5-1, also Fig. 5-2).

Alkalinity (pH > 7) – acidity (pH < 7)

Electrochemical parameters which are easy to determine in situ without too much expenditure (pH and conductivity; Fig. 5-1(2), (3) and Fig. 5-2(2), (3)) and the easily analysed guide parameter chloride (Figs 5-1(4) and 5-2(4)) enable an unambiguous distinction between the deposits. The leachates from the industrially non-contaminated waste react in a neutral to weak alkaline way ($7.1 < pH < 8.3$ in the base discharge, and up to $pH = 9$ inside the landfill in the case of a retarded degradation process). Leachates from industrially contaminated deposits can be separated into two clearly distinguished ranges: approx. 65% of the leachates from the highest deposit covered with soil and containing high contamination (Lys. 9, base) react in a weakly acidic way: $6.3 < pH < 6.9$. Neutral to weak alkaline ($7.0 < pH < 7.7$) leachates were only discharged in dry periods over a short time. A rising trend was observed even 12 years after deposition, despite a very small layer thickness (approx. 1.5 m). Within the same time interval the pH values of all other leachates of the base discharges, also those from the highly contaminated aerobic stage (Lys. 6/10, $\frac{1}{1}$h), increased to $8.3 < pH < 8.8$ and within the waste bodies to $8.1 < pH < 9.4$. The weak alkaline range $7.1 < pH < 8.4$, typical for all discharges from the non-contaminated deposits tested, is almost completely missing. Low pH values, as in the discharge of the contaminated wastes of Lys. 9, were only measured inside the poorly stabilised and chemically non-contaminated deposits (Lys. 3). The various

120

acidic and/or basic reactions of the leachates indicate that even low contamination by industrial wastes obstruct biochemical stabilisation to the same extent as a long-term acidic phase caused by an unfavourable installation technology. Purposeful biochemical degradation (in this case, 'aerobic decomposition') may compensate in part for the delay even after a high contamination, but the industrial contamination clearly impairs stabilisation in comparison to the non-contaminated waste. If wastes containing high industrial contamination are deposited using an unfavourable landfill technology, extremely long delays occur in the biochemical stabilisation process.

Chloride content

When comparing the conductivity and chloride contamination, it has to be noted that chlorides make only a minor contribution to the conductivity in leachates from wastes. In the initial phase of the degradation, e.g., the organic acids are important for the conductivity, and chloride is an indicator for the degradation of organic substances. The chloride content and conductivity are only proportional to each other in leachates from extensively degraded wastes. In this comparison the leachate of the anaerobic deposit with maximum industrial contamination (Lys. 9, base) contains the lowest chloride concentrations and is the only deposit to exhibit a decreasing trend in chloride contamination (from 400–700 mg Cl/l to 100–300 mg Cl/l). The low chloride concentration may not have been the cause of the high conductivity (6000–8000 μS/cm, decreasing to 1000–3000 μS/cm). Comparing the leachates of the highly contaminated aerobic comparison bodies (Lys. 6/10, $\frac{1}{1}$h), (300–3300 mg Cl/l, 2000–15 000 μS/cm, synchronous fluctuations of chloride content and conductivity) it can be concluded that degradation is inhibited in the anaerobic deposit (Lys. 9). Chloride contamination and conductivity of the deposits with a low industrial contamination (Lys. 2 and 6/10, $\frac{1}{2}$h and $\frac{3}{4}$h in each) and the non-contaminated deposits (Lys. 3, 5 and 7) are within the same range. The ratio of conductivity to chloride content shows in all stabilisation stages: increasing chloride content with constant or decreasing conductivity indicates an increase in mineralisation. This process can be observed despite industrial contamination in the discharges of the aerobic deposits (Lys. 6/10, base). Stabilisation was only proved unambiguously in the leachate of permanently aerobic sewage sludge waste deposits with no industrial contamination (Lys. 5), where conductivity was at a low level and chloride content changed synchronously (conductivity = 1000–4000 μS/cm, individual peaks up to $8000\,\mu S/cm$; $Cl = 100–600\,mg/l$, peaks up to $1800\,mg/l$).

Sum of organic contamination

The comparison of the parameters of the organic contamination (BOD_5, DOC, COD and organically bound nitrogen org. N) (Fig. 5-1(5)–(8) and Fig. 5-2(5)–(8)) confirms the indirect conclusions from the electrochemical parameters and the guide parameter chloride.

High contamination with easily degradable materials were also measured as BOD_5 in the leachates of the industrially highly contaminated wastes: maximum approx. $9000\,mg\ O_2/l$ after about 7 years of anaerobic deposition (Lys. 9, base). This condition ($BOD_5/COD \approx 0.8$), in comparison to the initial phase, is a clear improvement of the biochemical degradation processes, since in the first years these leachates were characterised by a very high chemical oxygen demand but restrained degradation ($BOD_5/COD \approx 1000/10\,000 = 0.1$) (cf. Ehrig, 1986). After about 2 years of decomposition (loose stockpiling with bottom ventilation), the relocation of the highly contaminated deposit of the aerobic mix (Lys. 10) on Lys. 6 substantially accelerated the biochemical degradation despite the extremely high final compaction, while the contamination of the highly contaminated anaerobic comparison body (Lys. 9, base) corresponded to the acidic initial phase even after 7 years of storage. The biochemical oxygen demand of the other deposits was insignificant but clearly distinguishable in the investigated stabilisation phase of the deposit. It was usually below the detection limit in the leachate discharges of industrially non-contaminated wastes but could be detected in the leached discharges of industrially contaminated wastes.

The sum parameters of organic materials (COD, DOC, org. N) behaved in a similar way to each other. High overall contamination ($DOC > 2000\,mg/l$, $COD > 5000\,mg/l$, org. $N > 30\,mg/l$) was associated with good degradability over the long term ($BOD_5/COD \approx 0.7–0.8$), i.e. the high degradation inhibition in the initial phase was overcome even in the most unfavourable, highly contaminated anaerobic deposit (Lys. 9, base). Conditions for a high residual COD contamination ($COD > 1000\,mg/l$, $BOD_5/COD < 0.1$) known from the methanogenic phase of operating landfills were only measured inside the anaerobic waste bodies ($600 < COD < 2000\,mg/l$, average approximately $1200\,mg/l$; $200 < DOC < 700\,mg/l$, average approx. $400\,mg/l$). Inside the waste bodies the residual COD of the leachates from weak to medium industrially contaminated deposits (Lys. 2, $\frac{1}{2}$h and $\frac{3}{4}$h) was the same as the leachate from an industrially non-contaminated municipal waste deposit with sewage sludge lenses and local stagnation (Lys. 7). The weak to medium industrial contamination therefore retarded stabilisation to the same extent as a temporary leachate stagnation.

With the exception of the highly contaminated anaerobic deposit (Lys. 9, 3rd stage of contamination, base), the parameters of the organic substance in the base discharges correspond to the discharges of an extensively degraded, permanently aerobic landfill (COD < 300 mg/l, DOC < 200 mg/l, org. N < 25 mg/l, $BOD_5 \rightarrow 0$). Nevertheless, the residual DOC (dissolved organic carbon in the case of $BOD_5 \rightarrow 0$) proves the influence of industrial deposits:

- without industrial contamination: residual DOC < 110 mg/l
- with industrial contamination: 100 mg/l $<$ residual DOC < 300 mg/l.

The contamination data from inside the waste body have to be used for the leachate discharges of common landfills and contaminated sites, since the consecutive degradation only occurs under oxygen influence in the base.

Mineral and organic nitrogen compounds

The compounds of mineralised nitrogen (Fig. 5-1(9), (10) and Fig. 5-2(9), (10)) prove the conclusions from the parameters of the organic contamination compared previously: degradation was still ongoing in the anaerobic deposit with high industrial contamination (Lys. 9, base) in the acidic phase 7 years after landfilling started (NH_4-N < 200 mg/l, peaks up to 300 mg/l, NO_3-N $\rightarrow 0$). Those anaerobic deposits with little or no industrial contamination (Lys. 2, $\frac{1}{2}$ h and $\frac{3}{4}$ h, base) in principle did not differ from one another in the waste body and corresponded to the final phase of methanogenesis with temporary aerobic degradation (sum NH_4-N $+ NO_3$-N $=$ approx. 500–1000 mg N/l). The nitrogen emissions, which were roughly twice as high from the anaerobic deposits without industrial contamination, may also have been caused by sewage sludge. The variant with industrial contamination (Lys. 6/10 with $60 < NO_3$-N < 110 mg N/l) from the simultaneous, permanently aerobic sewage sludge waste mixes (NH_4-N $\rightarrow 0$) emitted twice as much nitrate-nitrogen than the non-contaminated variant (Lys. 5 with $10 < NO_3$-N < 50 mg N/l).

Sulphates and carbonates

If the easily degraded substances are extensively degraded, the sulphate (SO_4^{2-}) (Figs 5-1(11) and 5-2(11)) and hydrogen carbonate (HCO_3^-)

minerals (Figs 5-1(12) and 5-2(12)) enable a more exact differentiation of the stability state than the sum of parameters from the organic contamination COD and DOC. All samples of the base discharge of all deposits without industrial contamination dropped below 600 mg/l HCO_3^-. Deposits without a long-lasting acidic phase were always below 350 mg/l HCO_3^-. However, only six samples from the base discharge of deposits with industrial contamination remained below 600 mg/l HCO_3^-, being the maximum contamination of non-contaminated wastes and they reached a maximum contamination of 3000 mg/l HCO_3^- (in peaks 4000 mg/l HCO_3^-). The maximum values inside the waste bodies of deposits with industrial contamination and no hydrated lime conditioning (Lys. 2: 9000 mg/l HCO_3^-) were also measured. But the lowest values were measured in the aerobic sewage sludge waste mix, which contained sewage sludge conditioned with lime at population equivalent mass ratios (Lys. 5: 200 mg/l HCO_3^-).

The amount of sulphate contamination is inversely proportional to that of hydrogen carbonate, i.e. the concentration of hydrogen carbonates decrease and the concentration of sulphates increase with increasing aerobic mineralisation. With a few exceptions, the base discharges of wastes with no industrial contamination, contained at least 200 mg/l and maximum 1000 mg/l sulphate (SO_4^{2-}). Peak values were detected in the leachate of the permanently aerobic sewage sludge waste mix (Lys. 5). For comparison: a maximum of 300 mg/l sulphate was contained in the leachates of the contaminated deposits with four exceptions. Similar conditions prevailed inside the waste bodies but at a lower level.

The carbonates (Figs 5-1(13) and 5-2(13)) show the differences between permanently aerobic sewage sludge waste mixes with industrial contamination (Lys. 6/10) and without (Lys. 5):

- contaminated: usually 10–60 mg/l, in exceptions ~0 mg/l carbonate (CO_3^{2-});
- non-contaminated: usually ~0, in exceptions up to 10 mg/l carbonate.

The higher carbonate content also indicates higher industrial contamination.

The results of the comparison can be summed up in the following statements:

- A high industrial contamination of the municipal solid waste in the composition tested (galvanic sludge, phenol sludge, cyanides, pesticides) may retard the anaerobic biochemical degradation by more than an order of magnitude (proof: an approx. 1.5-m high waste deposit, whose non-contaminated comparison body had reached the stable methanogenic phase in about 1 year (see Lys. 1 in Ehrig, 1986), was still unchanged in the acidic phase 7 years after deposition, and there was yet no sign of transition to the methanogenic phase).

- If the stabilisation of the wastes is substantially accelerated by biochemical degradation – e.g. intentional aerobic degradation – the above industrial contamination may substantially retard or inhibit the degradation processes (high COD at a low BOD_5). The long-term organic residual contamination is markedly increased.
- Moderate industrial contamination of the above composition has no significant influence on the anaerobic or aerobic degradation, if one considers the usual parameters of the organic contamination and the nitrogen compounds as well as the electrochemical parameters for the assessment.
- In the mineralisation phase, clear differences emerge even in the lowest grade of contamination. At the same time, all comparison bodies with industrial contamination were still in the intensive hydrogen carbonate mobilisation phase while the non-contaminated deposits had long passed the phase of hydrogen carbonate mobilisation and entered that of sulphate discharge.
- One can conclude from the usual criteria of leachate contamination that the final phase of mineralisation of the municipal solid waste, both in anaerobic and aerobic deposits, is markedly retarded by industrial wastes (combined effect of cyanide, phenol sludge, galvanic sludge, pesticides) even at small doses, e.g. in the range of tolerance of AbfAblV (2001).
- The negative consequences of industrial contamination on the degradation of natural materials can considerably be reduced but not fully avoided by intentional aerobic degradation of a primarily non-contaminated bottom layer.

It can be clearly seen that the degradation performance of difficult-to-degrade organic substances, which are already considerably high in the bottom zone of non-contaminated waste bodies, are even higher in the contaminated waste bodies than one would expect based on the non-contaminated waste bodies. The permanently aerobic degradation reduces these contaminants by about an order of magnitude at maximum and even then only when the base layer is kept in an obviously aerobic state. After extensive anaerobic stabilisation inside the landfill, only a small amount of oxygen is necessary, which can penetrate by diffusion.

5.5 Influence of recycling on leachate contaminated from natural origins
Peter Spillmann

5.5.1 Objectives of the investigation and selection of landfill types

One would expect from recycling that the mainly separate collection of biowaste reduces the transported masses of organic substances and salts. In the test case the influence of organic substances was increased by including in the investigation a material variant with markedly higher organic content (Lys. B and C) than that of the total waste (Lys. A) and residual waste (Lys. D and E) (see Fig. 2-1). This material variant was produced by extensive recycling of valuable materials and omitting the separation of biowaste. One goal of the investigation was to see if and to what extent recycling can reduce the long-term emissions in comparison to aerobic stabilisation.

5.5.2 Assessment method

It can clearly be seen from the sums of the leachate discharges (Chapter 3, Fig. 3-4) that obvious phase shifts of the discharges between the anaerobic deposit (Lys. A, B and E) and intensive aerobic degradation (Lys. C and D) occurred within the first 2 years after the beginning of deposition in the storage phase (increase and saturation of the very different storage capacities) and the discharge sums of the residual wastes (Lys. D and E) were clearly behind those of the other comparison bodies (anaerobic: discharge $\sum$ Lys. E < discharge $\sum$ Lys. A < discharge $\sum$ Lys. B; aerobic: discharge $\sum$ Lys. D < discharge $\sum$ Lys. C). However, if one compares the discharge sums of the long-term trends after the end of the construction work and saturation of the storage capacities (Chapter 3, Fig. 3-4), the discharge sums of the anaerobic deposits fully coincide, and the considerably reduced sums of the aerobic deposits also fail to show large differences. Due to these hydraulic differences one should consider the distribution and cumulative curves of the transported masses instead of the concentrations of the materials in the assessment of leachate contamination. To characterise the long-term trends – separated from the overall trans-ported masses – the trends of the transported masses should be compared after the end of the construction and storage phases. The characteristic long-term trends refer to deposits of about the same dry waste masses.

126

5.5.3 Selection of parameters

Restriction of the parameters was justified in comparing the investigation of the overall waste (Section 5.3) because the COD sufficiently characterises the entire organic contamination. The Kjeldal nitrogen failed to provide substantial additional information in the investigation on the main plant and the transition from carbonate to sulphate phase could not be expected within the short observation periods chosen. Calcium was used as an indicator to estimate calcium hydrogen carbonate. Iron was added as a second indicator for incrustation risk. The temperature of the waste was also displayed because of the influence of decomposition.

In the case of large contamination differences the curves of the transported masses are represented separately according to landfill technology:

* Lys. A, B, E = anaerobic
* Lys. C, D = aerobic

5.5.4 Illustration and explanation of the results

Waste temperature

The aerobic degradation is characterised by heat emission and anaerobic by the formation of degradation products with retained high calorific values (e.g. methane and organic acids). Having identical waste masses and boundary conditions and methane emissions with negligible energy content, the path and extent of degradation and concomitantly the potential contamination of the leachates can be characterised by the heat emission (see Section 2.8.3.2 in Spillmann and Collins, 1979, and Chapter 11, Fig. 11-16 to prove that energy emission depends on leachate contamination).

Since identical heat-insulated compressible cylinders were selected to model various landfill types, the exterior surfaces were identical, therefore the same heat emission can be deducted from the same medium interior temperatures, so that the heat emissions refer to about the same dry waste masses within the aerobic deposits. After placing other wastes on the aerobic deposits, the deposited dry waste masses were approximately identical for each landfill type, so that it suffices to simply consider the temperature curve to characterise the heat emission.

The temperatures of the anaerobic deposits followed the medium air temperatures. The high initial temperature of the anaerobic residual waste can be traced back to air ingression through rough pores

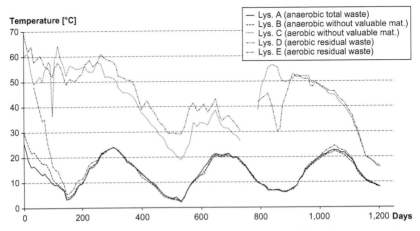

Fig. 5-3 Temperature curves of landfill sectional cores (Lys. A–E) with various recycling and operational influences (anaerobic/aerobic), Wolfsburg landfill

(Fig. 5-3). Anaerobic conditions in the residual waste had to be imposed by applying a cap of loamy soil. This proves that the intended anaerobic degradation in Lys. A, B and E was really achieved. The good agreement in decomposition temperatures is noticeable between municipal wastes after the separation of valuable materials, thus organically enriching the waste (Lys. C) and the remainder after extensive separation of biowaste (Lys. D). The temperature break-down in the residual waste after placing other wastes on Lys. D was due to water deficiency, which was compensated for by artificial irrigation. Both waste types needed 2 years until heat emitted to the air approached zero, both in the basic arrangement and in the expanded version. The identical energy emission per 1 tonne dry waste substance proves that the waste fraction classified as 'organic' in the separation analysis is not characteristic for the aerobically degradable mass.

Alkalinity (pH > 7) – acidity (pH < 7)

Regardless of the extent of recycling, the anaerobic deposits needed about 3 years until stable neutral to weak alkaline conditions developed (Fig. 5-4(1)). As expected, the deposit with the highest content of organic substance needed the longest time to achieve unmistakable alkaline conditions. However, the longest constant weak acidic phase (pH < 7) occurred in the overall waste (Lys. A). Unlike anaerobic land-fill types, aerobic deposits immediately developed an unambiguous alkaline environment once decomposition had begun. Characteristic,

128

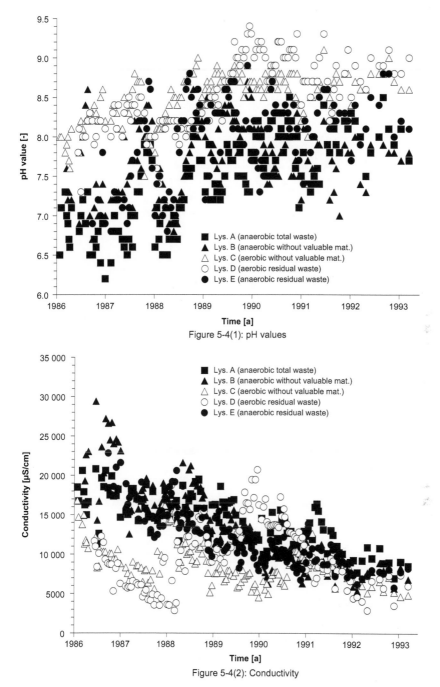

Figure 5-4(1): pH values

Figure 5-4(2): Conductivity

Fig. 5-4 Selected leachate parameters of landfill sectional cores (Lys. A–E) with different recycling and operational influences (anaerobic/aerobic), Wolfsburg landfill

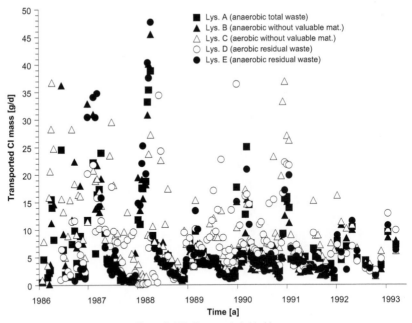

Figure 5-4(3): Transported chloride masses

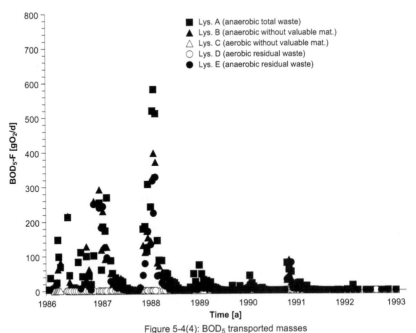

Figure 5-4(4): BOD$_5$ transported masses

Fig. 5-4 Continued

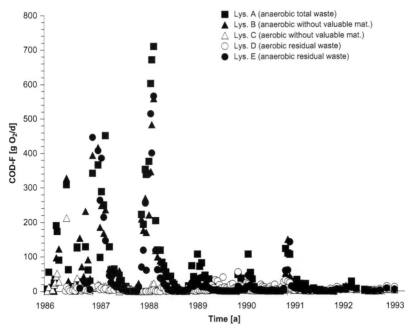

Figure 5-4(5): COD transported masses

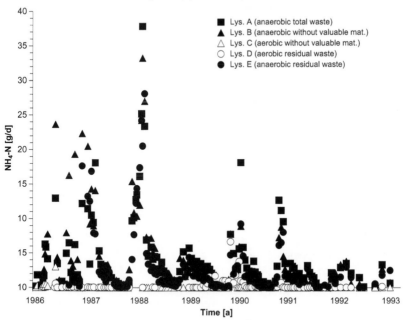

Figure 5-4(6): Transported NH₄⁺-nitrogen mass

Fig. 5-4 Continued

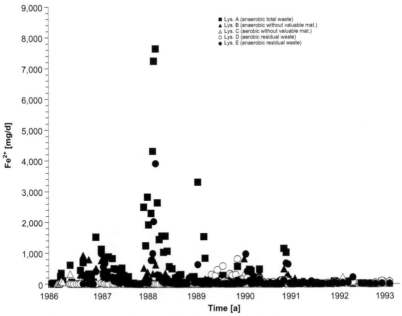

Figure 5-4(7): Transported dissolved iron mass

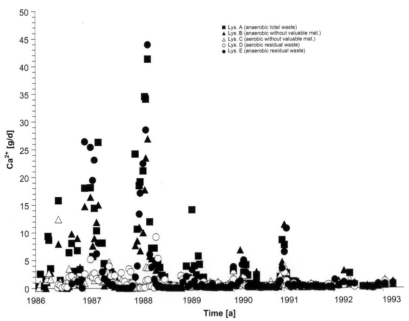

Figure 5-4(8): Transported calcium mass

Fig. 5-4 Continued

clear differences were detected between aerobic and anaerobic deposits regardless of recycling.

Conductivity

Conductivity indicates the presence of both salts and organic acids (Fig. 5-4(2)). A characteristic difference can be proved almost everywhere between anaerobic deposits (Lys. A, B and E) and aerobic waste (Lys. C and D). Placing more waste on the decomposition lysimeter (Lys. C and D) after approx. 650 days noticeably increased conductivity in the leachate of the decomposition lysimeter within as little as 200 days. Under anaerobic deposition conditions, even extensive separation of biowaste (Lys. E) failed to reduce the conductivity of leachates. Under aerobic conditions (Lys. C and D), the residual waste leachates (Lys. D) corresponded temporarily to the leachates of anaerobic deposits despite the removal of the compostable fraction, the 'biowaste'. Recycling failed to bring about a reduction in conductivity.

Chloride content

Chloride is a reliable indicator of the degradation of organic substance, since, as an element, it cannot be degraded and usually its adsorption remains at very limited levels. The curves in Fig. 5-4(3) prove that the fluctuations of chloride concentration in the anaerobic and aerobic deposits had similar profiles and the aerobic deposits differed from the anaerobic ones by phase shifts and additional emission peaks (e.g. 1300 to 1500 days). Accordingly, the aerobic deposits exhibited the greatest sums of the transported chloride mass (Fig. 5-5(1)). Within the same deposit the wastes supplied the largest transported chloride mass after separation of valuable materials due to the relatively enriched organic fraction. However, if only the longer-term trends are considered and the first 1200 days are disregarded (different masses and different storage processes (cf. Chapter 3, Fig. 3-4) per test body in the initial phase), then identical transported chloride masses are obtained for all anaerobic deposits and an approximately 1.5-fold increase of the transported chloride mass by the ongoing aerobic degradation. The effect of the biowaste was only a short-term increase of the transported chloride mass. Ongoing trends were not affected by recycling.

Organic and related mineral contamination BOD_5, COD, NH_4-N, Fe, Ca

The range of the biochemical and chemical oxygen demand (BOD_5 (Fig. 5-4(4)) and COD (Fig. 5-4(5)) and of ammonium (Fig. 5-4(6)) depend directly on each other. The iron (Fe) (Fig. 5-4(7)) and calcium

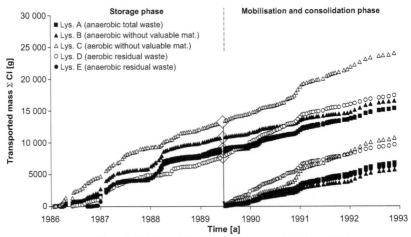

Figure 5-5(1): Cumulative transported mass of chloride content

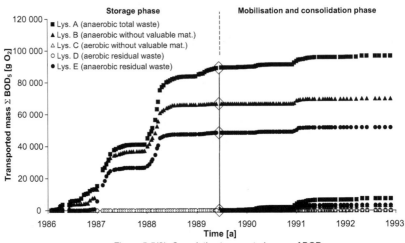

Figure 5-5(2): Cumulative transported mass of BOD$_5$

Fig. 5-5 Cumulative transported mass of selected leachate parameters of landfill sectional cores (Lys. A–E) with various recycling and operational influences (anaerobic/aerobic), Wolfsburg landfill

(Ca) minerals (Fig. 5-4(8)) are directly affected by the degradation processes. Therefore these parameters must be considered simultaneously.

The curves of the five transported masses of leachate contamination from anaerobic deposits compared here run synchronously: peaks in transported mass in the winter half-year (= high run-off per unit area, moderate oxygen influence in the base) and minima in transported mass in the summer half-year (low run-off per unit area, trickling filter

134

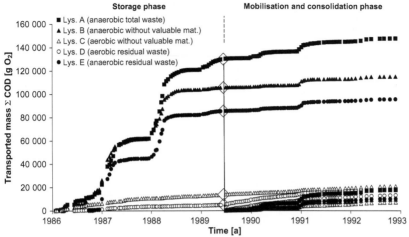

Figure 5-5(3): Cumulative transported mass of COD

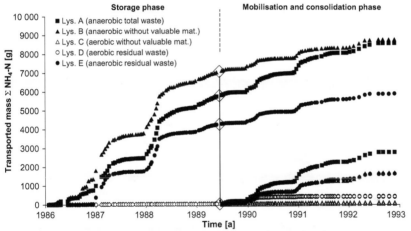

Figure 5-5(4): Cumulative transported mass of ammonium-nitrogen

Fig. 5-5 Continued

effect in the drainage gravel). The peaks in transported mass coincide with the minima of the pH values pH < 7 (except basic ammonium). In addition, the conditions from $BOD_5/COD >$ approximately 0.7 during the transported mass peaks indicate a good degradability of the organic contamination.

The maxima of the transported masses of all five parameters were measured with the exception of an individual calcium peak in the leachate of the total waste (Lys. A). The peaks of the transported ammonium mass, being indicative of an initial anaerobic stabilisation

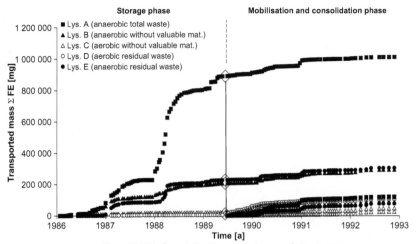

Figure 5-5(5): Cumulative transported mass of dissolved iron

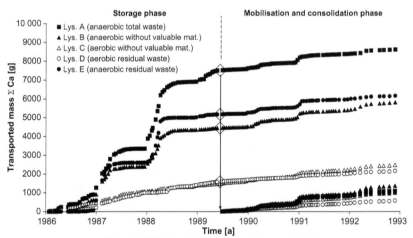

Figure 5-5(6): Cumulative transported mass of calcium

Fig. 5-5 Continued

in the sense of the degradation of organic acids in the waste body, emerged first but not to its full extent, in the waste with valuable material separation (= relative enrichment with organic substance, Lys. B). The peak of transported iron mass was measured, as expected, in the leachate of the total waste (Lys. A), while that of calcium in the discharge of the residual waste (Lys. E). After a steady increase in pH of all leachates to pH > 7 (clearly alkaline), the transported iron and calcium masses of all leachates also decreased to insignificant concentrations regardless of the recycling ratio.

As expected, the contamination parameters of the leachates from the aerobic deposit fell below those of anaerobic deposits to such an extent that they can hardly be recognised in the overall display of the curves of the transported masses. Only the increase in chemical oxygen demand, and the minerals in the leachate of the residual waste which contained weak chemical contamination and was compacted with an earth capping (Lys. D), are clearly detectable.

The curves of the transported masses allow the following conclusions to be drawn:

- Suitability of the test equipment: the test equipment enabled the construction of anaerobic waste bodies maintaining an acidic phase for several years despite the small dimensions.
- Influence of recycling on leachate contamination: landfill technology is a decisive factor which influences the transported masses of leachate contamination and not the recycling ratio.

Sums of transported Cl, BOD$_5$, COD, NH$_4$-N, NH$_4$, Fe and Ca masses
The cumulative transported masses provide a more exact proof for various influences, including the distinction of short-term landfill reactions from long-term trends, than the concentration curves (Fig. 5-5(1)–(6)).

Chloride content
The transported chloride mass (Fig. 5-5(1)) is a reliable indicator for the degradation of fresh organic substances. Having identical organic content and identical degradation, the transported chloride masses are identical. In the case of rapid aerobic degradation the recycling of organic substances (Lys. D) halves the transported chloride mass in this test in comparison to the waste with a commercial recycling of valuable materials (glass, sheet metal, plastic, composite materials, paper) (Lys. C). Over the long term, their transported chloride masses are about the same with a clear, increasing distance of the transported masses of the anaerobic deposits. An influence of recycling on anaerobic deposits cannot be detected, i.e. the preservation processes have a greater influence on the transported chloride mass in the anaerobic deposits than the large differences of the content of organic wastes between total waste (Lys. A), commercial recycling of valuable materials (Lys. B) and extensive separation of organic substance (Lys. E) do.

As an indicator for easily dissolved environmental pollutants, these transported chloride masses prove that easily dissolved pollutants are

137

stored both in the deposited anaerobic residual waste and in the total waste in such a way that they can be re-activated even after saturation of the storage capacity.

Easy-to-degrade organic substance

The cumulative transported masses of easily degradable organic substances, characterised by BOD_5 (Fig. 5-5(2)), were determined by landfill technology alone. Regardless of the content of organic wastes, the cumulative transported masses from aerobic deposits were so low that they were difficult to display at the same scale together with those from anaerobic deposits.

The total waste from anaerobic deposits (Lys. A) emitted more degradable organic material than the landfill type with an increased content of organic waste (Lys. B). As expected, only the residual waste (Lys. E) emitted the smallest transported mass of degradable organic materials. However, if one considers the long-term trends alone, the emissions of wastes enriched in organic substances (Lys. B) coincide with those of the residual waste (Lys. E). The total waste emitted a larger amount of transported mass. It can be concluded from these measurements: when the emissions of organic substances per 1 tonne of dry waste substance from wastes enriched in organic substances (BOD_5) are higher than those from the total waste, the selection of 'biowaste' cannot possibly have had any influence on the emission of easily degradable substances over the leachate path. If one compares the discharge of these transported masses with the curves of the pH values, then a direct relationship is obtained between degradation phases in the range $pH \leq 7$ and the emissions of easily degradable organic substances (BOD_5). From these considerations the following statements can be derived as conclusions:

- The separation of organic wastes did not affect the easily degradable substances in the leachate under aerobic conditions.
- In anaerobic deposits the poor compaction of the residual waste extended the initial aerobic phase. This influence reduced the acidic phase of the residual waste (Lys. E) and indirectly reduced the emissions. Stability conditions of the organic substances and not their mass are decisive for the long-term organic emissions in terms of humic substances.

Difficult-to-degrade organic substances

Extremely low long-term transported masses of the degradable organic substance BOD_5 (Fig. 5-5(2)) with short peaks in times of high discharges were emitted from the base of anaerobic deposits. It can be concluded from this that the slight, unavoidable oxygen influence with progressive stabilisation of the waste was sufficient in the lower region of the waste and in the base of the total waste (Section 5.2) to extensively degrade the easily degradable materials. This process takes place on the large scale in purification plants or in porous aquifers near the surface (cf. Spillmann *et al.*, 1995). Therefore, the COD measured here (Fig. 5-5(3)) refers to the transported masses of substances particularly difficult to degrade biologically. This is the group of organic substances decisive for long-term environmental impact. It corresponds to the residual COD after a chiefly aerobic degradation in the aerobic waste body or in an aerobic porous aquifer.

If one compares the sums of the transported masses of all organic materials in the initial phase, a similar picture emerges for the easily degradable organic substances:

- Clear reduction of the transported masses by aerobic degradation.
- Reduction of the transported masses by shortening the acidic phase.
- No direct effect of recycling.

The long-term trend of the residual COD is approximately the same in all deposits – regardless of recycling and deposition conditions. The temporary increase in the transported masses of the residual COD from the rotted residual waste coincides temporally with the pollutant spiking. It can be assumed that this caused the inhibition of degradation, as proved for the total waste (Section 5.3), and this along with other parameters, must be investigated.

Ammonium-nitrogen

High ammonium transported masses are emitted in the leachate of anaerobic deposits, when organic acids are further degraded to methane and carbon dioxide (transition of the acidic phase to the methanogenic phase). In the case of homogeneous waste composition and identical anaerobic degradation down to methane, the cumulative transported masses of the ammonium-nitrogen are about the same (Fig. 5-5(4)). The temporal process is primarily determined by the length of the acidic phase. High ammonium transported masses are only emitted in the initial phase in the leachate of overwhelmingly aerobic deposits, as long as the oxygen supply is sufficient for the nitrification of the ammonium-nitrogen (usually $BOD_5 < 50\,mg\ O_2/l$). Disturbances in degradation within the aerobic range and/or oxygen shortage can be recognised by an increase in ammonium-nitrogen.

139

The entire transported masses of ammonium-nitrogen between the anaerobic total waste (Lys. A) and the anaerobic organically enriched waste (Lys. B) coincide. The slow increase of the transported masses of the total waste corresponds to the longer acidic phase of the total waste. The cumulative transported mass of the anaerobic residual waste (Lys. E) is about 30% lower. A retardation due to a longer acidic phase cannot be seen and the long-term trends between waste with organic enrichment and residual waste are in agreement. A subsequent compensation of the cumulative transported masses between residual waste and other wastes is not expected.

As long as nitrogen losses did not occur at the base of the residual waste (nitrification and denitrification due to changes in oxygen influence), extensive recycling managed to reduce the transported masses of ammonium-nitrogen by about 30%. This difference is a clearly measurable advantage. However, this reduction is negligible in comparison to the nitrification performance of more than 90% of ammonium-nitrogen in an aerobic deposit.

Dissolved iron

The dissolved iron is one of the key elements, which becomes precipitated under anaerobic conditions in the case of a pH jump from an acidic to alkaline range or under aerobic conditions due to oxidation and cause incrustation in the drainage system. Based on this behaviour, dissolved iron is also an indicator of biochemical activities.

A comparison of the cumulative transported masses (Fig. 5-5(5)) from the entire test period suggests that the separation of the valuable material 'iron-containing metals' can reduce the cumulative transported masses by about 80% under anaerobic landfill conditions. Under aerobic conditions only, disturbances in oxidation can produce clearly measurable transported masses. However, if one considers the long-term trends, a reduction of the transported masses due to separation cannot be observed. The emission of dissolved iron proceeds proportionally to the pH value produced by the acidic phase of the deposit. It depends primary on the pH value (Fig. 5-4(1)), not on the separation.

Dissolved calcium

The bulk of calcium emissions is discharged as hydrogen carbonate during the acidic phase. If the organic acids are degraded by biochemical activity, a jump in pH occurs at the bacteria, which causes a local precipitation (Ramke and Brune, 1990). The effect of calcium precipitation can cause incrustation leading to a complete clogging in the landfill drainage layer. Oxidative processes do not affect carbonate precipitation. At the same content of the deposit, the calcium content of the leachate is an indicator for pH fluctuations during degradation caused by disturbances in acid degradation (see Fig. 5-4(1) and (8)).

A comparison of the cumulative transported masses over the entire test period indicates that the separation of valuable materials alone noticeably reduced the calcium emission within the anaerobic range (Fig. 5-5(6)). However, if one compares the increase in the transported masses with the change in the pH values in a differentiated way, it can be seen that the key increase in transported masses coincide with the reductions in pH at the end of the winter terms (Fig. 5-4(1)). The calcium emissions in the clear basic range are low from all deposits regardless of the oxygen influence (parallel slow increase of all cumulative curves). This relationship is particularly easy to recognise over the long term. A reduction of calcium emissions by recycling cannot be observed.

Effect of contaminant spiking in the residual waste (Lys. D)

An increase in the transported calcium masses from Lys. D immediately after pollutant spiking cannot be observed. Since the oxidation-dependent contamination such as difficult-to-degrade substances, soluble iron and ammonium were increased by spiking but the pH-dependent calcium remained constant, cyanide – initial material for HCN at low pH values (respirative poison) – most probably temporarily disturbed the degradation by inhibiting the respiration activities. The anaerobic degradation of organic acids was not measurably disturbed by the altogether moderate chemical spiking.

Comprehensive assessment of the individual results

The comparative parameters used here (pH, conductivity, Cl^-, BOD_5, COD, NH_4^+-N, Fe^{2+}, Ca^{2+}) are considerably influenced by the biochemical degradation processes, with the exception of the residual COD in the case of base aeration. Relating to 1 tonne of dry waste substance, no clearly measurable differences were seen between municipal total waste and residual waste from a comparable collection area either under anaerobic or aerobic landfill conditions. This was the case when a careful separate collection of the primarily compostable fraction and an intensive sorting of industrial valuable materials enabled to be recycled 60% by weight of the dry waste mass.

Concerning emissions on the flow path, the advantage of an extensive recycling including biogenic materials is that the emissions per inhabitant decrease to approximately the same extent as the mass of the residual waste per inhabitant. Recycling does not substitute material stabilisation of the residual wastes to be deposited, however, as an effective additive action – per inhabitant – primarily reduces the emissions of easily soluble salts and the extremely persistent soluble humic-type substances, characterised by the residual COD.

The disadvantage of poor compactibility of the residual waste can be used as a demonstrable biochemical advantage. If the area of the landfill is kept large, despite a reduction of the waste masses, atmospheric oxygen can affect the upper residual waste layer through the coarse cavities, which, together with a slower building process, reduces the acidic phase of the degradation process and the high initial emissions. The medium- and long-term emissions per 1 tonne dry waste substance are, however, not reduced.

If the tipping surface is reduced as recycling progresses and residual wastes delivered in a dry state are deposited without pretreatment, the characteristic emissions over the flow path change over time towards saturation of the additional storage capacity as compared to the total waste. The cumulative transported mass per 1 tonne dry waste substance cannot be reduced without targeted biochemical stabilisation.

6

Transportation of industrial contamination in the flow path

6.1 Typical residues of industrial production
Hans-Hermann Rump, Wilhelm Schneider, Heinz Gorbauch,
Key Herklotz and Peter Spillmann

6.1.1 Deposits in the total waste

6.1.1.1 Form of illustration

Cyanide, galvanic sludge and phenol sludge were investigated as typical contaminants from industrial residues which had been co-deposited with municipal solid waste before the introduction of hazardous waste landfills (see Chapter 2 for details of intentional contamination). The concentrations in the leachates exhibited large fluctuations in the level of contamination over the short term and considerable differences over the long term. In order to make the effect of landfill technology on the discharge of chemicals clearly visible, the 'curve envelope' of the measured values is illustrated, i.e. the analysis values were not connected in the order of the measurement. Instead, in each case maxima and minima were shown as the upper and lower limit of an area which contained all measured values. The fluctuations of this area illustration were compared to the different phases of leachate generation. In a second step, analysis results of mobile substances (e.g. nickel, cyanide), including all single fluctuations, are represented and compared to the short-term fluctuations of leachate discharges in order to determine whether the fluctuations of the discharge volumes cause the fluctuations in the concentrations.

6.1.1.2 Transportation of non-degradable elements

The environmental impact of heavy metals can be avoided if these elements are discharged only at a level in which they occur in nature. The galvanic sludge investigated mainly contained nickel, a particularly mobile heavy metal. Chromium contamination was also significant. Zinc is also displayed for validation of the measured trend. This vitally necessary heavy metal is less important from the toxicology point of view. The measured values in Figs 6-1–6-3 should be considered in connection with storage, mobilisation and consolidation processes of the water regime (see Fig. 3-3).

Leachate from the highly contaminated aerobic sewage sludge waste mix (Lys. 6/10, $\frac{1}{1}$ h, cf. Fig. 2-7) has already become grossly contaminated in the decomposition process. The extreme nickel concentrations decreased with increasing consolidation (pore water is squeezed out by pressure until the pore water pressure approaches zero). In the equilibrium phase, the concentrations are low and insignificant in comparison

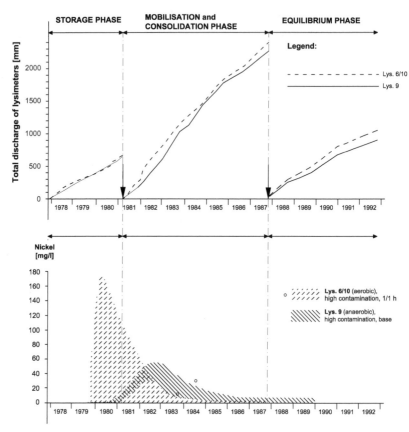

Fig. 6-1 Relationship between water regime, degradation processes and leachate contamination from industrial deposits with high nickel contamination; comparison between anaerobic operation (Lys. 9) and a permanently aerobic deposit (Lys. 6/10); Braunschweig-Watenbüttel landfill

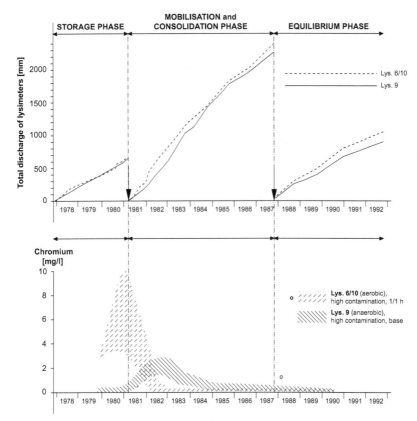

Fig. 6-2 Relationship between water regime, degradation processes and leachate contamination from industrial deposits with high chromium contamination; comparison between anaerobic operation (Lys. 9) and a permanently aerobic deposit (Lys. 6/10); Braunschweig-Watenbüttel landfill

to landfill contents (Fig. 6-1). Nickel concentrations in the leachate of the highly contaminated anaerobic deposit (Lys. 9) are characterised by a steep rise in the 'mobilisation phase' of the waste body (considerably more leachate leaves than corresponds to the climatic water balance) (Fig. 6-1). The high levels of contamination occur only after exhaustion of the storage capacity and the beginning of intensive degradation in the mobilisation phase (water-saturated storage of organic substance is degraded). The contamination peak only achieves 40% of that of the aerobic mix since the galvanic sludge was anaerobically deposited in 'lenses'; nevertheless it lasted substantially longer. The contamination remained high even after 9 years.

Chromium (Fig. 6-2) is contained in the leachate of both deposits at a considerably lower concentration than nickel (Fig. 6-1). The

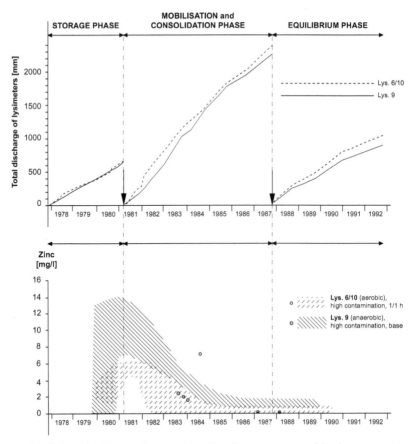

Fig. 6-3 Relationship between water regime, degradation processes and leachate contamination from industrial deposits with high zinc contamination; comparison between anaerobic operation (Lys. 9) and a permanently aerobic deposit (Lys. 6/10); Braunschweig-Watenbüttel landfill

characteristic of the contamination sequence is, however, still more clearly pronounced: after high initial peaks, the emissions from the highly contaminated rotted material decrease after 3–4 years to insignificant concentrations toward the end of the primary consolidation, while the anaerobic deposit releases concentrations >1 mg/l in a pronounced mobilisation phase (re-increase of contamination) over a period as long as 6 years. The zinc contamination curve (Fig. 6-3) confirms the measurements for nickel and chromium. Mobilisation is even more pronounced and the difference of long-term emissions more obvious: only the emission from the aerobic stabilised waste body is insignificant over the long term.

146

6.1.1.3 Emission of potentially biochemically degradable industrial contamination

Cyanide originating from hardening salts is also produced and degraded in nature. It was therefore possible that biochemical degradation also decomposed this waste. As far as the environmental impact is concerned, easy-to-relocate cyanide, whose behaviour (Fig. 6-4) in the total waste corresponds to that of heavy metals, is decisive. However, it was mostly available in a complexed form. Initial and consolidation peaks from aerobic waste roughly occurred during the first 2.5 years. Later, cyanide contamination was negligible except for a few clearly measurable individual values. Emissions from the anaerobic

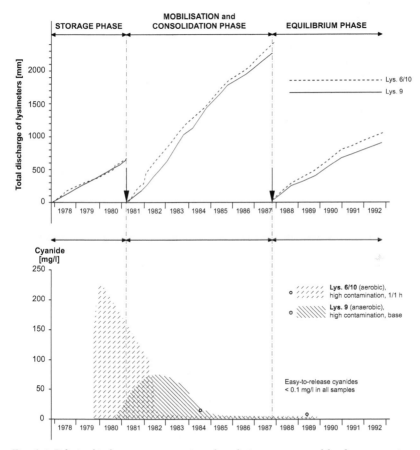

Fig. 6-4 Relationship between water regime, degradation processes and leachate contamination from industrial deposits with high cyanide contamination; comparison between anaerobic operation (Lys. 9) and a permanently aerobic deposit (Lys. 6/10); Braunschweig-Watenbüttel landfill

body only started decreasing after 5 years, around the end of the mobilisation phase, and became insignificant after 7 years. Individual easily measurable contamination also occurred here, so that a marked contamination potential remained up to the end of the test.

Phenol is also produced as an anaerobic degradation product in nature and can be degraded in an aerobic microbial way. The aerobic body (Lys. 6/10) exhibited a short initial peak (Fig. 6-5) and, as expected, degraded the contamination. The highly contaminated anaerobic waste body (Lys. 9) emitted increasing phenol concentrations with a large volatile phenol fraction, from the beginning of the mobilisation phase until the end of the observation time. A decrease within the 'stabilisation phase' was not detected. The minimum values exhibited a rising trend.

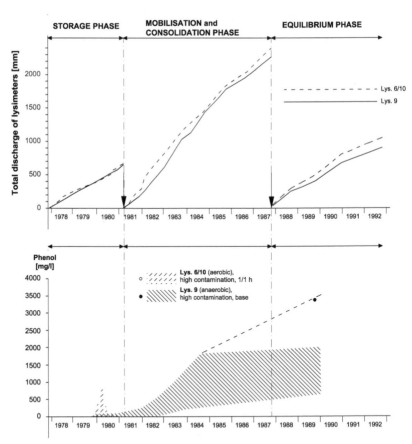

Fig. 6-5 Relationship between water regime, degradation processes and leachate contamination from industrial deposits with high phenol contamination; comparison between anaerobic operation (Lys. 9) and a permanently aerobic deposit (Lys. 6/10); Braunschweig-Watenbüttel landfill

6.1.1.4 Checking of hydraulic influences on emissions

It can be concluded from the agreement between emissions and high leachate discharges that the long-term intensity of the flow affects the emissions considerably. The brief concentration jumps allow the assumption that there is a direct hydraulic influence, since the discharge can vary by more than an order of magnitude. The discharge hydrograph from the anaerobic waste body was therefore compared to the pronounced hydrographs of the obviously hydraulically influenced heavy metal and cyanide contamination (Figs 6-6 and 6-7). When discharge and concentration are compared, it must be understood that each sampling for analysis purposes was combined with a discharge measurement. If contamination increases due to an increase in discharge, it is therefore easy to detect.

The same tendency, but no synchronous events, can be observed for both heavy metals and cyanide within the mobilisation phase. This applies both to the relationship between discharge and contamination of the individual parameters and to the individual contamination parameters among each other. However, the influence of stabilisation by biochemical degradation could be clearly detected: the discharge level remained constant over a 1200-day observation period. But contamination, particularly cyanide, continued to decrease. Following an initial peak of 30 mg/l, easy-to-release cyanide was very low at 1–3 mg/l after 600 days and negligible after 1000 days. However, discharge rarely

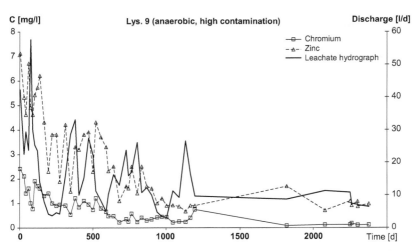

Fig. 6-6 Comparison of individual fluctuations of chromium and zinc with leachate discharge hydrograph of a chemically highly contaminated anaerobic deposit (Lys. 9); Braunschweig-Watenbüttel landfill

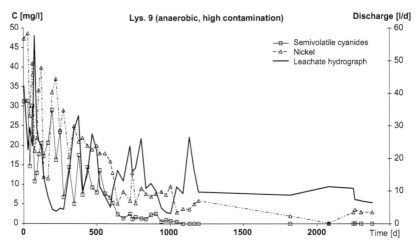

Fig. 6-7 Comparison of individual fluctuations of cyanide und nickel with leachate discharge hydrograph of a chemically highly contaminated anaerobic deposit (Lys. 6/10, 1/1h); Braunschweig-Watenbüttel landfill

decreased to values below 10 l/d. Hence, it follows that, although a high leachate flow rate removes large amounts of mobile substances, the extent of mobility essentially depends on biochemical stabilisation.

The long-term emission trends are displayed in Table 6-1 in accordance with the criteria of the Waste Water into Waters – Waste Water Ordinance (Direkteinleiterverordnung, German Wastewater Charges Act, AbwV 2002, Annex 51). This comparison particularly clearly shows that the intentionally spiked industrial waste only significantly exceeded the limiting values of direct waste water discharge within the leachate of the highly contaminated anaerobic deposit and only for nickel: nickel contamination $Ni \leq 3$ mg/l >nickel limiting value $Ni = 1.0$ mg/l. A high phenol emission of this deposit – not available as a single criterion for discharge – however, caused the cumulative parameters of the organic contamination to greatly exceed the limiting values. Ammonium nitrogen (NH_4-N) and organically bound chlorine (AOX) also considerably exceeded the discharge criteria. A tendency to adhere to the discharge criteria was not detected in the leachate of the highly contaminated anaerobic deposit.

The heavily contaminated stage of the highly compacted, permanently aerobic deposited sewage sludge waste mix (Lys. 6/10, 1/1h) fulfils all conditions for direct discharge concerning spiked industrial wastes, also those for particularly mobile nickel. The aerobic degradable phenol can hardly be detected in the leachate. In contrast to the highly contaminated

150

Table 6-1 Long-term trends of emissions of highly contaminated deposits (Lys. 9 and 6/10) compared with the respective limiting values of direct discharge (AbwV 2002)

Parameter	Dimension	Lys. 6/10 permanently aerobic deposit (Non-contaminated base)	Lys. 6/10 permanently aerobic deposit (High contamination, 1/1h)	Lys. 9 permanently anaerobic deposit (Highly contaminated base)	Limiting value for direct discharge (AbwV, Annex 51, 2002)
pH	–	8.3–8.7	8.2–9.1	6.2–7.3	–
Conductivity	μS/cm	3500–7500	3500–14 000	1000–5000	–
TOC	mg/l	<200	<1300	<4000	–
BOD_5	mg/l	~0	<500	<7500	20
COD	mg/l	<500	<3000	<12 000	200
Phenol index	mg/l	<0.2	<0.25	<3500	–
As	mg/l	<0.02	<0.02	<0.02	0.1
Pb	mg/l	<0.1	<0.2	<0.1	0.5
Cd	mg/l	<0.001	<0.001	<0.005	0.1
Total Cr	mg/l	≪0.1	<0.2	<0.15	0.5
Cr VI	mg/l	<0.1	n.d.[a]	n.d.	0.1
Cu	mg/l	<0.1	<0.15	<0.1	0.5
Ni	mg/l	<0.3	<0.5	<3	1.0
Hg	mg/l	<0.01	<0.01	<0.01	0.05
Zn	mg/l	<0.05	<0.05	<1	2.0
NH_4-N	mg/l	<10	<50	<150	70 N_{total}
Total CN	mg/l	<2.5	<3	<1	–
Easy releas. CN	mg/l	<0.03	<0.1	<0.04	0.2
AOX	mg/l	<1	<1.5	<2.5	0.5

(a) not determined

anaerobic deposit, aerobically deposited phenol does not increase the organic contamination. Nevertheless, organic contamination and its degradation products greatly exceed the limiting values for direct discharge. The indirectly influenced organic contamination (TOC, COD, BOD_5) only approaches the orders of magnitude of discharge limiting values after the passage through the two low-contamination stages and the non-directly contaminated base. Ammonium nitrogen (NH_4-N) and organically bound chlorine (AOX) approximately meet the permitted limiting values (Lys. 6/10, uncontaminated base). It can be concluded from the trend comparison of the tested contamination that only the mainly biologically stabilised material can approximately meet the discharge criteria over the long term. Therefore, stability of the permanently aerobic deposit was tested after approx. 17 years deposition in all contamination stages according to the current assessment criteria (AbfAblV 2001) (Table 6-2).

Table 6-2 Solid analyses of the deposited material after 18 years permanent aerobic deposition (Lys. 6/10) compared with the deposition parameter of TASi (1993) and AbfAblV (2001)

Parameter	Solid dimension	Lys. 6/10 Solid parameters of the deposited material after 18 years permanent aerobic deposition				Lys. 6/10 Eluate values after about 17 years deposition under permanent aerobic conditions [mg/l]				Eluate limiting values [mg/l] according to AbfAblV (2001)		
										Annex 1		Annex 2
		High contam. ¼h	Medium contam. ¾h	Low contam. ½h	Non-contam. base	High contam. ¼h	Medium contam. ¾h	Low contam. 4/4h	Non-contam. base	Landfill Class 1	Landfill Class 2	MBT
Ign. loss	% by wt	21	28	28	24					≤3	≤5	–
TOC	% by wt		Not determined			No eluate criteria				≤1	≤3	–(c)
EOX	% by wt	4.7	6.0	9.2	4.6					<0.4	<0.8	<0.8
COD solid(a)	mg/kg											
pH	–					7.8–8.3	8.1–8.3	7.9–8.2	7.8–7.9	5.5–13	5.5–13	5.5–13
Conduct.	µS					8600	20 700	17 900	17 900	≤10 000	≤50 000	≤50 000
TOC	mg/kg	53	50	90	53	168	156	98	82	≤20	≤100	250
Phenol index	mg/kg					<0.02	<0.02	<0.02	<0.02	≤0.2	≤50	50
As	mg/kg	2.7	4.4	6.0	4.8	<0.025	<0.025	<0.025	<0.025	≤0.2	≤0.5	0.5
Pb	mg/kg	329	516	809	262	0.007	<0.005	<0.005	<0.005	≤0.2	≤1	1
Cd	mg/kg	<10	<10	<10	<10	0.0011	0.0008	0.00011	<0.005	≤0.05	≤0.1	0.1
Total Cr	mg/kg	710	350	780	390		Not determined		0.0008	No criterion		
Cr VI	mg/kg		Not determined			<0.025	<0.025	<0.025	<0.025	≤0.05	≤0.1	0.1
Cu	mg/kg	452	201	223	177	1.2	0.2	0.2	0.1	≤0.05	≤0.1	0.1
Ni	mg/kg	2385	1330	1630	219	2.6	0.8	0.5	0.1	≤1	≤5	5
Hg	mg/kg		Not determined, since insignificant here				Not determined, since insignificant here			≤0.2	≤1	1
Zn	mg/kg	889	1224	1225	1088	<0.1	0.4	<0.1	<0.1	≤0.005	≤0.02	0.02
										≤2	≤5	5

Parameter	Unit	Solids				Eluate				Limiting value	Limiting value	Value
F	mg/kg	Not determined in solid, since contamination insignificant (cf. elements)				<2	<2	<2	<2	≤2	≤5	5
Cl	mg/kg					66	138	113	136	(≤500[b])	(≤500[b])	–
SO_4	mg/kg					79	335	317	253	(≤500[b])	(≤1400[b])	–
NH_4-N	mg/kg					5.4	18.6	19.6	24.5	≤4	≤200	200
NO_3	mg/kg					17.0	22.5	21.5	7.8	No limiting value		
NO_3-N	mg/kg					3.8	5.1	4.9	1.8			
Easy-to-release CN	mg/kg					0.27	<0.02	<0.02	<0.02	≤0.1	≤0.5	0.5
AOX	mg/kg	740	320	480	940	0.5	0.6	0.8	0.1	≤0.3	≤1.5	1.5
Evapor. residue	% by wt	No solids criterion				0.8	1.6	1.4	1.2	≤3	≤6	6
RA_4(d)	mg O_2/g	Not determined								–	–	<5
GG_{21}[8]	litre CH_4/kg	Not determined								–	–	<10
H_0	kJ/kg	Not determined								–	–	<6000

(a) Solids COD organic substance oxidisable with potassium dichromate, calculated as carbon fraction org. C % by weight DS

(b) Limiting value TASi draft August 1992

(c) No limiting value when lower calorific value H_0 is adhered to

(d) RA_4 or alternatively GG_{21} are adhered to

*GG_{21}: gas generation in 21 days

153

To characterise the stability of the organic substance, similar to the organic leachate contamination, the Institute Fresenius determined the mass fraction oxidised with potassium dichromate and compared it with ignition loss. Only about 5–9% by weight of DS waste of oxidised fractions analogous to COD can be compared with the high ignition losses of between 20 and 30% by weight of DS waste. This explains why the low-contamination stage meets the eluate limiting value according to TASi 1993 for soluble organic substance (TOC < 100 mg/l), all stages meet the limiting value for biological treatment and all stages fall far below the limiting value for ammonium nitrogen (NH_4-N). Easily soluble chlorides and sulphates released by degradation even fall below the eluate limiting value of landfill class I, TASi (TI Municipal Waste: Third General Administrative Provision to the Waste Avoidance and Waste Management Act: Technical Instructions on Recycling, Treatment and Storage of Municipal Waste of 14.5.1993) draft 92. The spiked phenol was also degraded in the extremely high dosage of the high contamination stage except for insignificant residues and the fraction of free cyanide was negligibly low despite a high dosage, even in the highly contaminated stage. The organically bound chlorine (AOX) parameter proves substantial contamination in the solid. However, it cannot be attributed to the spiking zones, i.e. the spiked chlorinated hydrocarbon compound lindane is not characteristic for the total contamination (maximum AOX = 940 mg/kg DS waste in the non-contaminated base). The eluates, however, closely adhere to the limiting value of landfill class I. The stability criteria according to Table 6-2 overwhelmingly prove that the material is biologically stabilised. In this basic mass, the extremely high spiked amounts of toxic elements are proved in the selected contamination stages.

The easily destabilised nickel (Ni) with a maximum 2.4 g/kg DS waste is very strongly represented from among the other elements. The elution behaviour of this toxic element with high mobility potentially provides information on the retardation of the toxic elements, which are considerably influenced by the stability of the organic substance (Frimmel and Weis, 1995). They transport heavy metals from the landfill in a soluble form (Spillmann, 1993a); i.e. they only bind metals in a biologically stable, insoluble form. Despite an extremely high initial contamination, the material of the highly contaminated stage only exceeded the limiting value of landfill class I for copper (Cu) and the limiting value of landfill class II for nickel. All other eluates met the limiting values of landfill class I despite an extremely high solid contamination. If one considers the fact that the spiked industrial wastes were distributed in the waste mix and were thus entirely exposed to the leachate flow and the elution process, the long-term reduction of leachate contamination concerning the elements can be explained by a better retention after stabilisation. Phenol was extensively degraded and cyanide disintegrated and/or complexed.

Summing up, it can be said that an extensive biological stabilisation substantially reduces the long-term emissions by degradation, complexation or simply by improved adsorption. This contamination however substantially increases the long-term organic emissions (residual COD) due to unfavourable influences on biological degradation in comparison to uncontaminated wastes.

6.1.2 Effect of recycling on mobilisation and transportation

It was particularly easy to prove the transportation of cyanide from the industrial residues tested in the total waste. A direct hydraulic relationship between cyanide contamination of the leachate and the discharge volume could be proved particularly easily in decomposing residual waste (Lys. D) (Fig. 6-8). Cyanide concentration and discharge volume are directly proportional to each other (more water causes higher contamination per litre). The hydrographs run – with a small phase shift – almost synchronously. It can be recognised from the cumulative curves of the transported masses (Fig. 6-9) that the

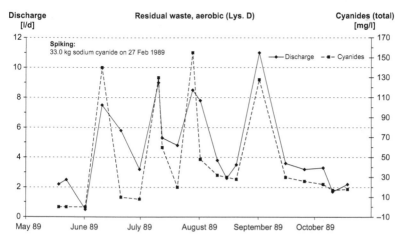

Fig. 6-8 *Relationship between leachate discharge and cyanide transportation of an aerobic residual waste (Lys. D, Wolfsburg landfill), hydrographs*

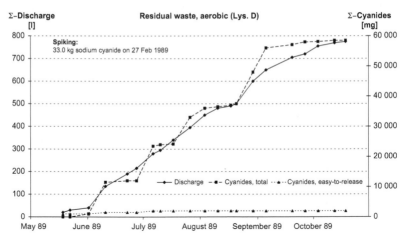

Fig. 6-9 Relationship between leachate discharge and cyanide transportation of an aerobic residual waste (Lys. D, Wolfsburg landfill), cumulative curves of discharge and transported masses

beginning of the higher discharge substantially increases the concentration (the transported mass curve is steeper), while dilutions can be observed at the end of high discharges (the transported mass curve is flatter). The sum of the easily released cyanide, however, remained insignificant within cyanide transportation. The same effect was proven in the anaerobic residual waste body at the lowest concentration level. Thus it can be concluded that, as expected, particles on the surface of hydraulically mobilised industrial wastes are directly dissolved by intensive precipitation through the large channels of residual waste.

Phenol and heavy metal contamination did not provide any significant results. It corresponded to the retention of the total waste in Braunschweig. The large channels did not have any accelerating effect on the release of compactly stored galvanic and phenolic sludge.

6.2 Pesticides simazin and lindane – examples of toxic industrial products

Henning Nordmeyer, Wilfried Pestemer and Key Herklotz

6.2.1 Model substances and relevant transportation mechanisms

Model substances

The pesticides lindane (insecticide, multiple-chlorinated hydrocarbon ring) and simazin (herbicide, simply chlorinated triazine ring), as

156

examples of organic industrial products, were intentionally spiked to estimate the retention capacity of waste materials for pesticides or similar reacting materials. The test results of short- and medium-term reactions were published in detail (Herklotz, 1985; Herklotz and Pestemer, 1986).

Transportation mechanism in solution

Herklotz (1985) has already described the characteristic of mass transport of this type of chemical in dissolved form. He found that both materials were extensively adsorbed by municipal solid waste, regardless of the influence of sewage sludge content. The degradation by about 50% of the organic substance does not reduce the sorption capacity of the deposit because the specific capacity of the humus-like products increases. However, it does not reach the sorption capacity of a peaty field, although the fraction of organic substance lies in the same order of magnitude. Decisive for the long-term effect is the proof that not only simazin, but also lindane can almost completely be desorbed over the long term, both from fresh waste including sewage sludge and from a sewage sludge waste mix extensively biochemically stabilised. It follows from these findings that, initially, adsorption prevents emissions from taking place. Then a step-wise desorption commences when water transport starts in the capillaries of the organic sorption substance due to increasing moisture penetration. Contamination peaks, like those of easily dissolved substances, e.g. cyanides, cannot be expected in the deposits because mobilisation is only possible within a limited solubility. Substances with moderate water solubility but good adsorption capacity will be re-adsorbed by non-contaminated organic substances along their route of transportation. During dry weather discharges, a decrease in the contamination concentration is to be expected due to this material behaviour because water movement only takes place in the large pores.

Transportation on suspended matter

Due to an extensive initial sorption, both substances can be directly discharged on suspended matter, as soon as the particles are mobilised in large vertical channels. This fraction is not subjected to chromatographic effects because wall contact is moderate in large channels. This transport system therefore has to be considered separately.

6.2.2 Influence of landfill technology and extent of contamination on the emissions from the total waste

6.2.2.1 Contamination of leachates from standard compacted landfills
(a) Rapid anaerobic degradation, low to moderate contamination

Lysimeter 2: non- to medium contaminated municipal waste, sludge-free, anaerobic

Figures 6-10 and 6-11 illustrate the concentration curves of the pesticides spiked in 1978 and 1979. They were detected in the base discharge of the anaerobic Lys. 2 beginning at the end of 1985 – partly due to higher detection sensitivity of analysis methods (0.05 µg/l for simazin and lindane). Thus, the first occurrence of the two pesticides was detected in the base discharge approx. 7 years after spiking. The measured concentrations for simazin and lindane after the breakthrough were 159 µg/l and 1.4 µg/l, respectively, in the transition from the low-contamination stage to the unloaded base ($\frac{1}{2}$h and 15 µg/l and 0.6 µg/l, respectively, in the base discharge.

In the later test period, lindane was detected in the base discharge up to 1990 and at the transition from the moderately contaminated stage

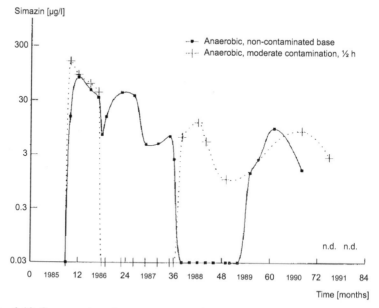

Fig. 6-10 Concentration of simazin in non-, low- and medium-contaminated anaerobic municipal waste (Lys. 2) (n.d. = non-detectable)

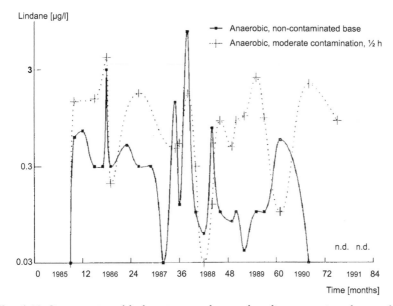

Fig. 6-11 Concentration of lindane in non-, low- and medium-contaminated anaerobic municipal waste (Lys. 2) (n.d. = non-detectable)

to the non-contaminated base until 1991. The highest concentration found at the transition was 3.9 µg/l in June 1986, while the highest concentration measured in the base discharge was 7.3 µg/l in January 1988. From August 1990, concentration in the base discharge was below the detection limit. This was tested and identified using mass-spectrometric investigations (GC/MS analysis).

The proven concentrations for simazin in the base discharge (base) and at the transition from the low-contamination stage to the non-contaminated base ($\frac{1}{2}$h) were somewhat higher than the values of lindane. The maximum concentration was 159 µg/l at the transition in October 1985 and 80 µg/l in the base discharge in December 1985. Strong concentration fluctuations were sometimes observed during the test period. Simazin concentrations at the transition were below the detection limit from July 1986 to December 1987 and in the base discharge from March 1988 to July 1989. This is probably due to very small amounts of leachate. The simazin concentration of 1.6 µg/l detected in August 1990 in the base discharge was confirmed by means of GC/MS. Sampling in March 1991 failed to find any simazin in the base discharge. However, simazin and lindane were found in the leachate at the transition to the non-contaminated base up to the last sampling (March 1991). It is therefore to be expected that further

159

pesticides will be released from the waste body. It can be concluded from the results that the anaerobic sludge-free waste body (Lys. 2) exhibits a substantial, but time-limited adsorption capacity for pesticides (7 years delay along a 2-m seeping distance). Thus the substances contained in the leachate could not have been prevented from penetrating the subsoil when the basal liner was missing or leaky. The release of pesticides from the waste proves that a complete chemical and biological degradation of pesticides did not take place in the municipal waste. The proven higher concentrations of simazin in the leachate indicate a higher mobility of this substance as compared to lindane. This is also proved by the adsorption constants determined (also see Herklotz, 1985). The K_d values for lindane are somewhat higher by about a factor 10.

The measurements were in full agreement with Herklotz's investigations (1985). Both substances were first completely adsorbed, as long as leachate passed through large cavities in the storage phase, in this case over 7 years for 5 m waste height. Desorption commenced in the mobilisation phase (exhaustion of storage capacity and mobilisation of stored water by degradation of organic storage material). Simazin was more intensively dissolved by more than an order of magnitude according to the different solubility. The influence of the large channels can be recognised by the fact that simazin was only partially retained in the non-contaminated base. After the end of the mobilisation phase, hardly any hydraulic transport of simazin took place in the predominantly large pores during the dry weather discharge. In the case where capillary water was mobilised, both simazin and lindane transports were detected over the long term. The difference to the transportation of highly soluble substances was that transportation of contamination peaks was limited due to a low water solubility and contamination sometimes decreased to below the detection limit due to high sorption.

(b) Extremely long acidic phase (acidic preservation), high industrial contamination

Lysimeter 9: highly contaminated municipal waste without sewage sludge, anaerobic

Neither simazin nor lindane was detected in leachate of the highly contaminated anaerobic contamination stage (Lys. 9, base) until 1985, although the waste body had only been 1.50 m high. Starting from September 1985, minor traces of lindane appeared in the leachate

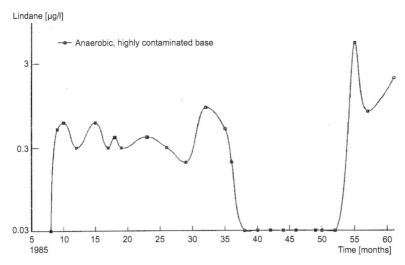

Lindane [μg/l]

Fig. 6-12 Lindane concentration in a chemically highly contaminated anaerobic municipal waste (Lys. 9)

near the detection limit (Fig. 6-12), while simazin was still not detected. No pollutant thrust, comparable to cyanide, was measured after the end of the storage phase.

Adsorption and desorption conditions rather unfavourable for simazin would have suggested the opposite result. Sorption measurements performed (Herklotz, 1985) showed a markedly higher mobility as opposed to lindane. No further lindane contamination was detected in the leachate in 1988. This was confirmed on selected samples using GC/MS.

Simazin was first detected in July 1989 (10 years after spiking) at 98.5 μg/l in the leachate of the only 1.50-m high waste layer. In a further test period, major concentration fluctuations were observed. The contamination was below the detection limit on individual sampling occasions. Lysimeter 9 exhibited a high retention effect despite extensive spiking. The delay was considerably greater than in the case of transport from a degrading material. The degradation of the storage material thus had a major influence on the transportation of semi-water-soluble contamination of industrial origin.

Effect of the reduction of natural organic substance due to recycling
The emission of the easily soluble simazin was accelerated in the initial phase of deposition (Lys. E) to such an extent that the sorption effect was not detectable despite a greater sorption constant (K_d value).

161

Substances of this type are therefore transported in large channels over the short term. Transportation of lindane was not detected even at about 4 years after deposition. A much stronger sorption and not the more favourable transportation was decisive for the removal of the semi-soluble substance.

6.2.2.2 Contamination of leachates from decomposition-landfills with extensive stabilisation

Extensively stabilised, permanent aerobic sewage sludge waste mix, two
initial, three contamination stages in the final phase on top of a
non-contaminated base (Lys. 6/10, cf. Fig. 2-7):
In contrast to the sludge-free waste (Lys. 2 and 9), pesticides were also used to spike the aerobic sewage sludge waste mix (Lys. 6/10). In October 1981, the high-contamination stage (Lys. 10) was placed on Lys. 6 in a highly compacted state (nomenclature: Lys. 6/10). Leachate contamination of the aerobic mixes contaminated in three stages can be described as follows: the breakthrough of the plant protection product was observed at the measuring points ($\frac{1}{4}$h, $\frac{3}{4}$h, $\frac{1}{2}$h and base) five years after the first spiking. In the later test period, a constant lindane and simazin discharge was detected at the sampling heights at all transitions of the three contamination stages and in the base discharge. Figures 6-13–6-16 show the concentration of simazin and lindane. It is conspicuous that simazin and lindane can be detected almost simultaneously at the measuring points. Relying on the stronger adsorption of lindane, a markedly stronger retention effect and thus a retarded discharge with the leachate could have been expected. The otherwise usual development of preferential paths in the waste cannot be blamed for this phenomenon. When the decomposed sewage sludge waste mixes were homogenised, they contained a large amount of fines and were extensively compacted. However, biochemical degradation removed about 50% of the organic substance and, thus, capillary storage capacity. The extreme compaction of the plastic, decomposed material and the application of new load on top of the lysimeter accelerated the consolidation (compressive stress squeezed out capillary water). These findings are based on simultaneous measurements performed after the high-contamination stage (Lys. 10) had been placed in an extremely compacted state on the three other contamination stages (base, non-contaminated to medium contaminated stages).

The anticipated adsorption and desorption was partly annulled by forced convection and the non-appearance of the time-delayed

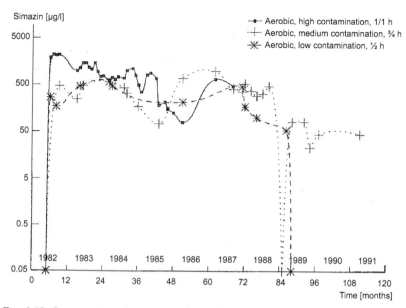

Fig. 6-13 Concentration of simazin in chemically low- to high-contamination permanent aerobic sewage sludge waste mix (Lys. 6/10, $\frac{1}{2}$h, $\frac{3}{4}$h, $\frac{1}{1}$h)

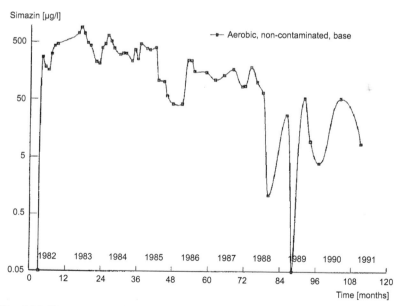

Fig. 6-14 Concentration of simazin in initially chemically non-contaminated permanent aerobic sewage sludge, municipal waste mix of the base (Lys. 6/10)

163

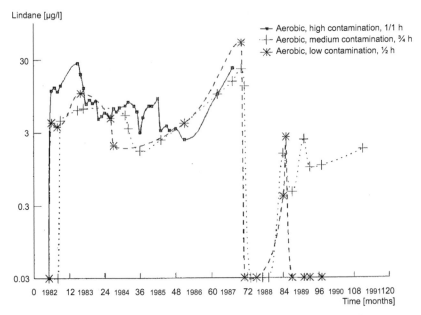

Fig. 6-15 *Concentration of lindane in chemically low- to highly contaminated, permanent aerobic sewage sludge, municipal waste mix (Lys. 6/10, $\frac{1}{2}$h, $\frac{3}{3}$h, $\frac{1}{1}$h)*

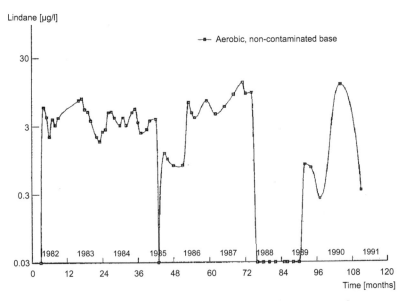

Fig. 6-16 *Concentration of lindane in initially chemically non-contaminated, permanent aerobic sewage sludge, municipal waste mix of the base (Lys. 6/10)*

break-through curves. It was particularly conspicuous that the initial occurrence of pesticides was simultaneously detected (between April and September 1982) at all transitions ($\frac{2}{2}$h, $\frac{3}{4}$h and $\frac{1}{1}$h) and in the base. 271 µg/l simazin and 5.7 µg/l lindane was detected in the base as early as April 1982. The maximum concentration for simazin, 910 µg/l, was reached in May 1983, while for lindane it was 13.0 µg/l in October 1987. However, it is possible that not all high concentrations were detected due to long measurement intervals (because of small amounts of leachate flow). As expected, both simazin and lindane exhibited markedly higher concentrations in the transition from the high- to medium-contamination stage ($\frac{1}{1}$h) than in the discharge of the non-contaminated base. This concentration reduction can be explained by different mechanisms.

In low-contamination stages the concentration of a contaminant is usually decreased by dilution (dispersion and diffusion), adsorption (reversible and irreversible) and degradation. The transportation of the pesticides used here also depended on the flow. The transportation only started after saturation of the storage capacity (after about 7 years without decomposition) and sometimes dropped below the detection limit in dry summer months. However, contrary to the highly soluble cyanides and heavy metals from the galvanic sludge, no pronounced mobilisation peaks emerged. The overall main difference to the waste without extensive biological stabilisation before deposition is that squeezing out the consolidation water accelerates transportation.

6.2.2.3 Comparison of transportation and relocation

The transported masses of pesticides discharged with the leachates were very small in comparison to the input (Table 6-3) and could not be explained alone by dissolution processes. The fraction of suspended

Table 6-3 Sorption of simazin and lindane by suspended particles (>2µm) of leachate at the base discharge of the permanent aerobic sewage sludge waste mixture (Lys. 6/10); AS = active substance

Substance	Municipal waste/sewage sludge, contaminated, aerobic (Lys. 6/10)		
	Leachate [µg AS/l]	Suspended matter (265 mg/l) [µg AS/l]	Sorption at suspended matter [%]
Simazin	100.0	0.32	0.32
Lindane	6.8	0.28	4.0

matter obtained by means of membrane filtration (>2 μm) from the leachate of stabilised waste (Lys. 6/10) at the sampling date in August 1987 was on average 265 μg/l leachate. Thus 1.1 μg lindane and/or 1.2 μg simazin were adsorbed per g suspended matter, i.e. 4% lindane and 0.3% simazin were discharged in the adsorbed form. Thus the emergence of pesticides in the leachate is not only explained by the 'chromatography effect' of adsorption and desorption but also by pesticides being adsorbed by mobile solid particles. These solid particles are thus to be regarded as carriers for pesticides. Less than 1% of the substances were discharged from 1982 to 1991.

The acceleration of discharge by biological stabilisation was insignificant in comparison to the inventory. Relocations determined by solid analyses and the measurements within reactivation tests are decisive for the final assessment of the long-term effect (Chapter 8).

To clarify the material relocation within the waste bodies, solid samples were taken from the anaerobic waste without sludge (Lys. 2) and the permanent aerobic sewage sludge waste mix (Lys. 6/10) at different depths and tested for pesticides in October 1989.

Figures 6-17 and 6-18 show the results of solids analyses for simazin and lindane of the first two anaerobic contamination stages (Lys. 2, non-contaminated base, low- to medium-contaminated stage) and all three aerobic contamination stages (Lys. 6/10, base, non- to medium-contamination stage). When interpreting the concentration differences,

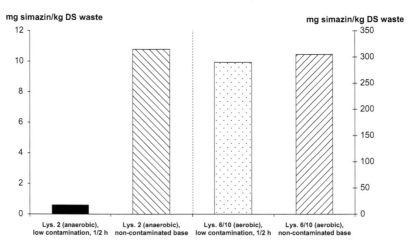

Fig. 6-17 Residues of simazin on solid material, comparison between an anaerobic municipal waste (Lys. 2) and a permanent aerobic sewage sludge waste mix (Lys. 6/10), 12 years after spiking

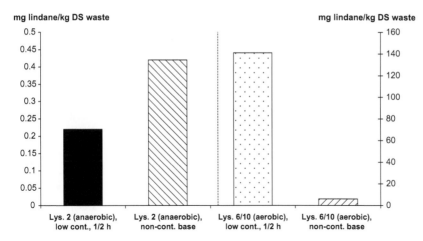

mg lindane/kg DS waste

mg lindane/kg DS waste

Fig. 6-18 Lindane residues on solid material, comparison between an anaerobic municipal waste (Lys. 2) and a permanent aerobic sewage sludge waste mix (Lys. 6/10), 12 years after spiking

it should be noted that the high-contamination stage was not tested in Lys. 2 but separately in the anaerobic material in Lys. 9, and the biocide was spiked in a liquid form. The material samples of the anaerobic waste therefore only cover an undefined partial flow of a cross-distribution. They are therefore less contaminated than in the aerobic, homogenised waste (Lys. 6/10).

The highest simazin concentration (302.4 mg/kg dry waste) was found in the bottom range (non-contaminated base) of the aerobic stabilised waste (Lys. 6/10) (Fig. 6-17). The highest lindane concentration (approximately 142 mg/kg dry waste) was found in the middle layer ($\frac{1}{2}$h) (Fig. 6-18). The relocation front of lindane obviously moves more slowly, so that the expected higher retention effect compared to simazin is indeed present. The relocation behaviour can also be confirmed by the sorption studies performed. The average distribution coefficients (K_d values) were markedly greater (factor >10) than for simazin (Herklotz and Pestemer 1986).

The highest simazin concentration (10.8 mg/kg) in the anaerobic waste (Lys. 2) was also found in the bottom range of the initially non-contaminated base (Fig. 6-17). The highest lindane concentration (0.42 mg/kg) was also found in the bottom layer (base) (Fig. 6-18). There were clearly less simazin and lindane residues in the anaerobic waste (Lys. 2) than in the aerobic waste (Lys. 6/10). Lindane can be degraded particularly well under the conditions of Lys. 2 (pH >7, anaerobic) (Adams, 1973). However, the concentration difference in

167

solid samples from wastes with different contamination is not sufficient to provide the necessary proof.

The analysis results of the solids tests were checked in various points using mass spectrometric methods (GC/MS) which confirmed the findings. Simazin degradation was found in the predominantly aerobic sewage sludge waste mix (Lys. 6/10): 0.3 mg/kg desethylsimazin was proven as a metabolite in 1989.

Even if one considers that the solid material was taken from waste bodies as punctiform samples, and pesticides were spiked to the anaerobic wastes (Lys. 2) in a lens-shaped manner, an extensive relocation in flow direction of the centre of contaminant plumes was detected both in aerobic and in anaerobic waste to initially non-contaminated areas.

6.2.3 Effect of recycling on the emissions

6.2.3.1 Change of sorption

Investigations into sorption (K_d values) of simazin and lindane with selected municipal waste resulted in K_d values of 227–802 for lindane and of 3.3–30.1 for simazin (Table 6-4), determined on samples of different sampling points from the different deposits (anaerobic = Lys. A, B and E; aerobic = Lys. C and D) and the different selecting stages (total waste = Lys. A, valuable material recycling = Lys. B and C, residual wastes = Lys. D and E). At the same time the content of organic carbon varied between 27 and 39% by weight. However, waste samples

Table 6-4 Sorption (K_d values) of simazin and lindane in different municipal waste samples

Lys.	Particle size [mm]	C_{org} [%]	K_d simazin [-]	K_d lindane [-]
B Anaerobic without valuable material	2.00	39.5	15.8	802
C Aerobic without valuable material	2.00	27.6	30.1	494
D Aerobic residual waste	0.25	35.3	13.2	370
	0.25	37.8	24.3	511
	0.25	34.2	28.7	576
	2.00	30.9	4.7	287
	2.00	30.4	3.3	511
	2.00	31.4	12.0	532
	2.00	33.2	15.1	552
E Aerobic residual waste	2.00	36.7	6.7	n.d.[a]
	2.00	33.9	3.3	227

(a) n.d. = non-detected

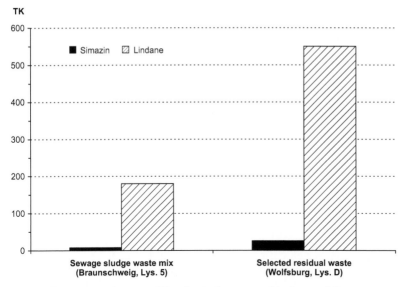

Fig. 6-19 Comparison of sorption (K_d values) of simazin and lindane in different municipal waste samples (Lys. 5, Braunschweig-Watenbüttel landfill (without selection) and Lys. D, Wolfsburg landfill (extensive selection)); average value, waste = 2 mm

with the highest C_{org} content did not exhibit the highest adsorption. It can be seen that the composition and quality of the organic fractions have a major influence on the extent of the adsorption. It must be noted that plastics, e.g. polyethylene, adsorb materials such as lindane very well and plastic containing composite packing materials remained in the residual waste according to the then state-of-the-art recycling.

Comparing the K_d values measured for Wolfsburg residual waste (wet waste = bio waste, glass, metal and paper selected, Lys. D) with the values of non-selected municipal waste from Watenbüttel (aerobic sewage sludge waste mix = Lys. 5), a markedly stronger adsorption (two- to threefold) can be observed in the residual waste from Wolfsburg at a similar particle size (1–2 mm) (Fig. 6-19). This confirms the assumption that the residual waste exhibits high adsorptivity for the materials tested. One of the reasons for the high adsorption may be the plastic content of the residual waste. Pestemer and Nordmeyer (1988) found high adsorption of pesticides by plastics.

The determined K_d values enable a much higher filter and buffer capacity of the Wolfsburg municipal waste for pesticides. Therefore, less contamination risk of the leachate was to be expected. In Watenbüttel, initial leachate contamination with pesticides was detected

169

approximately 4 years after spiking (homogenised municipal waste (anaerobic) = Lys. 2, sewage sludge waste mix (aerobic) = Lys. 6/10). Therefore it was expected that no major leachate contamination would occur in Wolfsburg in the investigation period of 1989–1991.

6.2.3.2   Transportation by leachate

The leachate was continuously sampled and tested for pesticide residues after the total waste (Lys. A), the residual wastes with decomposition (Lys. D) and without decomposition (Lys. E) were spiked with pesticides. Simazin was detected in the trace range as early as in 1990 3 months after spiking. To confirm the results a punctiform mass spectrometric check of the contaminated water samples was performed using GC/MS with SIM modus (select ion monitoring). The February 1990 sampling provided a leachate contamination of 24 µg simazin per litre in the anaerobic total waste (Lys. A) and 1.2 µg per litre in the anaerobic residual waste (Lys. E). When evaluating these results it must be taken into account that the total waste from Wolfsburg contained substantially less organic substance than that of Braunschweig. A concentration increase was observed in the consecutive period.

The fast breakthrough of the simazin front (3 months after spiking) stands in contradiction to the adsorption measurements and the long-term investigations of total waste in Braunschweig (main plant). The somewhat obvious higher K_d values, compared to the total waste in Braunschweig suggested a higher retention effect in Wolfsburg. However, due to a lower organic content in the municipal waste in Wolfsburg, the fraction of bulky components had increased relative to the total waste in Braunschweig, so that preferential flow paths increased. In such transportation paths the leachate can flow off very rapidly and annul the usual adsorption and desorption processes. A different form of seeping in the lysimeter is also confirmed by water content measurements. Moist and dry zones were found.

As expected, lindane was not simultaneously detected in the leachate at any sampling time. That met the expectation, since the adsorption of the semi-soluble lindane was about an order of magnitude greater than that of simazin.

6.2.3.3   Investigation of solids for relocation

In December 1990, solid samples were taken from the total waste contaminated with simazin and lindane (Lys. A) and from residual

Table 6-5 *Solids analyses of anaerobic total waste (Lys. A), aerobic residual waste (Lys. D) and anaerobic residual waste (Lys. E). Residual contents of simazin and lindane at different sampling depths*

Lys.	Layer	Simazin [mg/kg]	Lindane [mg/kg]
A Anaerobic total waste	top (0–0.8 m)	1.1	n.d.[a]
	middle (0.8–1.6 m)	0.1	n.d.
	bottom (1.6–2.4 m)	1.3	n.d.
D Aerobic residual waste	top (0–0.8 m)	47.5	n.d.
	middle (0.8–1.6 m)	1.4	n.d.
	bottom (1.6–2.4 m)	13.7	n.d.
E Anaerobic residual waste	top (0–0.8 m)	15.9	n.d.
	middle (0.8–1.6 m)	0.5	n.d.
	bottom (1.6–2.4 m)	0.2	n.d.

(a) n.d. = non-detectable

waste (Lys. D and E). These samples were intentionally extracted below the contaminated levels and tested for lindane and simazin to determine possible relocations. The waste body was divided into three layers: top, middle and bottom. The results of solid analyses are summarised in Table 6-5.

It was shown that simazin was detectable both in total waste and in residual waste at all sampling depths. The highest residue, 47.5 mg/kg, was found in the top layer in the aerobic residual waste (Lys. D). As expected, relocated lindane was not detected.

This enables the conclusion to be drawn that the increase of preferential seeping paths in the residual waste had accelerated the discharge of hydraulically mobilisable pesticides, even if adsorption of the waste has increased by an increase in plastic content. Only in the case of high adsorption and low solubility (e.g. lindane) do the larger channels in the residual waste not accelerate the discharge.

6.2.4 Conclusions

The discharge from the waste body in the flow path was only a matter of time both in the anaerobic and in the aerobic system. An effective degradation was not found when contamination was high.

The difference between the model substances lindane and simazin was that lindane was relocated more slowly due its better adsorption and lower solubility in solution than simazin. However, there was no difference when suspended matter adsorbed the substances.

In the case of high chemical contamination, the difference between anaerobic deposition and extensive aerobic stabilisation prior to a compacted deposition is that the transportation from the stabilised waste comes to an end faster. The possible biochemical degradation of lindane under anaerobic conditions was – against all expectation – moderate in the final development (see Chapter 8) under anaerobic alkaline conditions (Lys. 2) and non-detectable under anaerobic acidic conditions (Lys. 9). When intensive, almost un-disturbed aerobic degradation processes took place (low to medium contamination stage, Lys. 6/10), the degradation corresponded to that in an arable soil.

The influence of the larger channels due to recycling was that the easily-dissolved substance simazin was relocated in an accelerated way despite a substantially increased sorption capacity, while, as expected, the relocation of the semi-water-soluble lindane was retarded proportionally to the higher sorption capacity of the residual waste.

7

Microbiological investigations to characterise stabilisation processes in landfills

Wolfgang Neumeier and Eberhard Küster

7.1 Objective

Test results from the initial phase up to a measurable biological stabilisation of the deposited municipal waste (1976–1991) have already been published (Filip and Küster 1979, Neumeier and Küster 1986). It was possible to conclude from those results early on in this phase that wastes of the same composition may be degraded in different ways according to whether deposition conditions encourage or inhibit degradation. On the other hand, similar emission characteristics were observed from largely different waste compositions (e.g. sewage sludge total waste mixture compared to sludge-free residual waste).

The intentional spiking of typical industrial wastes (galvanic sludge, phenol sludge, cyanide or pesticides) increased the long-term natural contamination of leachates. However, the extent of the increase within the investigation time interval remained below the effect of the deposition conditions and was only detectable by the direct comparison of parallel non-contaminated and contaminated cases. The reason can be found in the differences of the biochemical degradation to be analysed in this chapter. According to the different extent of influences, first wastes deposited in different ways without any addition of industrial wastes and then the effect of chemicals was investigated.

The estimation of long-term environmental effects in future landfills of the influence of recycling on the stabilisation of deposited municipal waste, which is influenced both by the selective collection of industrial products and the selective re-use of natural organic substances, also needed to be investigated. Therefore, both effects were investigated in a supplemental test programme.

173

7.2 Materials and methods

7.2.1 Preliminary comment on criteria selection and their description

Long-term investigations started in 1976. The methods used at that time had to be applied later because this was necessary to enable comparison of results from medium-term investigations. They were supplemented by other methods as the investigation continued. In order to facilitate an interpretation of the results for the reader, the methods used will be briefly described once again.

7.2.2 Sampling the waste bodies

The samples were taken from the bottom (60 cm), middle (120 cm) and the top waste layer (190 cm) of the respective construction stage through lateral openings (Fig. 2.5) in two- or four-month distances up to 49 months after construction of the lysimeters. As soon as a stability trend was detected, the sampling intervals were increased, in order to investigate the long-term behaviour of the waste bodies at reduced costs. The waste samples for biological investigation were carefully dried to a constant weight and ground in a hammer mill to a grain size of ≤ 1 mm.

The following landfill types with an intermediate earth cover and one of the aerobic landfill sectional cores were excavated in May and June 1991 (Fig. 2.1):

- Lys. 1: municipal waste, anaerobic, 2-m stages
- Lys. 9: municipal waste with sewage sludge lenses, anaerobic, 2-m stages
- Lys. 4: sewage sludge waste mixture, anaerobic, 2-m stages
- Lys. 5: sewage sludge waste mix, aerobic, 2-m stages to decomposition start.

Individual samples were taken at additional heights and at defined distances from the edge. These additional sampling points will be described in detail in the discussion of the individual lysimeter results.

7.2.3 Microbiological test methods

The quantitative determination of the physiological microorganism groups in the lysimeters was carried out using the most probable number (MPN) estimates in a liquid media (McCrady, 1918) in culture tubes and/or according to Koch's pour-plate method. Since July 1997, microtitre plates were used for all microbial groups. A ten-fold dilution series was usually produced from the waste samples (see below).

Culture tubes
1 ml of each dilution stage was pipetted into 9 ml of a nutrient solution. Five duplicates were set up which were assessed after a relevant incubation period according to De Man's tables (1975). Sulphate, iron and nitrate reducing bacteria as well as cellulolytic bacteria were determined using this method.

Koch's pour-plate method
In three repetitions, approx. 10–15 ml of nutrient agar was inoculated with 1 ml of waste suspension, diluted to 10^{-9}, in a Petri dish. After the incubation, the microorganism colonies (colony-forming units (CFU)) were counted. This method was used for aerobic and anaerobic cultivable saprophytes and for aerobic spore-forming fungi. The anaerobic incubation was performed under a helium atmosphere.

Microtitre method
0.1 ml of each dilution stage was simultaneously filled into ten recesses of a microtitre plate. Then 0.1 ml of a special double-concentrated nutrient solution (see below) was added to each recess and the plates were incubated. The anaerobic incubation of the microtitre plates took place in Bohlender mini desiccators, the anaerobic environment was produced by Anaerocult A (Merck).

The following physiological microorganism groups were continuously quantitatively determined during the investigation period:

Aerobic microorganisms:

- aerobic cultivable saprophytes (standard I nutrient broth, Merck)
- acid generating bacteria (Difco BTB broth base with glucose addition, maltose and bromothymol blue as an indicator)
- cellulolytic bacteria (Imschenezki's nutrient solution with a filter paper strip) – microscopic fungi (malt agar)
- Actinomycetes (glycerin-nitrate agar with casein (Difco) according to Küster and Williams (1964) and Lab-Lemco agar (Oxoid) with 2 g/l yeast extract additive).

Anaerobic microorganisms:

- anaerobic cultivable saprophytes (standard I nutrient broth, Merck)

- sulphate reducing bacteria (sulphate API broth, Difco)
- iron-reducing bacteria according to Ottow (1969)
- nitrate reducing bacteria (nitrate broth, Difco)
- denitrifying bacteria (nitrate broth, Difco).

The incubation took place at 25°C and/or 27°C for mesophylic micro-organisms, at 49°C for thermophylic fungi and at 55°C for thermotolerant Actinomycetes. It took 5 to 14 days depending upon the group of organisms. The evaluation was performed according to the relevant VDLUFA guidelines (VDLUFA *Methods Manual*) in the individual dilution stages (De Man's MPN method, 1975).

7.2.4 Biochemical test methods

Respiration test
To assess the biochemical activity of the aerobic microflora of waste samples, Novak's (1972) respiration test was performed. The CO_2 delivery of the waste samples wetted to 60% of maximum water holding capacity (WHC) (sample diminution from 100 g DS waste to 10 g DS waste, results related to 100 g DS waste) was determined titrimetrically using Isermeyer's method (1952). To estimate the relative efficacy of individual nutrient additions (N (nitrogen), G (glucose), NG (nitrogen and glucose) and P (peptone)) the quotients of potential and basal respiration activity were calculated as required (N/B, G/B, etc.).

Potential methane generation (Neumeier and Küster, 1981)
100 g of re-wetted municipal waste (see respiration test above) was weighed in 1-litre Erlenmeyer flasks and incubated for 3 days in a $CO_2/H_2 = 50/50$ atmosphere at 25°C. A mixture of pyrogallol/K_2CO_3/diatomite was added to the Erlenmeyer flasks to ensure an anaerobic environment during the incubation. Methane detection was performed using gas chromatography and FID (Perkin Elmer 900) using a Porapak Q column. The results were given in μMol CH_4/(100 g DS waste × 72 h).

7.2.5 Chemical and physico-chemical test methods

The following analyses were carried out to investigate the waste materials according to the methods recommended by Rolle *et al.* (1970).

Ignition loss (IL)
Determination of the organic total mass as an ignition loss: municipal waste finely ground to a grain size ≤ 1 mm was incinerated at 600°C 3 h (heating up 0.5 h to 300°C).

Degradable organic substance (DOS)
To determine the degradable organic substance, 20 ml of sodium dichromate solution was added to a sample finely ground to 0.5 g, then 20 ml of sulphuric acid was slowly added after an hour and filled up with distilled water to approximately 200 ml. After adding 10 ml of phosphoric acid, 0.2 g of sodium fluoride and 30 drops of indicator, the non-used potassium dichromate was re-titrated with ammonium iron(II)-sulphate solution. Blank values were determined in the same way without any waste sample (according to the method suggestion of Rolle *et al.*, 1970).

Waste humic acids (WHA)
The humic acid-like substances (waste humic acids = WHA) were extracted with a mixture of 0.1 M $Na_4P_2O_7$ and 0.1 M NaOH from the waste samples (1:10, W:V – weight to volume) under a N_2 atmosphere for 24 hours. The WHA were precipitated from the colloid solution by acidification to pH = 1.5 of the alkaline extracts applying 37% HCl, then separated by centrifugation from the supernatant liquid, washed again with distilled H_2O, dissolved in 0.02 N NaOH and, finally, determined gravimetrically as a dry substance in an equivalence volume (Filip and Küster, 1979).

7.2.6   Enzymatic activities
Methods for the measurement of enzymatic activities in municipal waste were largely missing in the literature at the beginning of the tests (Grainger *et al.*, 1984). Therefore, it was necessary to adjust the relevant methods of soil microbiology to the needs of waste investigations.

Dehydrogenase activity in solid samples (Thalmann, 1967, amended)
1 g of the waste sample ground to 1 mm was treated in three parallel tests with 10 ml tris-TTC (triphenyl tetracolium chloride) solution in 100-ml test tubes (0.5 g TTC in 100 ml tris-buffer solution); after wetting the sample by careful agitation it was closed with a silicone plug and incubated in the dark at 30°C for 24 h. The reaction was interrupted by the addition of a 25 ml acetone-tetrachloromethane mix (9:1) and the

colour complex produced was extracted. The filtration took place after 2 h (Schleicher and Schüll 595.5*); the tubes were flushed with another 25 ml acetone-tetrachloromethane mix into the same filter. The filtrates were filtered again (Schleicher and Schüll 512*) and immediately photometrically measured at 546 nm compared to the blank value (10 ml tris buffer and 50 ml acetone-tetrachloromethane mix). Reference samples without TTC solution (tris buffer pH 7.6) and sterile samples, i.e. 1 g of sterile municipal waste (5 h at 140°C in a drying steriliser) and 10 ml tris TTC solution served as blank chemical or waste values. TTC is reduced by dehydrogenase to TPF; accordingly the results were indicated in mg TPF/g DS waste.

Alkaline and acidic phosphatase in solid samples (according to Hoffmann, 1969)

In 100 ml measuring flasks 1 g of waste ground to 1 mm was spiked with 10 ml substrate solution (6.75 g disodium phenyl phosphate/1000 ml distilled water) and 20 ml borate buffer (pH 9.6; alkaline phosphatase) and/or acetate buffer (pH 5.0; acidic phosphatase) and afterwards incubated for 3 h at 37°C. It was filtered after filling up with warm distilled water at 39°C (Schleicher and Schüll 512*). 10 ml borate buffer or acetate buffer was added to 5 ml filtrate and topped up with distilled water to 25 ml and coloured with 1 ml colouring substance solution (200 mg 2,6-dibromine quinone chlorimide dissolved in ethyl alcohol to form 100 ml). It was topped up to 100 ml after 30 minutes and measured photometrically at 600 nm and compared to the blank value. The reference sample was set with 10 ml distilled water instead of the substrate solution; the substrate was only added after the incubation and analysed according to the above guidance. The results were indicated in mg nitrophenol (NP)/g DS waste.

Phosphatase in leachate samples (according to Reichard, 1978)

1 ml sample was spiked with 1 ml 1 mM of p-nitrophenyl phosphate and 1 ml 0.3 M tris buffer (pH 9.5) to determine the activity of alkaline phosphatase. Citrate buffer (pH 5.6) was used for the measurement of acidic phosphatase instead of tris buffer. The reaction was stopped with NaOH/EDTA after 24 h and measured in a photometer at 419 nm. Heat-inactivated samples served as blank samples. The results were indicated in international units (U/ml).

Note: * Manufacturer and filter size.

Glucosidase in leachate samples

Reichard and Simon's (1972) method was modified for leachate tests. A leachate sample (2 ml) was incubated with 1 ml p-nitrophenyl-α-D-glucopyranosid (5 mM) and/or p-nitrophenyl-β-D-glucopyranosid (5 mM) and phosphate buffer pH 7.0 24 h at 25°C. Subsequently, the reaction was stopped by 1 ml 1 N NaOH in 0.05 M EDTA and the nitrophenol produced was measured in a photometer at 420 nm. Heat-inactivated samples served as blank samples. The results were indicated in ng nitrophenol (NP)/ml.

7.3   Classification and presentation of the results

Based on the results of the medium-term biological investigations (cf. Spillmann, 1986b), as well as Neumeier and Küster, 1986) the model landfills tested can be classified into three groups with a maximum of three types

2-m lifts, immediately compacted with earth cover (silt containing sand)

- Lys. 1 and 9: municipal waste
- Lys. 8: municipal waste with sewage sludge lenses
- Lys. 4: municipal waste intensively mixed with sewage sludge and extremely highly compacted.

0.50-m lifts, immediately compacted, without any earth cover

- Lys. 2: municipal waste, homogenised with water addition and compacted
- Lys. 7: municipal waste with sewage sludge lenses
- Lys. 3: municipal waste intensively mixed with sewage sludge and extremely highly compacted.

Municipal waste and sewage sludge intensively mixed and placed for decomposition on an air-permeable base with natural draught ('chimney draught' principle)

- Lys. 5 (base): sieving of compost-like material for the remediation of construction sites ('Gießen model', GufA, 1970), extremely high compaction of the residues
- Lys. 5 (1st and 2nd addition of waste), Lys. 6 and 10: extremely high compaction after extensive biochemical stabilisation.

Since the construction was carried out so slowly that the follow-up stage was only built after the end of the acidic phase of the preceding stage, the most favourable degradation and stabilisation conditions prevailed in all landfill sectional cores, which can also be achieved under similar conditions in practice.

The influence of spiked chemicals of industrial origin must be considered in three contamination stages: in the sludge-free anaerobic municipal waste (Lys. 2, non-contaminated, low and medium contaminated; Lys. 9, grossly contaminated) and in the extensively biochemically stabilised sewage sludge waste mix (Lys. 6, non-contaminated, low and medium contaminated; Lys. 10, grossly contaminated). Therefore these results will be presented separately.

As far as the time sequence of the results is concerned, first the short- and medium-term processes (5 years) will be described to such an extent that conclusions can be drawn from them for the long-term processes. These investigations have already been reported by Neumeier and Küster (1986) in Spillmann (1986b) in detail, so that only a brief version of the results will be given here. Diagrams and tables will only be repeated to such an extent as is needed to indicate typical changes. Deviations within the variants of the same types of landfill are only described verbally in this repetition.

7.4 Stabilisation and degradation processes in the total waste, short and medium-term processes (5 years)

7.4.1 Population dynamics, metabolic activity and humification effect without the influence of spiked industrial wastes

7.4.1.1 Landfill construction in 2-m stages with earth cover (silt containing sand)

Municipal solid waste without sewage sludge (Lys. 1)

The most common form of the early type of regular waste disposal, 2-m stages with earth cover, does not usually contain any sewage sludge (Lys. 1). The results of the measurements on this type of landfill will therefore be described in detail and the deviations of the variants with sewage sludge will be added.

Fresh municipal waste delivered for disposal was characterised by very high microorganism numbers as aerobic bacteria in particular (Fig. 7-1).

The microorganism numbers markedly decreased after compaction. This decrease was strongest among aerobic bacteria and microscopic fungi which is obviously due to a reduced oxygen supply. However, the microorganism numbers rose again between the 2nd and 8th

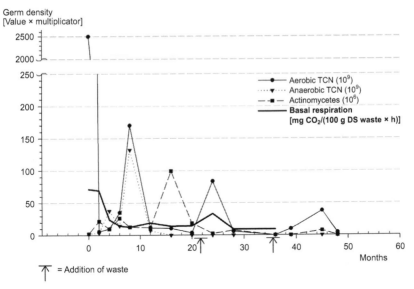

Fig. 7-1 *Stabilisation tendencies in the municipal waste without sewage sludge (Lys. 1), in a fresh condition, construction stages approximately 2 m high with earth cover. (Reading: unit × multiplicator = measured value): Example 8th month after test start: Aerobic TCN: $10^9 \times 170 = 1.7 \times 10^{11}$ total colony number, aerobics. Anaerobic TCN: $10^8 \times 130 = 1.3 \times 10^{10}$ total colony number, anaerobics. Actinomycetes: $10^6 \times 10 = 10^7$ total colony number, Actinomycetes. Basal respiration: $10\,mg\ CO_2/(100\,g\,DS \times h)$*

month after compaction and reached their maximum in the 6th (Acti-nomycetes, fungi) and 8th months. This maximum, however, was only significantly above the level of the initial material for the Actinomycetes and anaerobic bacteria. Between the 8th and 12th month after compac-tion, the microorganism numbers decreased strongly. This condition of reduced colonisation largely prevailed in the subsequent period despite an intermediate addition of waste to the municipal waste landfill.

The remaining physiological groups of microorganisms (such as cellulolytic bacteria and sulphate reducers) exhibited a similar trend (not discussed here). The separate incubation of the cultures at 55°C furnished the proof that most of the microorganisms in the deposited municipal waste were thermotolerant.

The metabolic activity of the complex microflora in the deposited municipal waste can be assessed by the respiration activity (CO_2 release) (Fig. 7-2). A basal and a potential respiration activity are distinguished. The basal respiration activity shows the CO_2 release which, under constant test conditions, depends both on the reliability of the tested samples and on the mineralising activity of the complex

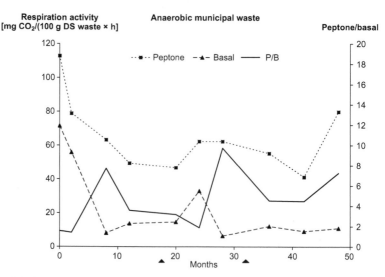

Fig. 7-2 Respiration activity of the microflora in compacted deposited municipal waste (Lys. 1), 'middle' sampling point in mg $CO_2/(100\,g$ DS waste $\times$ h); addition of waste after 22 and 34 months

microflora. Which of these two factors for the momentary CO_2 release is decisive, can only be decided after determining the potential respiration activity. The CO_2 release is considered as potential respiration activity that occurs after the addition of easily metabolised C and N sources alone or in combination. In the event of addition of a complex nutrient, conclusions on relative usability can be drawn or, conversely, on the stability of the original organic substance in the waste sample (Filip, 1983).

The basal respiration activity decreased progressively during the test (Fig. 7-2) and was five times lower after 6 months of deposition than at the beginning of the test. The CO_2 release dropped most strongly between the 2nd and 6th month after compaction (Table 7-1). The rise in microorganisms observed at the same time should not mislead one (Fig. 7-1): it was determined under optimum cultivation conditions

Table 7-1 Relative decrease in basal respiration activity in the anaerobic municipal waste (Lys. 1), 'middle' sampling point in %; addition of waste after 22 and 34 months

	Test duration [months]											
	0	2	4	6	8	12	16	20	24	28	36	42	48
Activity decrease [%]	100.0	78.2	40.9	21.8	11.5	19.3	21.8	20.6	46.2	8.9	17.1	12.9	15.4

Table 7-2 *Waste humic acid (WHA) content in compacted deposited anaerobic municipal waste (Lys. 1), in mg WHA/100 g DS waste.*

Sampling point	Test duration [months]			
	0	2	6	12
Top	271	281	283	236
Middle	–	300	405	260
Bottom	–	307	383	379

and was based to a large extent on the germination of resting micro-organisms. The rising values of the quotient peptone/basal (P/B) (Fig. 7-2 and Table 7-1) also showed that the native organic substance of the tested waste samples had already become rather resistant to microbial degradation after an 8-month test duration. The changes in values of the entire organic substance (ignition loss) and the degradable organic materials almost suggested that the stabilisation process indicated by the respiration test was only due to an increase in chemically resistant organic materials. However, the cause of stabilisation might also have been a biologically produced material change which could be characterised by the generation of microbially semi-degradable, humic-substance like compounds.

The values in Table 7-2 show that humic-substance like materials (waste humic acids (WHA)) were already present in the deposited municipal waste on delivery and were further generated after deposition. The maximum increase in WHA amounted to about 30% in the middle municipal waste layer within 4 months. The maximum was already achieved after 6 months. In addition, it was observed that the yield of WHA increased in the lysimeter from top to bottom.

In order to draw a qualitative comparison among waste humic acids (WHA), spectrophotometric techniques were used. The extinction curves of WHA took an ever steeper shape with increasing test duration. The material extracted from the municipal waste stored for 12 months was very similar to a soil humic acid, however it exhibited a deviating quality in the sedimentation test. The tested podsol humic acid was precipitated by 16 meq $CaCl_2$ from a weak (0.02 M NaOH) alkaline solution, while the best waste humic acid sedimented only after the addition of 20 meq $CaCl_2$.

The gradual degradation of the model landfill (Lys. 1, cf. Fig. 2-1) provided an opportunity in May 1991 to take solid samples over the entire profile other than from the used sampling points. However, no information beyond that already known was gained. Stabilisation already proved in the samples during the current tests also agreed with mass losses determined by weighing: >20% by DS waste weight.

Influence of population equivalent deposition of sewage sludge in 'lenses' (Lys. 8)

Colony (CFU) counts have not resulted in characteristics other than those of deposition without sewage–sludge addition (Lys. 1). A timely limited, strong rise of aerobic bacteria in the 24th month was only observed after an addition of waste in the 20th month. When the lysimeter was dismantled in May 1991, values were obtained for all parameters that were substantially lower in the bottom layer than in the middle one. The clearly recognisable stabilisation of municipal waste was about similar to that of the sludge-free deposit. It corresponded to the mass losses proved in the mass balance.

Effect of an intensive mixing of municipal waste with population equivalent sewage sludge and extremely high compaction (Lys. 4)

The extremely high compaction of the sewage sludge waste mix resulted in marked differences as opposed to common placement technique. The mix contained less aerobic microorganisms before the compacted deposition than the fresh municipal waste alone, while the numbers of anaerobic and/or facultative anaerobic microorganisms were often higher (Fig. 7-3). The number of aerobic microorganisms and other microbial settlements were generally stabilised after 16 months. Only the addition of waste in the 33rd month made a significant change in the colony numbers (rise in aerobic bacteria and Actinomycetes). However, anaerobic sulphate reducers behaved differently: their values were nearly three orders of magnitude above the initial number after 29 months due to an increasing deterioration in oxygen supply in the extremely compacted sewage–sludge waste mix. The maxima of aerobic microorganisms in the top layer (3rd construction stage) of the deposited material was usually an order of magnitude above those of the middle layer (2nd construction stage), while the bottom layer (1st construction stage) hardly differed from the middle layer.

Based on the extent and trend in all three layers (construction stages) of the anaerobic sewage sludge waste mix (Lys. 4), the basal and potential respiration activities (Fig. 7-4) indicate an unfavourable change in the physical conditions in the tested material. The ratio of the potential respiration activity to the basal respiration activity (quotient P/B)

184

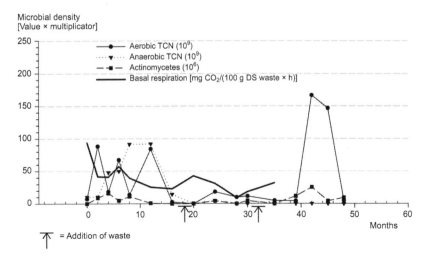

Microbial density
[Value × multiplicator]

$\underline{\mathbf{T}}$ = Addition of waste

Fig. 7-3 Stabilisation trends in a population equivalent sewage sludge municipal waste mix (Lys. 4), compaction in fresh condition, construction stages approx. 2 m high with earth cover. Reading: see Fig. 7-1

remained relatively constant up to the 20th month (decreasing trend). It increased in all three layers after the addition of waste and then stabilised again at a low level. In the case of a biochemical stabilisation, the ratios should have increased because of degradation processes and humic substance generation (see municipal wastes in 2-m lifts, Lys. 1).

A marked predominance of bacteria was recognised during the entire test period compared to Actinomycetes and fungi. However, this predominance was reduced by an increase of Actinomycetes compared to both the bacteria and fungi over increasing time. Bacteria and fungi therefore played a prominent role in the initial mineralisation of easily usable organic materials, while Actinomycetes increased in the following test period because they are capable of degrading more resistant substances. After exhaustion of the easily degradable organic substance, about 60% of the WHA was mineralised microbially without new ones being produced.

Lysimeter 4 was also dismantled in May 1991 and samples were taken at different depths and different distances from the edge. Micro-organism counting confirmed the preceding samplings. It was proved that degradation processes had not been affected by atmospheric oxygen either in the boundary region or in the crushed stone. The gravel was as clean as at the time of placement, it was only populated by anaerobes.

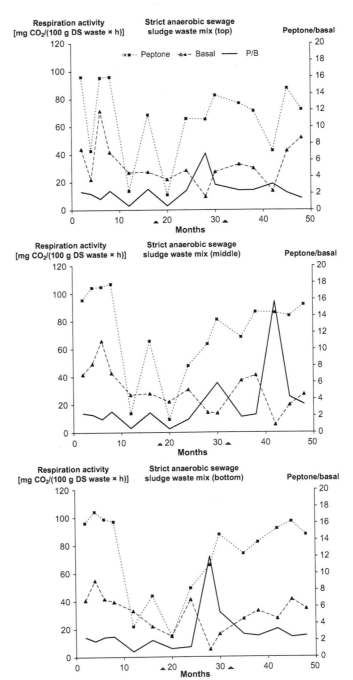

Fig. 7-4 Basal and potential respiration activity as well as the quotient potential/basal respiration in the strictly anaerobic sewage sludge waste mixes (Lys. 4)

186

The conclusion that the proved stabilisation in the anaerobic sewage sludge waste mix is due to a conservation and not to a degradation of the organic materials can be drawn from the condition of the material (no detectable degradation) and the mass balance (mass constancy) in connection with the proof of worsening degradation conditions. Thus, it is indirectly proven that the test equipment was adequate to produce and maintain anaerobic conditions over several years.

7.4.1.2 Landfill construction in 0.50-m layers without cover (thin layer placement)

Municipal waste with the addition of population equivalent sewage sludge in 'lenses', 0.50 layers, 4 lifts/year (Lys. 7)

Waste disposal with earth cover was replaced in practice by thin layer placement without an earth cover (onion skin layering) because the soil cover was inclined to form unwanted perched water horizons. In addition, it was felt that atmospheric oxygen could affect the open surfaces and thus degradation processes are encouraged more favourably. At the same time the deposition of sewage sludge also increased usually applied in compact bodies ('cassettes') or – in this model – 'lenses'. The investigation into thin-layer placement without chemical spiking was therefore focused on the deposition of population equivalent waste and sewage sludge masses in 'lenses'. Deviations due to other deposition techniques (without sewage sludge and sewage sludge waste mix) will also be discussed.

In comparison to sludge-free placement in 2-m lifts (Lys. 1), an increase in the inhomogeneity of the deposited material and the irregularity of the nutrient level was not the only effect of sludge lenses (Lys. 8). The frequent replenishment in short time intervals created an additional factor. The respiration values started to decrease only after the last addition of waste and did so until the end of the test. The long-lasting high fungus settlement was particularly conspicuous.

Microorganism populations and respiration values decreased after the end of the construction period to the same extent as in similar deposits built in 2-m stages. In addition, the occurrence of aerobic fungi indicated that the thin section placement had achieved the desired effect of extending the aerobic initial phase (Fig. 7-5).

Sludge-free thin layer placement, 0.50 m/lift, 4 lifts/year (Lys. 2)

Municipal waste was homogenised and wetted before compaction in a rotating body waste-collection vehicle to increase density and encourage

187

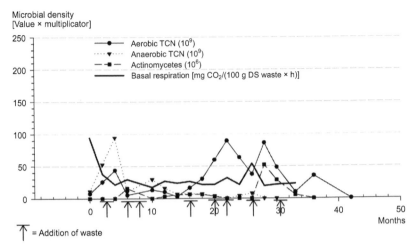

Fig. 7-5 *Stabilisation trends in municipal waste with added sewage sludge lenses (Lys. 7), compaction in fresh condition, construction stages approximately 0.5 m high without sand cover. Reading: see Fig. 7-1*

biological activity. Simultaneously to the co-disposal of municipal waste and sewage, the layers were placed at a distance of 6 weeks up to the 4th lift without any additional industrial contamination (5th to 9th lift = low-contamination stage, 9th and 10th lift = medium-contamination stage, see Fig. 2-7). The samples were taken from the bottom zones of the 2nd and 3rd lifts since the basal drainage (crushed stone) here had no influence and early stabilisation was expected. The results are very similar to the simultaneous deposition with sewage sludge lenses.

Municipal waste mixed with sewage sludge, 0.50-m lifts, 4 lifts/year,
extremely high compaction (Lys. 3)
After a thorough mixing (5 hours in a rotating body waste-collection vehicle), municipal waste and sewage sludge were intensively compacted. The microbiological investigations took between 45 and/ or up to 90 months to complete. Samples were only taken from the middle and bottom layers in this case to achieve better stabilisation. Strong fluctuations in the colony numbers of all microorganism groups were also observed in this variant due to frequent additions of waste (Fig. 7-6). In comparison to moderate compaction without sewage sludge and with sewage sludge lenses the stabilisation started later. The conservation effects resembled those of the anaerobic sewage sludge waste mix in 2-m stages and an earth cover (Lys. 4). Thus, stabilisation was caused in this case also by conservation of deposited waste and not degradation.

188

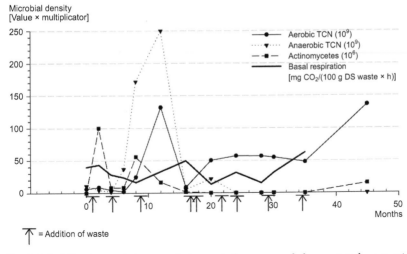

Microbial density
[Value × multiplicator]

$\top$ = Addition of waste

Fig. 7-6 *Stabilisation trends in a population equivalent sewage sludge municipal waste mix (Lys. 3), compaction in fresh condition, 0.50 m lifts, 4 lifts/year, extremely high compaction. Reading: see Fig. 7-1*

7.4.1.3   Extensive biochemical degradation before compacting the waste

Municipal waste, mixed with sewage sludge in population equivalent ratio, loosely placed for decomposition according to the 'chimney draught method', sieved after decomposition and the filter remainders deposited in a heavily compacted state (Lys. 5)

The aerobic treatment had a markedly stabilising effect on the numbers of aerobic mesophylic microorganisms (Fig. 7-7). The number of aerobic bacteria experienced a decrease after 2 months and remained constant from the 8th to the 20th month. An increase of these aerobic organisms was observed in the 24th and 33rd months due to the additions of waste 2 months previously. While the colony numbers of the anaerobes increased strongly after the 2nd and 8th months in the 1st year of investigation, they always remained at a low level in the subsequent months. The colony numbers of microscopic fungi remained unchanged apart from a decrease between the 8th to 12th test months.

A particularly conspicuous, rapidly progressing stabilisation process was found based on the curve of the basal respiration (Fig. 7-7). As early as in month 2, CO_2 release dropped to values that were only reached at 29 months in the sludge-free arrangement with 2-m stages (Lys. 1), i.e. in the landfill type with the fastest stabilisation within anaerobic deposits (see Fig. 7-2).

189

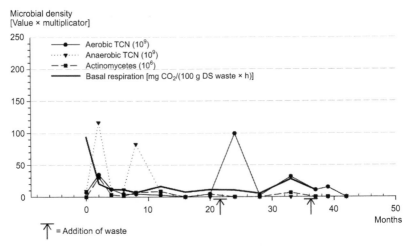

Fig. 7-7 *Stabilisation trends in a population equivalent sewage sludge municipal waste mix with aerobic degradation (Lys. 5), sieving after 20 months, placement of sieve residues in a highly compacted state and addition of waste with a new, decomposing mix on top of sieve residues, layers approximately 2 m high at the beginning of decomposition. Reading: see Fig. 7-1*

The rapid stabilisation seen in the basal respiration is unambiguously confirmed by comparing the population dynamics of the microorganisms between thin-layer placement (Lys. 3, Fig. 7-6) and the intentional aerobic degradation (Lys. 5) as shown in Fig. 7-7. The initial materials were identical in both deposits. The thin-layer placement (0.5-m layers) only contained approximately 0.3 t dry waste mass (approximately 0.5 t wet waste) per 1 m² surface area and was exposed to air contact for 3 to 4 months. Even after placing the subsequent layer, it was not possible to exclude the influence of air. Only three layers together contained the same dry waste mass per 1 m² of surface area (0.9 t DS waste/m² corresponds to 1.5 t wet waste/m²) after 6 to 9 months operational time, which was immediately put for intentional aerobic decomposition (Lys. 5). Although the thin layer placement (Lys. 3) achieved the intended stimulation, the microorganism populations and the values of basal respiration remained at a level about twice that of the intentionally aerobically degraded material (Lys. 5) after 3 months of decomposition time. As far as the population dynamics of microorganisms allows conclusions to be drawn, substantial conservation effects can also be expected from the thin layer placement, while the intentional aerobic degradation produces rather resistant humic-substance like matter.

The investigation of waste humic acids showed that their material composition was about the same in the entire aerobic mix regardless of the yield fluctuations and corresponded to a preliminary stage of natural soil humic acids.

The continuous hot decomposition had an effect on the whole waste body. Thermophilic fungi outweighed the mesophylic ones more than 20 times. Boundary influences due to test conditions were not observed.

A special test served for the always present potential degradation activity of the microflora and a direct check of material stability was proved by means of respiration tests. Eighteen months after the start of the intentional aerobic degradation, a sample of 2 kg was taken from the stabilised mix (Lys. 5), which was placed in a net bag into a freshly compacted layer of fresh municipal waste. A check after 2 months revealed that the supply of fresh organic substance, which corresponds to the supply from a domestic waste layer to an old municipal waste landfill, may even result in the reactivation of the microflora there, where the deposited municipal waste had proven extensively stable. However, microorganisms in a latent state were re-activated; the stabilised material suffered no further change.

Municipal waste, mixed intensively with population equivalent sewage sludge, loosely placed for decomposition according to the 'chimney draught method' and deposited after decomposition without sieving in an extremely highly compacted state (Lys. 6)

The sewage sludge waste mix formed a parallel during the decomposition phase to Lys. 5. Lys. 6 had been filled from the beginning in the same way as Lys. 5. The content was only treated differently after the end of the hot decomposition: it was not sieved, but placed on the same lysimeter surface in extremely highly compacted layers. The second lift was placed on the compacted 1st lift for decomposition. Colony counts performed in the material of Lys. 6 in the first months were similar to those in the material of Lys. 5. This was expected as both lysimeters contained the same aerobic material. These results will be used as a basis for comparison in Section 7.4.2.2 for the determination of the influence of industrial wastes (Lys. 10).

7.4.1.4 *Effect of different landfill technologies on the reduction of potential methane generation*

Neumeier and Küster's (1981) test method of the potential methane generation differs from the technique currently used in practice in as much as the landfill conditions are not simulated in the laboratory, rather – similarly to the measurement of the potential aerobic

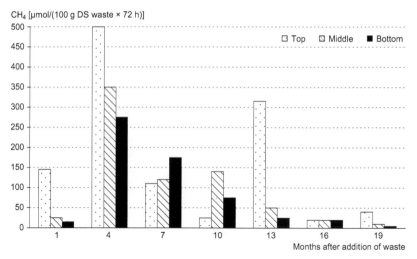

CH$_4$ [µmol/(100 g DS waste × 72 h)]

Fig. 7-8 Potential methane gas generation in the anaerobic municipal waste (Lys. 1) (Top = 1.80 m from bottom, Middle = 1.20 m from bottom, Bottom = 0.60 m from bottom)

respiration activity – extremely favourable laboratory conditions are provided by the abundant addition of hydrogen (H$_2$) and carbon dioxide (CO$_2$) (see Section 7.2.4).

The comparison of the potential methane generation failed to indicate any characteristic differences in the process itself, all the more in methane quantities. The main difference in a conventional deposit (2-m layers with earth cover, Lys. 1, Fig. 7-8 and Lys. 4, Fig. 7-9) was that the potential gas generation increased faster and dropped again under favourable degradation conditions (municipal waste, Lys. 1) in the extremely compacted sewage sludge waste mix (Lys. 4), although the water-saturated mix of the overlying waste lift provides nutrient-rich leachate to the base, which could have encouraged a process activation. One can conclude from this that the reactivation proceeds with retardation in the extremely compacted sewage sludge waste mix.

The extremely compacted thin-layer placement (0.5 m each layer) of anaerobically built sewage sludge waste mix (Lys. 3) differed from the 2-m lifts in a somewhat greater maximum methane production.

The difference to the material intentionally degraded biochemically (Lys. 6, Fig. 7-10) was the reduction by about an order of magnitude of the potential methane production in the compacted layer. The fresh material placed for decomposition was similar to the basal layer of a freshly contaminated, previously biologically stable old waste of a conventional landfill with regard to potential methane production.

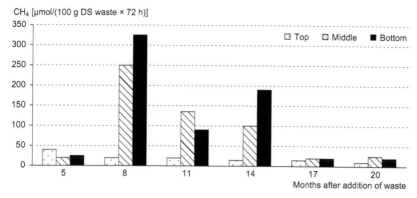

CH₄ [μmol/(100 g DS waste × 72 h)]

Fig. 7-9 Potential methane gas generation in the strictly anaerobic sewage sludge waste mix (Lys. 4) (Top = 1.80 m from bottom, Middle = 1.20 m from bottom, Bottom = 0.60 m from bottom)

Hence, it follows that though a sewage sludge waste mix placed for decomposition based on the 'chimney draught method' is mainly degraded in an aerobic way, anaerobic microorganisms including obligate anaerobic methanogenic bacteria are also available and can be activated.

7.4.1.5 Influence of site-related high precipitation on microbial activity
The Braunschweig landfill is in an area of moderate rainfall with respect to the average for the Federal Republic of Germany. Therefore, the rainfall was artificially increased by 50% in November 1982 simultaneously with the actual precipitation. Solid samples were taken before and after intensifying the irrigation (see Fig. 3-3).

The analyses showed an almost quadruple dehydrogenase activity (Fig. 7-11) both in anaerobic (Lys. 2) and aerobic waste deposits

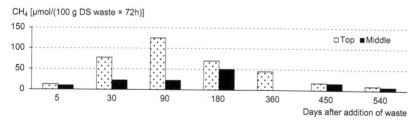

CH₄ [μmol/(100 g DS waste × 72h)]

Fig. 7-10 Potential methane gas generation in stabilised sewage sludge waste mix (Lys. 6) (Top = freshly placed for decomposition, Bottom = old decomposed material, highly compressed)

193

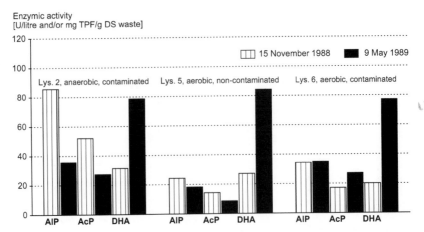

Fig. 7-11 Comparison of enzymic activity between chemically contaminated anaerobic municipal waste (Lys. 2) with permanently aerobic sewage sludge waste mix, non-contaminated (Lys. 5) and chemically contaminated (Lys. 6) before and after intensifying irrigation (AlP = alkaline phosphatase, AcP = acidic phosphatase, DHA = dehydrogenase activity)

(Lys. 5 and 6) after the irrigation period, i.e. the microbial activity increased due to the higher moisture content regardless of the type of landfill. The changes of alkaline and acidic phosphatase activity were insignificant and failed to exhibit any uniform trend. It follows for an average German municipal waste landfill that the degradation processes described here can proceed faster under equally favourable boundary conditions. This conclusion applies under the condition that higher precipitation does not cause leachate stagnation and have a conserving effect.

7.4.2 Change of the population dynamics of microorganisms due to spiked industrial wastes in high contamination stage

7.4.2.1 Municipal waste, non-mixed, placed in a highly compacted state in a 2-m stage with earth cover (Lys. 9)

The anaerobic municipal waste deposit with high industrial contamination, to which the chemicals were added in compact 'lenses', had been built as a separate lysimeter without any lateral sampling holes to ensure strict anaerobic conditions. The contamination was arranged in a contaminated level in the upper third of the deposit (see Fig. 2-7). In order to achieve a reliable understanding of the effect of volatile cyanide

Table 7-3 Colony numbers after cyanide addition in chemically highly contaminated anaerobic municipal waste (Lys. 9)

		MW, aerobic, highly contaminated (Lys. 9) Months after cyanide addition		
		0	2	4
Aerobes	$[\times 10^9]$	2.5	4.5	0.3
Anaerobes	$[\times 10^6]$	95	1.5	4.5
Acid-producers	$[\times 10^7]$	150	115	0.04
Cellulose decomposers	$[\times 10^5]$	2.5	0.075	2.5
Sulphate-reducing bacteria	$[\times 10^6]$	4	0.45	0.25
Mould fungi	$[\times 10^5]$	300	1	1
Actinomycetes	$[\times 10^6]$	<0.0002	0.001	0.2

from the compact inliers, addition of cyanide was coordinated with the sampling schedule.

The figures compiled in Table 7-3 prove that the populations of the microorganisms were sustainably reduced by at least an order of magnitude, with the exception of Actinomycetes whose number was already low. Therefore substantial delays can be expected in the biochemical stabilisation over the long term as a consequence of industrial contamination. It may represent the cause of the long-term leachate contamination detected.

7.4.2.2 Municipal waste intensively mixed with sewage sludge and extensive chemical addition and loosely arranged for decomposition (Lys. 10)

Comparing the colony numbers of different groups of microorganisms (Lys. 6, none to medium contamination) after 9 and 12 months with those of the about equal old mix with high chemical contamination (Lys. 10) (Table 7-4) shows that the chemical contamination significantly reduced the colony numbers, sometimes by several orders of magnitude, in most groups of organisms. Aerobic bacteria form an exception. Over the course of time this strongly inhibiting effect became somewhat balanced. The 'middle' samples originated directly from the highly contaminated municipal waste mixture. The comparison 'top' and 'middle' failed to show any clear trend. Only the respiration activity was constantly reduced in the contaminated centre. The comparison of the basal and potential respiration (Table 7-5) allows the conclusion that the respiration activity and thus the aerobic degradation was inhibited in the contaminated

Table 7-4 Colony numbers and CO_2 release in the aerobic municipal waste with medium chemical contamination (Lys. 6) and high chemical contamination (Lys. 10)

Physiolog. group	Sampling point	Factor	MW, aerobic, medium contamination (Lys. 6)		MW, anaerobic, high contamination (Lys. 10)	
			8 months	12 months	10 months	12 months
Aerobes	Top	$[\times 10^9]$	0.5	0.5	2.5	25.0
	Middle	$[\times 10^9]$	3.0	0.8	1.5	25.0
Anaerobes	Top	$[\times 10^9]$	0.25	0.45	0.000095	15.0
	Middle	$[\times 10^9]$	2.0	0.08	0.00075	75.0
Acid-producers	Top	$[\times 10^7]$	1.6	0.25	0.011	30
	Middle	$[\times 10^7]$	4.5	0.45	0.025	0.3
Cellulose	Top	$[\times 10^7]$	0.005	0.030	0.0000035	95.0
decomposers	Middle	$[\times 10^5]$	2.0	7.5	0.095	2.0
Sulphate-reducing	Top	$[\times 10^6]$	4.5	0.9	0.00045	2.50
bacteria	Middle	$[\times 10^6]$	9.5	1.5	0.0011	2.50
Mould fungi	Top	$[\times 10^5]$	1.0	8.0	0.001	21.0
	Middle	$[\times 10^5]$	6.0	10.0	n.d.	0.40
Actinomycetes	Top	$[\times 10^6]$	1.0	3.0	0.010	50.0
	Middle	$[\times 10^6]$	4.0	1.0	0.0001	0.050
CO_2 release	Top	$[\times 1]$	6.4	21.1	4.53.7	28.4
[mg O_2/100 g DS]	Middle	$[\times 1]$	5.5	12.8		11.9

n.d. = not detected

centre over the long term. Despite peptone addition, the respiration activity remained far below that in the low-contamination cover layer 'top' even after 12 months of decomposition time. Since the respiratory poison cyanide was assumed to be the main cause of inhibition, its effect was investigated in detail in a special test.

Table 7-5 Basal and potential respiration in the aerobic sewage sludge municipal waste mix with high chemical contamination (Lys. 6/10, $\frac{1}{1}$h), in mg CO_2/(100 g DS waste $\times$ h)

Variant [mg CO_2/(100 g DS waste $\times$ h)]	10 months		12 months	
	Top	Middle	Top	Middle
Basal respiration	4.5	3.7	28.4	11.9
On adding N	11.9	6.4	25.7	11.9
On adding C	3.6	0.9	47.7	7.3
On adding C + N	8.3	6.4	49.6	9.2
On adding peptone	57.7	4.6	66.0	10.1

7.4.3 Investigating the influence of cyanide on biochemical degradation and decontamination processes

7.4.3.1 Objective

Out of the intentionally spiked typical industrial contaminants (see Fig. 2-7), phenol and cyanides also occur in nature, and they can in principle be regarded as degradable. As expected, phenols were well degraded under aerobic conditions and not significantly degraded under anaerobic conditions. Cyanides, however, were removed from all deposits, and it was assumed that high residual COD values in the high contamination stage of the aerobic degradation (Lys. 10) had been caused by disturbing the respiration (cyanide is a respiratory poison). Therefore the medium-contamination stage of the anaerobic, sludge-free waste (Lys. 2) and the high-contamination stage of the aerobic sewage sludge waste mix (Lys. 10) were intentionally contaminated with sodium cyanide again 5 years after the start of waste placement in order to test the influence of cyanide on biochemical stabilisation.

7.4.3.2 Execution of the investigation

The cyanides were spiked in the following form:

- Sludge-free municipal waste, stored under anaerobic conditions after homogenisation, medium contamination stage (Lys. 2): placement of 100 kg hardening salt (sodium cyanide) in lenses at a depth of about 1 m. Subsequently, placement of municipal waste in the spiking point in a highly compacted state.
- Aerobic sewage sludge waste mix, high contamination stage (Lys. 10): the rotted material was removed from Lys. 10, 100 kg of hardening salts was added and then placed on the medium contamination stage of the aerobic stabilised sewage sludge waste mix (Lys. 6) in an extremely compacted state (nomenclature in the following: Lys. 6/10, cf. Fig. 2-7).

The investigation into the effects distinguished between the effect in the point of placement and that in the surrounding area in the case of placement in 'lenses'.

7.4.3.3 Influencing the population dynamics of microorganisms

Sludge-free, anaerobic municipal solid waste (Lys. 2)

Samples were taken before the addition of cyanide and in bimonthly intervals thereafter both directly from the cyanide repositories and at

197

a distance of 1.5 m from these above and at the base of the lysimeter. To quantitatively determine the total cyanide content and the easily released cyanide content, part of the samples were sent to the Fresenius Institute to be analysed.

Samples taken directly from the cyanide repository (Fig. 7-12) two months after cyanide addition showed a decrease of the colony numbers in contrast to the samples, which were taken at a lateral distance of 1.5 m (Fig. 7-13); however, the microorganisms were not completely destroyed. The cell number of aerobic microorganisms decreased by an order of magnitude in the cyanide repository, however, they rose again to the same values as before the cyanide addition at the next two investigation dates (5th and 9th month after addition).

Twelve months after a renewed spiking, another decrease and a renewed increase were experienced, however, only after 24 to 35 months. Similar developments were observed for anaerobes, sulphate-reducing bacteria and cellulose decomposers. The acid-producers showed an increase, which lay markedly above the initial value. The occurrence of Actinomycetes was decreased by the pollutant effect over a longer period. Their number over a period of 30 months was about three orders of magnitude lower than before pollutant addition. Subsequently, it rose again but failed to reach the initial value even after 50 months. Besides, the diversity of species was substantially impaired. Thus, only one kind of Actinomycete was found directly in the cyanide repository in relatively high numbers after 2 months. This organism was isolated and identified as the genus Micromonospora.

Among the microscopic fungi the thermophylic species decreased more strongly than the mesophylic fungi, proving they were more resistant. After 5 months, the first ones were no longer detectable. No unambiguous effect of cyanide addition on the colony numbers was found at a lateral distance of 1.50 m from the cyanide repository (Fig. 7-13). The anaerobic numbers showed a decreasing tendency up to 16th month after cyanide addition.

Sewage sludge waste mix, aerobic (Lys. 6/10)
Results from the aerobic stabilised sewage sludge waste mixes (Lys. 6/10) directly in the cyanide layer (Fig. 7-14) compared poorly with those of the anaerobic waste (Lys. 2). Thus, 2 months after cyanide addition the aerobes, acid-producers and mesophylic bacteria initially showed an increase in the colony numbers then the values decreased markedly. This is similar to the case of cellulose decomposers, sulphate-reducing bacteria and Actinomycetes immediately after

198

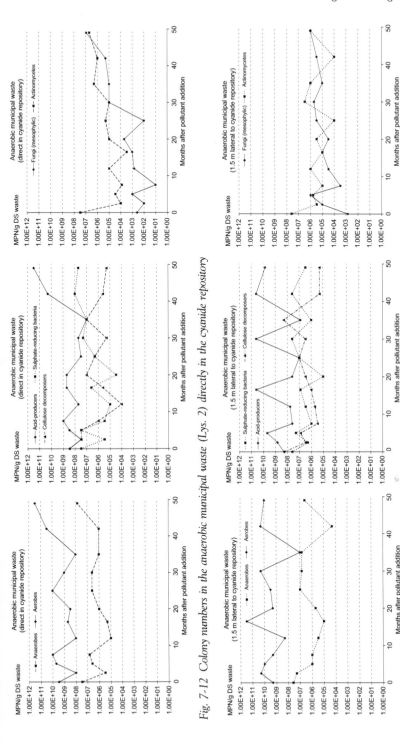

Fig. 7-12 Colony numbers in the anaerobic municipal waste (Lys. 2) directly in the cyanide repository

Fig. 7-13 Colony numbers in the anaerobic municipal waste (Lys. 2) at a distance of 1.50 m from the cyanide repository

199

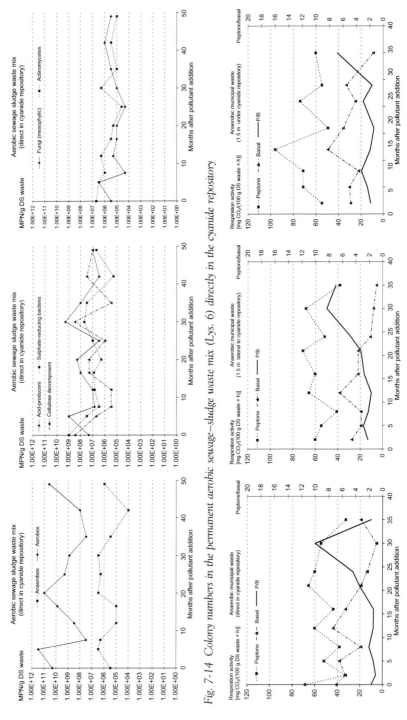

Fig. 7-14 Colony numbers in the permanent aerobic sewage–sludge waste mix (Lys. 6) directly in the cyanide repository

Fig. 7-15 Basal and potential respiration activities as well as the quotient potential/basal respiration in the anaerobic municipal waste (Lys. 2) after addition of the pollutants in different sampling points

cyanide addition. The trends, however, showed only a moderate decrease or even relatively constant colony numbers for nearly all investigated groups of microorganisms during the entire investigation period of 50 months after spiking.

7.4.3.4   Effects on metabolic activities of microorganisms

Sludge-free, anaerobic municipal waste, 0.50-m layers (Lys. 2)
The aerobic metabolic activities (measured as CO_2 delivery, Fig. 7-15) were first reduced in the cyanide repository of the anaerobic municipal waste (Lys. 2), but increased again after a year. Nutrient addition (peptone) was not able to increase the activity of the microflora at the beginning of the pollutant input, i.e. the metabolism was impaired by cyanide. Starting at the 20th month it was again possible to increase the aerobic metabolic activity by nutrient supply (up to 20-fold). The effect of the pollutants on the microorganisms had obviously become insignificant due to concentration reduction and immobilisation of the pollutants as well as adaptation of aerobic organisms. The limiting factor for their metabolism was rather the availability of nutrients.

Methane generation remained very low in the pollutant layer as opposed to the pollutant-free deposits such as Lys. 1 (see Section 7.4.1.4). The inhibition only had an effect on the aerobic degradation in the initial phase, while the anaerobic degradation was inhibited over the long term. One would have expected rather the opposite effect from a respiratory poison.

The contamination delayed and slightly weakened similar effects on the CO_2 delivery laterally from the cyanide repository.

Aerobic sewage sludge waste mix (Lys. 6/10)
The values of the basal respiration decreased in the aerobic sewage sludge waste mix (Lys. 6/10) immediately after the cyanide input and reached their lowest value after 9 months (Fig. 7-16). The potential respiration values first increased markedly to be followed by a decrease up to the 10th month. Then an increase set in up to the 24th month and, finally, the values decreased again up to the end of the test. These developments conclude that an inhibition occurred in the first 12 months which was caused by the pollutant input.

It can be concluded from the intermediate increase from the 10th to the 24th month that the inhibition carried on for at least 10 months and the following reduction after the 24th month could be attributed to stabilisation.

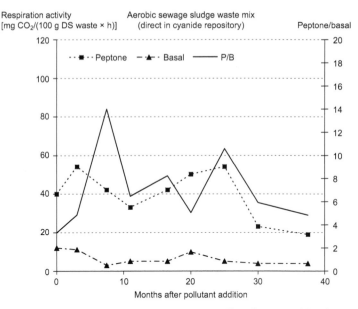

Fig. 7-16 *Basal and potential respiration activities as well as the potential/basal respiration quotient in the permanent aerobic sewage sludge waste mix (Lys. 6) after pollutant addition, sample taken directly from the cyanide repository*

Classification of metabolism

Combined measurements of oxygen uptake in 'Sapromates' (= closed respiratory vessel, produced by Voith) and CO_2 release were carried out according to Isermeyer (1952) to classify the metabolism under cyanide influence (Table 7-6). While aerobic respiration (respiration quotient RQ = 1) prevailed in the aerobic lysimeter without pollutants (Lys. 5), oxygen uptake was rather moderate under pollutant influence (Lys. 6/10). Prussic acid (HCN) is known as a blocker of final oxidation. The developing CO_2 from the contaminated wastes was thus of anaerobic origin, although sufficient O_2 was available in the test container.

Table 7-6 *Comparison between O_2 uptake and CO_2 release in aerobic sewage sludge waste mixes (high chemical contamination: Lys. 6/10; no chemical contamination: Lys. 5) and the respiration quotients (RQ)*

Variant	O_2 uptake [mMol O_2/h]	CO_2 release [mMol CO_2/h]	RQ [-]
With pollutants (aerobic Lys. 6/10)	0.023	0.270	About 0.1
No pollutant (aerobic Lys. 5)	0.235	0.230	About 1.0

7.4.3.5 *Pollutant tolerance of microbial isolates from sludge-free,*
anaerobic municipal waste (Lys. 2) and from an aerobic sewage
sludge waste mix (Lys. 6/10)

A number of Actinomycetes and bacteria strains which were isolated
during the research project from the pollutant repository of the
contaminated anaerobic municipal waste deposit (Lys. 2) (cyanides,
phenols and pesticides), were tested for their growth ability at a high
cyanide concentration. After a set of preliminary tests it turned out
that a hole or test well, as for the investigation of the inhibition effect
of antibiotics, is well suited for use as a screening method for cyanide-
tolerant microorganisms due to its easy handling. Although tolerance
limits cannot be detected by this method, comparisons of cyanide tolerances between the individual isolates can be drawn from the size of the
zone of inhibition. It turned out after preliminary tests that 0.1 ml of a
2 M KCN solution, pipetted into a hole of 9 mm in diameter, supplied
the best results. The diameters of the zones of inhibition measured
are displayed in Table 7-7. It was found that the growth of 11 isolates
($= 11.6\%$) out of 95 Actinomycete strains isolated from the cyanide
repositories, were not inhibited at these concentrations, while five of
them exhibited a strong inhibition (zone of inhibition 90 mm). Thus,
the inhibition effect was identified as selective.

Six of these strains, which proved CN^--unsusceptible as expressed,
were included in further investigations for CN^- tolerance. A relatively
susceptible strain (No. 44) served as a comparison.

Since the whole test does not allow any explanation about the
inhibiting KCN concentration, further tests with a highly contaminated
different nutrient solution were performed for this purpose. After
preliminary comparative tests with different nutrient solutions, the
Bacto Trypton yeast extract broth proved particularly suitable. 2 ml of
differently concentrated KCN solution was added to 19 ml of nutrient
solution. The KCN solution was prepared with distilled water which
was brought to pH 12 with NaOH. The culture containers were
sealed with gas-proof rubber plugs and incubated on a shaker table
at a temperature of 27°C for 24, 48 and 72 hours after inoculation
with a spore suspension of the relevant strains. Before adding the
KCN solution, a 24-hour growth phase in a pure nutrient solution
was arranged. The pH value only changed marginally (Fig. 7-17)
during the duration of the test (up to 120 hours). It dropped slightly
in the first days and then rose again. The renewed increase was due
to an autolysis of the microorganisms' cells, where traces of ammonia
are released.

Table 7-7 Cyanide tolerance of selected Actinomycete strains marked by numbers, from anaerobic municipal waste contaminated with cyanides (Lys. 2) and the permanent aerobic sewage–sludge waste mix (Lys. 6/10); concentration of the cyanide solution: 2 M KCN = 130 g/l

Actinomycetes						Bacteria	
Number	Zone of inhibition [mm]	Number	Zone of inhibition [mm]	Number	Zone of inhibition [mm]	Number	Zone of inhibition [mm]
1	30	33	–	76	–	113	35
2	42	34	52	78	90	114	32
3	60	35	64	79	–	115	37
4	–	36	48	80	38	Ps. aeruginosa	42
5	30	37	45	81	52	E. coli	53
6	n.t.[a]	38	38	83	58	Bac. pumilus	–
7	53	39	50	85	90		
8	40	40	90	86	53		
9	40	41	45	88	49		
10	30	42	41	89	58		
11	23	43	62	90	90		
12	–	44	67	91	43		
13	30	45	72	91	34		
14	26	46	34	93	33		
15	n.t.	49	38	94	–		
16	37	50	43	97	45		
17	32	51	33	98	74		
18	36	52	45	99	57		
19	–	55	57	100	60		
20	34	56	30	101	49		
21	40	59	30	102	42		
22	44	61	30	103	35		
23	35	62	30	104	41		
24	34	63	30	105	–		
25	43	65	39	106	52		
26	29	66	43	108	41		
27	39	67	57	109	34		
28	26	68	90	110	74		
29	26	71	–	111	37		
30	32	72	52	112	41		
31	39	73	48				
32	–	75	–				

[a] not tested

Seeing that pH started increasing as early as on the 2nd day, but mycelium weight only decreased on the 3rd day, this indicates that the cell death phenomena may have emerged in an early growth phase (2nd day). However, pH fluctuations were so small that they did not play any role in CN volatility. When the pH value was as low

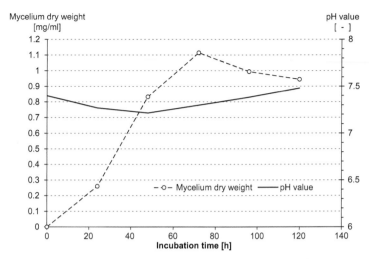

Fig. 7-17 Mycelium dry weight and pH value in the cultivation of the Streptomycetes strain 12 on the base nutrition medium

as 7.0–7.5 and cyanide was able to change into its gaseous phase in the form of HCN, the gas-proof rubber plugs prevented it from escaping from the containers. Thus, an equilibrium developed between the gaseous phase and the solution. However, gas analysis showed that only very small amounts of CN^- changed into the gaseous phase under these conditions. Thus it was clear that the spiked amount of KCN was indeed able to exert its effect.

The selected Streptomycetes strains were inoculated into the selected nutrient solution at increasing KCN concentrations and the percentage growth in comparison to the mycelium weight was determined after a 24-hour growth phase (Table 7-8). Negative percentages indicate that cell death (autolysis) took place at certain concentrations and after a certain time. The values showed that the strains reacted very differently to KCN in growth and autolysis, i.e. their tolerance was very different. The growth of the CN^- susceptible strain 44 was already markedly inhibited at the lowest KCN concentration, 2.5 mg/l, and negative values occurred at concentrations of 7.5 mg/l and above. The strains 32 and 33 were less susceptible. After 48 hours the growth values were still around 50% at 2.5 mg/l KCN. The strain 33 proved more tolerant at 20 mg/l KCN and above. The strains 71, 94 and 105 can be considered as another group. Even here a mycelium increase was observed during the growth period at low KCN concentrations (up to 12.5 mg/l) which indicates different growth rates particularly under the given stress conditions. This was especially conspicuous in the case

Table 7-8 Growth of Streptomycetes isolates after KCN addition. Mycelium weight [%] related to the control mycelium weight (24 h after KCN addition = 100%)

| | Streptomycetes strain | | | | | | | | | |
| | 44 | | 12 | | 32 | | | 33 | | |
CN⁻ [mg/l] h	24	48	24	48	24	48	72	24	48	72
0	100	275	100	154.5	100	100	85.7	100	94.3	77.1
2.5	17.8	3.75	45.5	110.9	41.4	50	72.8	52	55.7	61.8
7.5	−71.4	−82.1	3.6	23.6	28.6	22.8	0	26	27.6	36.9
12.5	−89.3	−85.7	0	14.5	7.1	8.5	−1.4	5.7	5.73	16.1
20	−78.6	−82.1	−0.9	0.9	1.4	−2.0	−38.5	−8.1	−8.1	−7.3
25	−89.3	−78.6	0	1.8	1.4	−48.6	−51.4	−5.7	−6	−7.8
50	−121.4	−157.1	1.8	−0.9	−22.8	−71.4	−77.1	−6	−7.5	−10
250	−117.8	−153.6	−2.7	−1.8	−57.1	−87.1	−130	−8	−10	11.2

| | Streptomycetes strain | | | | | | | | |
| | 71 | | | 94 | | | 105 | | |
CN⁻ [mg/l] h	24	48	72	24	48	72	24	48	72
0	100	94.9	86.9	100	111.2	96	100	100	97.2
2.5	87.7	100	82.8	40.3	106.7	111.2	47.5	97.5	95.1
7.5	5.1	44	56.6	11	13.4	40	−12.3	12.3	34.4
12.5	−1.71	0	26.9	−2.2	−2.2	25.3	−16.4	−18.8	35.2
20	−4	−4.6	−5.7	−2.2	−9.7	−11.9	−18.9	−20.5	−27.9
25	−2.3	−18.8	−19.4	−0.5	−15.7	−20	−17.2	−22.1	−27
50	−20	−25	−30.3	−4.3	−18.5	−22.8	−22.1	−21.3	−22.1
250	−30	−40.2	−41.5	−4.5	−20.6	−28	−22.95	−24.6	−31.2

of strain 12, which showed a better mycelium growth under CN addition after 48 hours. Strain 71 also behaved similarly up to a concentration of 12.5 mg KCN/l. The example of mycelium weights of strains 44 and 12 is illustrated in another form to provide better clarification (Fig. 7-18). The various KCN tolerances of the tested strains can possibly be explained by the origin of the sample from which these strains had been isolated. The strains 12, 32 and 33 were directly isolated from the CN⁻ block in an anaerobic municipal waste (Lys. 2), strains 71, 79 and 44 at a distance of 1.5 m from this block in the same material. Isolates 94 and 105 originated from the aerobic sewage sludge waste mix (Lys. 6).

7.4.3.6 CN⁻-degradation tests on isolates from anaerobic municipal waste (Lys. 2) and an aerobic sewage sludge waste mix (Lys. 6/10)

Actinomycetes which are able to use CN⁻-compounds as a C and/or N source are described in the literature (Winter, 1963). On a mineral salt

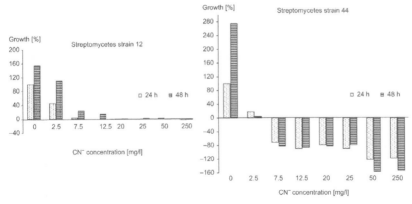

Fig. 7-18 *Pollutant tolerance of the Streptomycetes strains 12 and 44 at increasing CN^- concentrations (growth in% related to the control mycelium weight, 24 h and 48 h after KNC addition)*

solution according to Grün (1994) the original C and N supply was replaced by KCN based on the following test arrangement (Table 7-9). CN^--concentrations of 2.5 mg/l, 12.5 mg/l, 25 mg/l and 250 mg/l were used.

No variant, in which KCN was the only C or N source, exhibited measurable mycelium growth. The control measurement of cyanide in the inoculated test variants and non-inoculated controls also failed to show any differences, so that a microbial CN^- degradation by the test strains could be excluded. The strains found were CN^- tolerant, but did not metabolise CN^-. The ability to degrade CN^- seems to be reserved for some specialist microorganisms.

Oi and Yamamoto (1977) described a Streptomycete, which was able to convert CN^- into thiocyanate in the presence of thiosulphate with

Table 7-9 *Test arrangement to replace the original mineral salt solution for C and N supply by KCN*

	Mineral salt solution	
C	N	CN
+	+	−
+	+	+
+	−	−
+	−	+
−	+	−
−	+	+
−	−	+

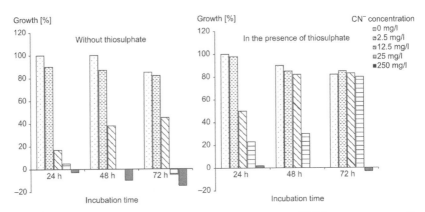

Fig. 7-19 Growth of the Streptomycetes strain 12 at various KCN concentrations with and without thiosulphate (growth in % related to the mycelium dry weight after a 24-hour growth phase)

the help of a Rhodanase. This finding should also be proved for our test strains. CN^--concentrations of 2.5 mg/l, 12.5 mg/l and 250 mg/l were added to a mineral salt solution inoculated with the strain 32 according to Oi and Yamamoto (1977) with and without sodium thiosulphate. KCN was also added after a 24-hour growth phase. Mycelium growth after the first 24 hours served as a scale for the following mycelium yield. The illustration (Fig. 7-19) allows the following conclusions:

1. Growth on KCN was somewhat retarded in the presence of thiosulphate, the optimum only occurred after 72 hours.
2. The mycelium yield was still very high in the presence of thiosulphate at a CN^--concentration as high as 25 mg/l.

The better growth may be due to a conversion into thiocyanate and an accompanied decontamination. This was also confirmed by the chemical analysis: the cyanide removed was converted almost completely into an innocuous, complexed form and only small quantities of the released cyanide were removed with the leachate.

7.5 Investigations into the long-term behaviour of the total waste

7.5.1 Microbial colonisation

Figures 7-20–7-24 illustrate the development of microbial colonisation in different types of landfill at the start of the test and after a 13- or

208

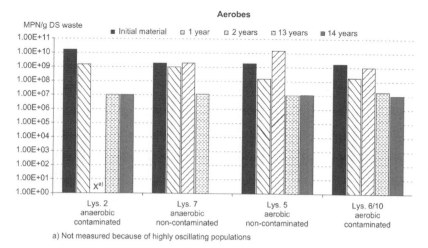

Fig. 7-20 Comparison of colony numbers of aerobic microorganisms between anaerobic municipal waste with moderate chemical contamination (Lys. 2), anaerobic municipal waste with sewage sludge lenses (Lys. 7), permanent aerobic sewage sludge waste mix (Lys. 5) and permanent aerobic sewage sludge waste mix, chemically contaminated in three stages (Lys. 6/10) at start and end of test

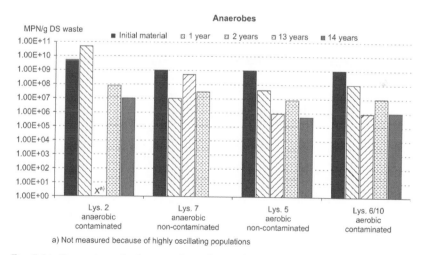

Fig. 7-21 Comparison of colony numbers of anaerobic microorganisms between anaerobic municipal waste with moderate chemical contamination (Lys. 2), anaerobic municipal waste with sewage sludge lenses (Lys. 7), permanent aerobic sewage sludge waste mix (Lys. 5) and permanent aerobic sewage sludge waste mix, chemically contaminated in three stages (Lys. 6/10) at start and end of test

209

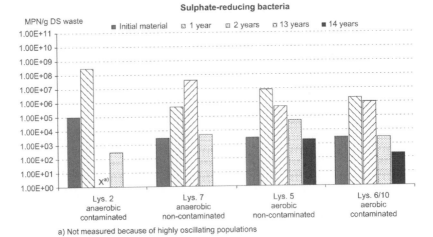

Fig. 7-22 Comparison of colony numbers of sulphate-reducing bacteria between anaerobic municipal waste with moderate chemical contamination (Lys. 2), anaerobic municipal waste with sewage sludge lenses (Lys. 7), permanent aerobic sewage sludge waste mix (Lys. 5) and permanent aerobic sewage sludge waste mix, chemically contaminated in three stages (Lys. 6/10) at start and end of test

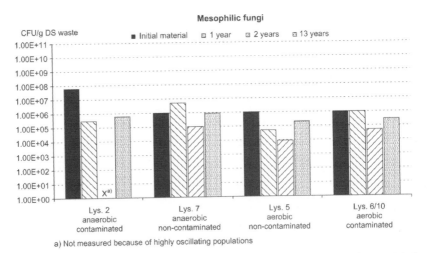

Fig. 7-23 Comparison of colony numbers of mesophilic fungi between anaerobic municipal waste with moderate chemical contamination (Lys. 2), anaerobic municipal waste with sewage sludge lenses (Lys. 7), permanent aerobic sewage sludge waste mix (Lys. 5) and permanent aerobic sewage sludge waste mix, chemically contaminated in three stages (Lys. 6/10) at start and end of test

210

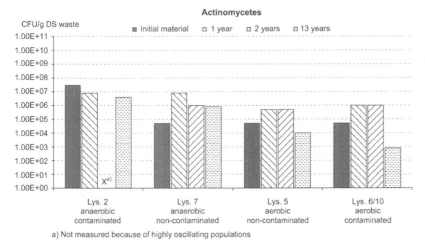

a) Not measured because of highly oscillating populations

Fig. 7-24 Comparison of colony numbers of Actinomycetes between anaerobic municipal waste with moderate chemical contamination (Lys. 2), anaerobic municipal waste with sewage sludge lenses (Lys. 7), permanent aerobic sewage sludge waste mix (Lys. 5) and permanent aerobic sewage sludge waste mix, chemically contaminated in three stages (Lys. 6/10) at start and end of test

14-year test duration (anaerobic landfill: municipal waste without sewage sludge (Lys. 2), municipal wastes with sewage sludge lenses (Lys. 7); aerobic landfill: extensive biochemical degradation before compaction, sewage sludge waste mix (Lys. 5). When interpreting the results it should be noted that the orders of magnitude are represented in a logarithmic scale.

The aerobic and anaerobic microorganisms (Figs 7-20 and 7-21) show the same tendency and decreased about 2 to 4 orders of magnitude within a decade. The obligatorily anaerobic sulphate-reducing bacteria, which were present only in small numbers in the initial material, also found quite good environmental conditions in the decomposing mixes. They were able to increase markedly in the first years. Toward the end of the investigation period they fell back approximately to the initial level (Fig. 7-22). The mesophylic fungi (Fig. 7-23) exhibited only a moderate decrease over the course of time, they even showed a slight increase in the 13th year. Actinomycetes (Fig. 7-24) hardly decreased during 13 years, which made the decreasing trend in the decomposition lysimeters more pronounced. The microorganism groups discussed previously however showed no differences between the various lysimeter types. Not only the colonisation strength, but the development of the colonisation was also fairly similar over the course of the years. As far as consequences can be drawn from the colonisation, the significant decrease depends on the reduction of the biological degradation.

7.5.2 Respiration activity and reactivation ability

Both CO_2 values of basal and potential respiration activities decreased in all landfill types during the test. Landfill technology with encouraged biochemical degradation (decomposition) only achieved a significantly higher basal and potential respiration activity during the initial phase. Reactivation ability, determined by the numerical ratio of the respiration stimulated by an abundance of easily degraded organic substance (peptone) to basal respiration, shows a similar picture (Fig. 7-25). The decomposition deposits exhibited markedly lower reactivation potentials from the 20th to the 90th month than the compacted 'anaerobic' deposits in Lys. 2 and 7. However, all values were on an equally very low level towards the end of the observation period, so that a high degree of stabilisation was reached in all lysimeters. Hence, it follows that stability, proven according to biochemical criteria, does not show the degree of degradation and does not differentiate between stabilisation by degradation and stabilisation by conservation.

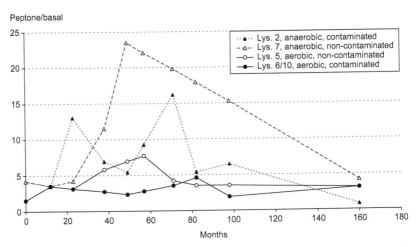

Fig. 7-25 Comparison of reactivation ability between anaerobic municipal waste with moderate chemical contamination (Lys. 2), anaerobic municipal waste with sewage sludge lenses (Lys. 7), permanent aerobic sewage sludge waste mix (Lys. 5) and permanent aerobic sewage sludge waste mix, chemically contaminated in three stages (Lys. 6/10) during the test period

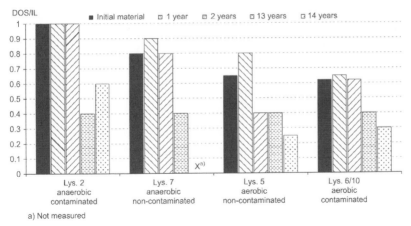

Fig. 7-26 *Comparison of the ratio of degradable to total organic substance (DOS/IL) between anaerobic municipal waste with moderate chemical contamination (Lys. 2), anaerobic municipal waste with sewage sludge lenses (Lys. 7), permanent aerobic sewage sludge waste mix (Lys. 5) and permanent aerobic sewage sludge waste mix, chemically contaminated in three stages (Lys. 6/10) at start and end of test*

7.5.3 Stability of organic substance

The absolute determination of the organic substance in the form of ignition loss (IL) (only a rough guide value for wastes) and as a degradable organic substance (DOS) gives little information about the actual stabilisation because of the inhomogeneity of the initial material. However, the ratio of degradable to total organic substance of the same sample is a measure of stability, because only inhomogeneity of degradation performance within a waste body is a measurement tolerance. Regardless of the deposit form and degree of degradation, almost the same ratio was found between degradable and total organic substance in all lysimeters over the years: it was 0.45 after 13 years and even 0.30 after 14 years in the decomposition lysimeters. Smaller ratios were more often found in decomposition lysimeters (after about 2 years) than in anaerobic landfill types (Fig. 7-26).

7.6 Influence of recycling on biochemical degradation

7.6.1 Investigated landfill types and recycling stages

The following types of landfill and recycling stages were investigated:

- Total municipal waste, unselected, highly compacted and anaerobically deposited (Lys. A).

213

- Municipal waste, highly compacted and deposited in an anaerobic way after metal, glass and paper recycling (Lys. B).
- Municipal waste, loosely arranged aerobically for decomposition after metal, glass and paper recycling (Lys. C).
- Municipal waste without compostable kitchen and garden wastes, loosely arranged aerobically for decomposition after metal, glass and paper recycling (Lys. D).
- Municipal waste without compostable kitchen and garden wastes, highly compacted and deposited in an anaerobic way after metal, glass and paper recycling (Lys. E).

The composition of wastes in the recycling stages of the individual test variants is plotted in Figs 2-2 and 2-3, Section 2.3.2.

7.6.2 Microbial colonisation

From the colony numbers of aerobes and anaerobes it can be concluded that the overall microbial colonisation is high (Fig. 7-27).

The aerobic deposits without valuable material but with biowaste (Lys. C) and residual waste without biowaste (Lys. D) exhibited the highest aerobic colony numbers at the beginning of the test. However, only small differences were observed in December 1990, while the cell number of aerobes decreased and that of anaerobes increased. Mesophylic and thermophylic mould fungi (Fig. 7-27) were readily present in the aerobic residual waste (Lys. D) in each test; their occurrence was usually higher in the aerobic waste without valuable materials (Lys. C) than in other types of wastes. Regardless of the intensity of recycling, a significant decrease in both groups of organisms was observed in the anaerobic deposits (Lys. A, B and E) until 1990, while remaining at same level in both aerobic wastes.

A strong presence of thermophylic Actinomycetes (Fig. 7-27) was observed in both aerobic deposits independently of the amount of biowaste (Lys. C and D) at the beginning. The colonisation of the anaerobically arranged and immediately highly compacted residual waste deposit (Lys. E) was at first similar to the aerobic types of landfill because residual waste is extremely difficult to compact and anaerobic conditions were only achieved after a subsequent cover of compacted loam was added. Thermophylic Actinomycetes are particularly involved in the degradation of polymeric substances – usually plastic bags.

The generic affiliation of fungi found in waste deposits was also determined. Species of the genuses *Penicillium* and *Aspergillus* were mainly identified. Representatives of the genuses *Mucor, Cephalosporium,*

214

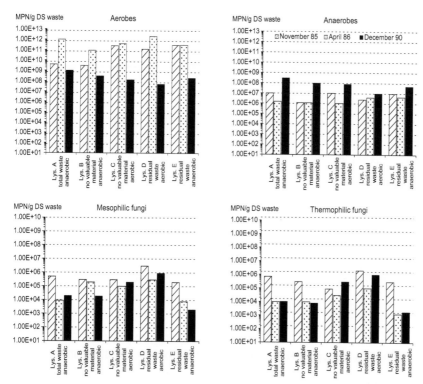

Fig. 7-27 *Comparison of microbial numbers of physiological microorganism groups in solid samples from waste deposits of various recycling influence and various operating technologies (anaerobic/aerobic) (Lys. A–E, Wolfsburg), sampling at 1 m depth in November 1985, April 1985 and December 1990*

Graphium, Trichoderma and in some cases *Paecilomyce* and *Stysanus* also occurred, and also some yeasts. Two species of the genus *Chaetomium* and one of the genus *Stachybotrys* were restricted to residual waste (Lys. E). Aromatics decomposers were found in all deposits from the outset, but their occurrence was somewhat lower in the residual waste (Lys. D and E) at the beginning. The cell numbers were very close to each other in all deposits in December 1990 and reached orders of magnitude of 10^7 to 10^9.

7.6.3 Parameters of biochemical activity

7.6.3.1 Respiration activity

Respiration activity was measured as O_2 uptake and as CO_2 release. O_2 uptake was measured in Sapromates (= closed respiratory vessel,

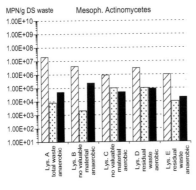

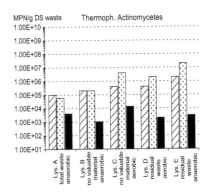

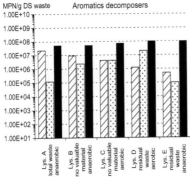

Fig. 7-27 Continued

produced by Voith) just on the November 1995 samples, since this method can only handle relatively small sample volumes due to measurement error issues. For this reason, suitable column equipment was developed to measure CO_2 release which permits the application of larger sample volumes. Figure 7-28 illustrates the results of O_2 uptake in Sapromates. Waste from the anaerobic deposit without valuable materials (Lys. B), which contained a larger amount of organic waste than the total waste, had the noticeably greatest O_2 demand. The extensively compacted residual waste (Lys. E) also had higher O_2 consumption than the aerobic residual waste (Lys. D). It has to be taken into account in the interpretation of the results that a material stored at reduced oxygen supply for a longer period of time experiences great activity increase when exposed to optimum conditions, in this case sufficient oxygen supply.

CO_2 measurements on large waste samples in 1985 yielded the same results (Fig. 7-29) as those of O_2 consumption. A further small increase in activity was observed in the anaerobic deposits of organic enrichment

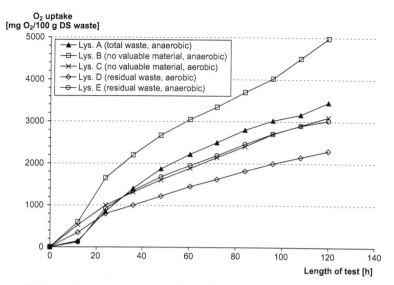

Fig. 7-28 Respiration activity (O_2 uptake) of solid samples from waste deposits with various recycling influence and various operating technologies (anaerobic/aerobic) (Lys. A–E, Wolfsburg), samples of November 1985 at 1 m depth

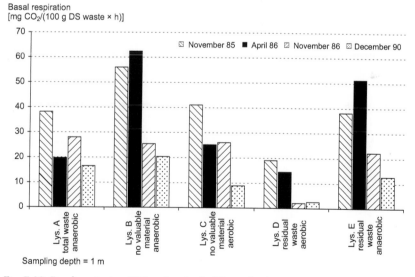

Fig. 7-29 Basal respiration (CO_2 release) of solid samples from waste deposits with various recycling influence and various operating technologies (anaerobic/aerobic) (Lys. A–E, Wolfsburg), at 1 m depth in 1985 to 1990

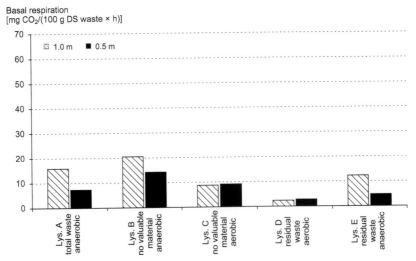

Fig. 7-30 Basal respiration (CO₂ release) of solid samples from waste deposits with different recycling influence and different operating technologies (anaerobic/aerobic) (Lys. A–E, Wolfsburg), at a depth of 0.5 m and 1 m in December 1990

in the next sampling due to the removal of valuable materials (Lys. B) and in the residual waste (Lys. E). However, both aerobic deposits (Lys. C and D) and the anaerobic total waste (Lys. A) exhibited a significantly lower CO_2 production. This decreased further in the aerobic residual waste (Lys. D) in November 1986, and also in the aerobic waste without valuable materials (Lys. C) in December 1990. The remaining deposits showed almost the same level (approx. 25.5 mg CO_2/(100 g DS waste × h)) in November 1986. All deposits produced only a moderate CO_2 release in December 1990.

CO_2 production of the samples from the top and bottom part of the waste bodies (Fig. 7-30) generated the same level in both aerobic deposits (Lys. C and D). However, the samples from the anaerobic deposits were different: those from the top half of the deposit produced less CO_2 than the ones from the bottom half.

7.6.3.2 Methane generation

Methane generation, a measure of anaerobic metabolic activities, only started to a small extent at the time of the second sampling in all variants (Fig. 7-31). In November 1986, however, the anaerobic deposits (Lys. A, B and E) exhibited very high methane values. In the material, following valuable material recycling and relative enrichment

218

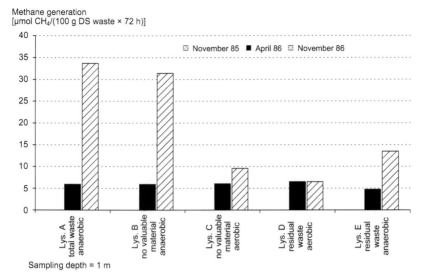

Methane generation
[µmol CH$_4$/(100 g DS waste × 72 h)]

☒ November 85 ■ April 86 ☑ November 86

Sampling depth = 1 m

Fig. 7-31 Methane generation capability of solid samples from waste deposits with different recycling influence and different operating technologies (anaerobic/aerobic) (Lys. A–E, Wolfsburg), at 1 m depth, in 1985 to 1986

of the organic substance (Lys. B), a higher methane generation capability was found than in the residual waste (Lys. E) (anaerobic, without metal, paper and compostable wastes). In the aerobic deposits (Lys. C and D) methane generation remained moderate as expected, regardless of the influence of recycling.

7.6.3.3 *Autogenous heating capability*

The autogenous heating capability has been used in composting technology for a long time to describe biochemical stability ('Rottegrad' – autogenous heating capability) (e.g. Niese, 1963). As an addition to the respiratory test and methane generation capability, it can also indicate the stability of residual waste. The further the biochemical stabilisation has progressed, the lower the autogenous heating capability. It was the same for the material of all deposits, regardless of recycling influence, and at 75°C very high in the initial phase (November 1985), i.e. the degradation degree was low (Fig. 7-32). Five months later (April 1986) samples were again taken from the three biochemically most active deposits (decomposition, Lys. C and D) and anaerobic residual waste (Lys. E) at two depths. The anaerobic residual waste (Lys. E) produced the highest temperature (= lowest

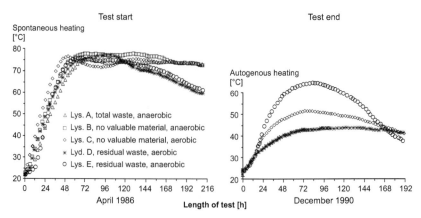

Fig. 7-32 Comparison of the change of spontaneous heating capability of the waste between different recycling and operating influences between test start in 1985 and test end in 1986 (total waste, anaerobic (Lys. A), valuable material recycling, aerobic (Lys. C), residual waste, aerobic (Lys. C), residual waste, anaerobic (Lys. D))

stability) with 65°C. The aerobic waste without valuable materials and thus relatively increased biowaste content had achieved 51°C. The aerobic residual waste reached a temperature of 44°C after 132 hours of test time and therefore had been stabilised considerably.

Autogenous heating also allows the conclusion that residual waste still has a substantial autogenous heating potential from the biochemically degradable materials. Intentional biochemical degradation reduces the autogenous heating potential of biologically degradable materials faster than selective collection. It is only the selective collection combined with intentional biochemical degradation which considerably reduces the autogenous heating potential of degradable materials.

7.6.3.4 Dehydrogenase activity

Dehydrogenase activity has proved in soil investigations to be a general parameter and an indicator of microbial activity. However, its information content is much lower for municipal waste because of a number of disruptive influences. For this reason, in order to separate biological reduction processes from chemical ones, a sterile sample was analysed additionally in each case. It turned out that chemical reduction processes covered those of microbial activity to a very large extent so that no 'biological activity' was detected in April 1986.

Dehydrogenase activity was high in all deposits at the beginning of the test series, being highest in anaerobic residual waste (Lys. E).

220

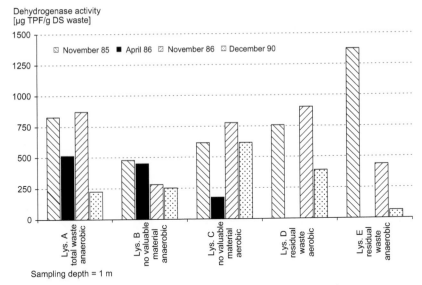

Dehydrogenase activity
[μg TPF/g DS waste]

Sampling depth = 1 m

Fig. 7-33 Dehydrogenase activity of solid samples from waste deposits with various recycling influence and different operating technologies (anaerobic/aerobic) (Lys. A–E, Wolfsburg), at 1 m depth, 1985–1990

However, higher activities were observed in both aerobic deposits (Lys. C and D) in November 1986, regardless of recycling influence (Fig. 7-33). This became completely clear in the investigation in 1990 (Fig. 7-34) where lower activities were found throughout rather than at the beginning of the investigations. However, samples taken at two depths within the last sampling campaign showed that the top layer of two anaerobic deposits, i.e. Lys. B and E exhibited higher dehydrogenase activity than the bottom layers (Fig. 7-34). The anaerobic residual waste (Lys. E) showed the highest activity in the top layer of all lysimeters. It was noticeable that the material with a relatively high content of organic waste in the bottom layer of aerobic deposits (Lys. C) showed higher activity than the residual waste without organic content (Lys. D). However, those deposits, both anaerobic and aerobic, without organic waste, exhibited the higher activities in their top layer.

7.6.3.5 Alkaline and acidic phosphatase
Neither alkaline nor acidic phosphatase were able to provide information on the influence of recycling or degradation path. These measurements will therefore not be presented.

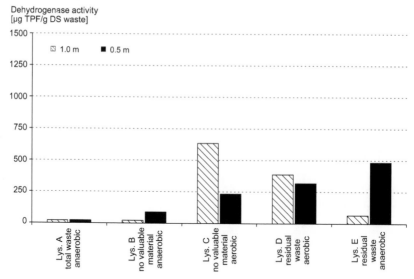

Fig. 7-34 Dehydrogenase activity of solid samples from waste deposits with different recycling influence and different operating technologies (anaerobic/aerobic) (Lys. A–E, Wolfsburg), December 1990, depths: 0.5 m and 1 m

7.6.4 Chemico-physical stability parameters of solids

7.6.4.1 Decrease in the organic substance
As already described for total waste, initially the absolute value of ignition loss and that of non-degradable organic substances can only provide low-reliability information due to large inhomogeneity and a high content of degradable organic substance in the waste. These results will not be used in the assessment because of their wide spread.

7.6.4.2 Alkalinity and conductivity
No significant acidic phase was observed in the solid of any deposits regardless of landfill technology and recycling influence. The pH values lay in the neutral to high alkaline range ($7.4 \leq pH \leq 9.1$). The trend in the aerobic deposits (Lys. C and D) indicated somewhat higher pH values than in other variants. Conductivity values from 119 to 193 $\mu S/cm$ in the sampling regimes from November 1985 to November 1986 also failed to indicate any variant-specific reaction. Conductivity decreased considerably toward the end of the investigation period, but it was not possible to detect any influence of various landfill technologies or different degrees of separation.

7.6.5 Leachate parameters to characterise stability

7.6.5.1 Investigation objectives

Taking solid samples is not only expensive (use of excavator with special drills) and time-consuming, but the representativeness of even large samples is problematic. Besides, sample preparation is connected with risks (drying process, grinding etc.). For this reason it was decided to characterise microbial processes in the waste bodies by analysing freshly discharged leachate (Küster *et al.*, 1989). For this purpose microorganisms removed with the leachate and selected typical enzymatic activities were determined quantitatively.

7.6.5.2 Enzymatic activities

Enzymatic activities, measured in terms of phosphatase and glucosidase, indicated a clear difference between aerobic and anaerobic deposits. No influence of recycling was detected (Figs 7-35–7-38). The activities of both aerobic deposits (Lys. C and D) were below the values of all four anaerobic deposits by one to two orders of magnitude.

The results of the first two sampling campaigns in September 1987 and May 1989 showed that the opposite behaviour of acidic phosphatase (Fig. 7-35) on the one hand and that of alkaline phosphatase (Fig. 7-36)

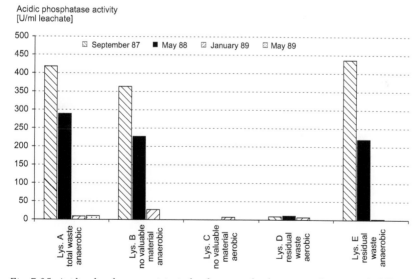

Fig. 7-35 Acidic phosphatase activity in leachate samples from waste deposits with different recycling influence and different operating technologies (anaerobic/aerobic) (Lys. A–E, Wolfsburg), 1987–1989

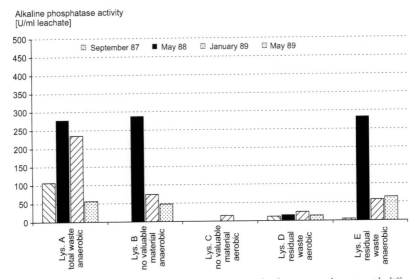

Fig. 7-36 *Alkaline phosphatase activity in leachate samples from waste deposits with different recycling influence and different operating technologies (anaerobic/aerobic) (Lys. A–E,
Wolfsburg), 1987–1989*

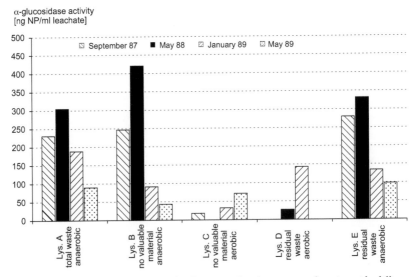

Fig. 7-37 *α-glucosidase activity in leachate samples from waste deposits with different
recycling influence and different operating technologies (anaerobic/aerobic) (Lys. A–E,
Wolfsburg), 1987–1989*

224

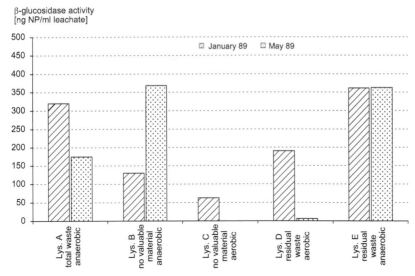

β-glucosidase activity
[ng NP/ml leachate]

Fig. 7-38 *β-glucosidase activity in leachate samples from waste deposits with different recycling influence and different operating technologies (anaerobic/aerobic) (Lys. A–E, Wolfsburg), 1987–1989*

on the other indicated a change from an acidic to alkaline phase within the waste bodies. The activity of acidic phosphatase was hardly measurable by 1989, the activity of alkaline phosphatase decreased from May 1988 to May 1989, the aerobic deposits (Lys. C and D) exhibiting the lowest values.

The activity of α-glucosidase was substantially lower in the sampling campaign of 1989 than that of β-glucosidase (Figs 7-37 and 7-38) which indicated a rapid stabilisation of the aerobic deposits. No influence of recycling was detected.

7.6.5.3 *Physiological groups of microorganisms*

The analyses of physiological groups of microorganisms in the leachate showed that initially the cell numbers of the anaerobes (1987 and 1989) usually exceeded those of aerobes (Fig. 7-39). A large increase of up to two orders of magnitude in the aerobes was observed in 1989, while the anaerobes only increased in the leachates of anaerobic deposits, and decreased markedly in the leachates of aerobic deposits (Lys. C and D). Spore-forming bacteria (not shown) mainly occurred in the leachate of aerobic residual waste (Lys. D) which had gone through a particularly intensive autogenous heating phase with temperatures exceeding 70°C. Nitrate-reducers and iron-reducers were represented the least in the leachate of both aerobic deposits (Lys. C and D). The number of

225

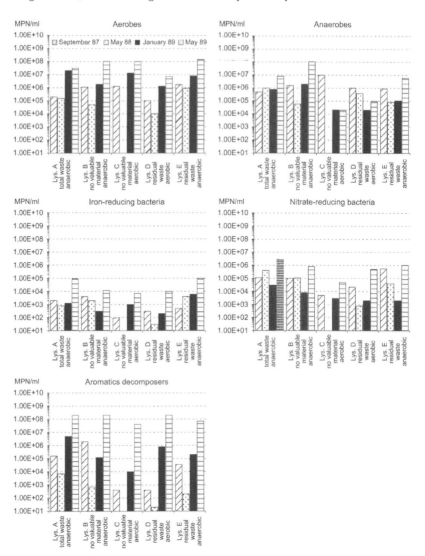

Fig. 7-39 Germ number of physiological microorganism groups in leachate samples from waste deposits with various recycling influence and various operating technologies (anaerobic/aerobic) (Lys. A–E, Wolfsburg), 1987–1989

aromatics decomposers in the leachate of all deposits increased as a function of time, with the exception of a minimum in the leachate of the anaerobic deposits in May 1989, in particular the anaerobic total waste (Lys. A). Aromatics decomposer increased with increasing ageing and variant-specific differences disappeared.

226

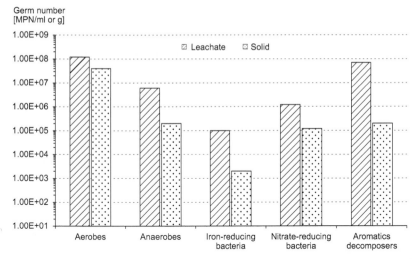

Fig. 7-40 Comparison of germ numbers of physiological microorganism groups in solid and leachate samples from waste deposits with various recycling influence and various operating technologies (anaerobic/aerobic) (Lys. A–E, Wolfsburg), May 1989

7.6.5.4   Comparison of microbial colonisation between solid and leachate

A comparison of microbial colonisation of a solid sample with that of a leachate of the same deposit and the same time shows similar trends (Fig. 7-40). This was also observed in samples of the total waste of the Braunschweig-Watenbüttel facility. The results determined in individual solid samples were confirmed by the tests on leachates.

7.7   Interpretation of the test results

7.7.1   *Transferability of the results to real landfills*

The investigations were performed on cylindrical sectional cores of flat landfills without leachate accumulation, whose content was identical to carefully operated municipal waste landfills with respect to waste composition, pre-treatment, compaction and seepage behaviour. An acceleration of the degradation processes, e.g. by increasing the internal surface due to a test-specific comminution of the wastes – typical for laboratory tests – was avoided. The sectional cores from aerobic flat windrows were, as expected, identical to the operational facility (e.g. Schwäbisch Hall landfill since 1976). The biochemical proof of a strict anaerobic environment during the entire test period in extensively compacted deposits is crucial for the transferability of the results of anaerobic degradation. This proof was successfully provided within this test. It follows from this finding that the measured influence of atmospheric oxygen on degradation and stabilisation processes occurs in the same way due to diffusion in moderately compacted wastes in large-scale facilities and is not caused by leakage in the equipment. The influence of atmospheric oxygen has to be calculated for higher compacted deposits according to the laws of diffusion (see details in Chapter 2).

It should be noted for the practical application of the results that the municipal waste was carefully compacted in several lifts, thus both water retaining horizons and coarse drainage paths were avoided and that a slow building technology was chosen. Therefore, more rapid and extensive stabilisation cannot be achieved in practical landfills of the same type and operating technology.

7.7.2 Conclusions from the detailed medium-term investigations

7.7.2.1 Waste without industrial contamination

Total waste
The effects of various landfill conditions can be best illustrated by comparing the extensive aerobic stabilisation (Lys. 5) with the most favourable anaerobic type (Lys. 1) and the most unfavourable anaerobic type (Lys. 3) of landfill technology. The results of the remaining types of landfill correspond to an interpolation between these types of landfill.

Stabilisation was detected in all types of landfill, based on the applied test criteria, in terms of a reduction in biochemical activity in the first 5 years of deposition despite a potentially active microorganism population. The difference between the operational types has a mainly time-related character. In an intentionally aerobic degradation of a population equivalent sewage sludge waste mix (Section 7.4.1.3) stability phenomena were found after 2 months that can only be achieved in other types after 29 months in the most favourable case (slow placement of municipal waste, Section 7.4.1.1, Lys. 1). The biochemical stability of the anaerobic thin layer placement of an identical material, i.e. sewage sludge waste mix (Section 7.4.1.2, Lys. 3) was considerably lower than the biological stability parameters of the intentional aerobic stabilisation of the same mix even after 5 years, e.g. double level of the population density and basal respiration in comparison with the intentional degradation after 3 months. Taking into account that a time interval of 2–3 months only makes up about 15–20% of the aerobic stabilisation time in which only about 25–30% of the produced heat energy of the intensive degradation phase is delivered, and the extensive stabilisation by humifi-cation is still in its initial phase, there is a difference of an order of magni-tude between the stabilisation times of the intentional aerobic degradation and the most favourable case of anaerobic degradation (Lys. 5 vs. Lys. 1) at the same deposit geometry. The long time delay due to the most unfavourable anaerobic deposit (Lys. 3) cannot be estimated using the medium-term measurement because of the extent of the delay.

Another differentiation can be derived from the potential methane generation (Section 7.4.1.4), as follows.

The intentional stabilisation (Lys. 5) reduced the potential by an order of magnitude in comparison to the most favourable standard placement (Lys. 1). The stability of the standard placement was primarily based on the conservation effect, which was clearly proved in the most unfavourable case (Lys. 3).

The humification of organic material proceeded in the intentional aerobic degradation successfully and uniformly, while it proceeded non-uniformly but successfully in the most favourable standard place-ment. However, the process broke down in the most unfavourable stan-dard placement. Similar differences were measured as mass degradation: approx. 20% by DS weight both by intentional degradation and by most favourable standard deposition, 0% by DS weight in the most unfavour-able case of the standard deposit.

It has to be taken into account when extrapolating over time that the material of the most favourable standard deposit changed into aerobic degradation to a large extent over 5 years (brown material, earthworms up to 1.5 m under the surface), while the most unfavourable standard variant remained anaerobic and was still in the acidic phase (no precipitation in the gravel, still grey colour in some places).

It follows from these observations that extensive stabilisation in the originally anaerobic municipal waste, favourable case (Lys. 1) only applies to the area where atmospheric oxygen may have an influence, i.e. near the surface and up to 4-m depth in a landfill without gas pressure. The same applies to the extensive degradation of the wastes in co-disposal including sewage sludge lenses. If these conditions are not fulfilled, stabilisation occurs in the sense of anaerobic conservation. The greater the content of stagnating pore water, e.g. in sewage sludge, the more extensive the conservation.

Influence of recycling

The investigations on residual waste from various intensive recycling up to a maximum recycling ratio of 60–65% by DS weight have proved that

229

degradation and stabilisation processes agreed with those of the total waste. Therefore, the long-term forecast for total waste also applies to residual wastes when the different water storage capacities and their saturation degree at the time of landfilling and the potential improvement of degradation by possibly having a longer aerobic initial phase in the standard landfill without earth cover are considered.

7.7.2.2 Effect of industrial contamination

The negative impact on biochemical stabilisation of industrial contamination recognised indirectly by the leachate analysis (Section 5.4) was clearly proven by the microbiological investigations.

The chemical contamination of the high contamination stage of the tested deposits of industrial residues (galvanic sludge, phenol sludge, cyanide and pesticides), reduced both aerobic and anaerobic biochemical stabilisation to such an extent that the parameters decreased by several orders of magnitude. The intentional aerobic degradation was thus retarded and disturbed but was still clearly measurable. The anaerobic degradation was inhibited to such an extent that, apart from a substantial, still acidic conservation, no trace of long-term stabilisation was detected after 5 years.

The disturbances in leachate contamination proven indirectly (Section 5.4) are thus clearly proven as degradation inhibition by industrial waste. This extensive inhibition had not been expected in a mixed landfill for the following reasons:

- The spiking of pesticides was not sufficient for the inhibition of the biochemical degradation.
- The toxic effect of heavy metals at high concentration on microorganisms is well-known. However, no major inhibition was expected at an adequate distance due to dilution in a compact deposit of galvanic sludge in 'lenses'.
- High-concentration phenol is a well-known, old disinfectant. However, its biochemical degradability under aerobic conditions is known which has also been confirmed in this test. The biochemical conversion to humic substances is known in soil science.
- Cyanides are highly toxic but are degradable materials which also occur in nature. However, their degradation in waste deposits was not proven.

The special test programme for the biochemical degradation of cyanides (Section 7.4.3.6) clearly showed that microorganisms convert cyanide into innocuous complexes under landfill conditions in the best case but do not degrade them. It has been proved unambiguously that industrial wastes have an extremely negative influence on the biochemical stabilisation process which is in agreement with the indirect conclusions from the increased leachate contamination. It follows from the increased contamination of the leachate in the low and medium contamination stages that moderate spiking also has a negative influence on the long-term stabilisation.

7.7.3 Checking the extrapolation of intensive medium-term measurements by extensive long-term measurements

Extensive measurements were used over a period of 13–14 years to investigate the biochemical stability of those landfill types whose wastes had clearly proved as being increasingly stable over the medium term (5 years): municipal waste without sewage sludge (Lys. 2), municipal waste with sewage–sludge lenses (Lys. 7), sewage sludge waste mixes after intensive aerobic degradation before compaction (Lys. 5). Colonisation, respiration activity, reactivation potential and reduction of the organic material exhibited equally low values after 13–14 years. According to these criteria all three deposits have achieved extensive biochemical stability regardless of landfill technology.

The influence of the intentional aerobic pre-treatment was recognised by the substantially faster degradation of the organic substance and the higher intensity of the degradation process. The intentional aerobic degradation achieved the same result within 2 years, which needed more than 10 years under unfavourable site conditions and at the same dimensions and depended on oxygen supply by diffusion and rain water. For conversion to deep landfills a time factor of >5 of the model lysimeter has to be multiplied with the time requirement of the stabilisation process which increases as a function of the square of landfill depth (cf. Section 2.8.3).

Since it has been proved that intentional aerobic degradation increased the intensity of the degradation processes in the same initial material, hence it follows that standard deposits must have a higher mobilisation potential. That contradicts the investigation results on samples that yielded the same reactivation potentials after 13 years. This statement on reactivation is crucial for the long-term forecast, therefore this contradiction must be clarified by a large-scale reactivation test (Chapter 8).

It also has to be noted for the assessment of compact depositions of sewage sludge that sewage sludge was very probably not considered representatively in the sampling. Again, reactivation tests can be used to check this effect (Chapter 8).

8

Checking biochemical stability using reactivation measures on selected deposits

8.1   Objectives

The measurements assessed in Sections 3–7 were performed on the entire test body (landfill sectional core) or on samples from the body. The measurements on the entire body infer indirect conclusions on stability, and the spread of the samples is very great because of the typical inhomogeneity of wastes. In addition, the measurements were carried out on a material at rest, so it was not possible to distinguish between reversible conservation and irreversible stabilisation. There-fore, to achieve a reliable assessment of material stability, it was necessary to perform investigations into the reactivation of degradation processes on the entire material of the landfill sectional cores (see Fig. 2-1) at least on the selected landfill types, in particular. The possibility to carry out these investigations arose in connection with the Deutsche Forschungsgemeinschaft DFG (German Research Foun-dation) key programme.

The interdisciplinary integrated project 'Mining – Re-using – Depos-iting Landfill Wastes from Old Sites', sponsored by the Volkswagen foundation Hanover, dealt with the measurement of those reactions and emissions from selected landfill types of the DFG key programme that, depending on the landfill technology used and the industrial contamination of the wastes, emerge when old waste dumps are emptied. The results will be evaluated with the aim of testing the long-term stability with regard to potential mobilisation and activation.

The wastes removed were homogenised in a mixing drum (waste collection vehicle) and arranged as a sectional core of a bottom-venti-lated flat windrow in bottom-ventilated lysimeters (air supply through one-way palettes) (cf. Section 11.4.1) in accordance with the principle of the 'chimney draught method' (Spillmann and Collins, 1981). Unlike the purely aerobic 'chimney draught method', the lysimeter was temporarily sealed with an aluminium polyethylene sandwich foil in order to alternate between aerobic and anaerobic degradation in the activation phase within short periods of time (see Fig. 2-1). One expected an increase of the degradation activity from this change, in particular a dechlorination of organic compounds. However, the BOD of the mined wastes was low but it was not possible to ensure strict anaerobic conditions due to the inevitable leakiness. Where anaerobic conditions are not expressly mentioned in the results, the biochemical processes were activated aerobically.

8.2   Criterion selection

The research project on landfill mining of old waste dumps was carried out in an interdisciplinary way. The results were published as reports and papers, and can be divided into the following subprojects:

A   Waste management: F. Brammer and H.-J. Collins (Report 1995)
B   Ecotoxicological and ecophysiological monitoring of landfill mining, intermediate storage and re-landfilling, control of in-situ activities and optimisation of the degradation processes: M. Kucklick, P. Harborth and H. H. Hanert (Report 1995)
C   Chemical-analytical subproject: J. Gunschera, J. Fischer, W. Lorenz and M. Bahadir (Report 1995)

The overall publication was published in 1997 (Brammer *et al.*, 1997). The following investigation criteria will be used for the assessment of the long-term stability of the deposits from the above programme:

Waste management
- State of the material, in particular, the recognisable degree of degradation.
- Mass changes and their densities.
- Activation of exothermic degradation processes.
- Change of sieve curves.
- Influence of reactivation on leachate contamination.
- Assessment of solids using landfill engineering criteria.

Chemical-analytical criteria
The chemical parameters for the assessment of the expected long-term emissions are identical to those of landfill mining. The analysis programme covered the following parameters:

234

- Very volatile chlorinated hydrocarbons and BTX in gaseous emissions.
- Very volatile and chlorinated compounds in leachate, eluate and solids.
- Organic acids, chlorine and alkyl phenols in leachate, eluate and solids.
- Semivolatile substances (phthalate, PAH, triazine) in leachate, eluate and solids.
- TOC content in leachate, eluate and solids.
- Element content in leachate, eluate and solids.

Biochemical conditions of stability
The following criteria have been used for the assessment of stability from the ecotoxicological and ecophysiological monitoring of landfill mining and control of in-situ activities:

- ecophysiological assessment
- ecotoxicological assessment
- checking the biochemical activation using microbiological methods
- extent and change of microbial numbers.

The advantage of using waste management criteria including the determination of landfill allocation criteria is that the measurements are performed on the entire material in an integrated way. They are therefore used as the prime parameters. The results of chemical analysis obtained from individual samples always have to be assessed in view of the large inhomogeneity of waste. Therefore, those significant results will only be quoted that yield information about stability and degradation of the basic substance and the intentionally spiked chemicals or give information on diffuse chemical base contamination.

Initial material, landfill conditions and chemical contamination form the basic environmental conditions for biochemical processes. Relying on special, additional tests, the working group 'Biochemistry' succeeded in clarifying contradictions and unclear results from waste-management and chemical investigations. They will therefore be used for the interpretation of the waste-management test results and chemical analyses.

8.3 Municipal solid waste without industrial contamination and without biological pre-treatment

8.3.1 *Selection of landfill type*
The population equivalent deposit of municipal waste with sewage sludge lenses was selected (Lys. 7) as a waste without industrial contamination

and without any pre-treatment because best reflected the current place-ment technique. On the basis of recent contamination of the basal leachate discharges (Chapter 5) but also on all other measurements of the entire waste mass, it was expected that an extensive stabilisation had occurred.

8.3.2 Landfill engineering assessment
Friederike Brammer and Hans-Jürgen Collins

Condition of material
Landfill mining exposed the sewage sludge lenses as discrete compact sludge masses (see Fig. 8-1). Certain sections of waste within the sludge ranges of the waste body were kept in such a good condition that a newspaper of 1978 from the top third of Lys. 7 was still easily readable after 14 years. The same observation was made in the simulta-neous test (Lys. 8) dismantled in 1981 after 5 years of test time (see Fig. 2-11(3)). The same applies to the water content (Table 8-1).

Mass changes
The total mass of the landfill sectional core (Lys. 7) was 42 t DS waste and sewage sludge at the time of installation but only 31.4 t DS waste

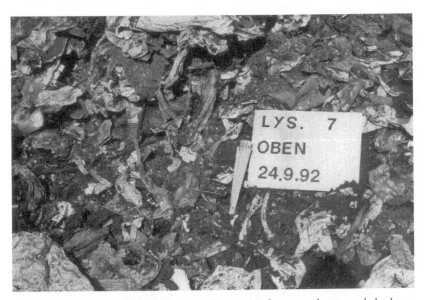

Fig. 8-1 Wastes on removal from the anaerobic municipal waste with sewage sludge lenses from lysimeter (Lys. 7, top) (OBEN = TOP)

Table 8-1 Composition of anaerobic municipal waste with sewage sludge lenses (Lys. 7, top) on removal from lysimeter in 1992

Material groups	Water content [% dry]	Weight	
		[% by weight, wet]	[% by weight, dry]
Paper/paperboard	200.31	5.68	3.68
Packaging material	113.14	1.12	1.02
Fe metals	29.53	1.57	2.36
Non-Fe metals	63.55	0.29	0.35
Glass	7.83	7.26	13.09
Plastics	90.21	9.42	9.63
Textiles	104.91	1.73	1.65
Minerals	8.53	2.98	5.34
Composite materials	53.54	1.84	2.32
Problematic wastes	2.17	0.40	0.76
Nappies	0.00	0.00	0.00
Wood	92.65	0.80	0.81
40–100 mm	118.06	3.70	3.30
8–40 mm	120.26	48.20	42.56
<8 mm	122.13	15.01	13.14
	$\overline{WC} = 94.5$	$\sum = 100.00$	$\sum = 100.00$

was removed in 1992. The total dry substance was thus reduced by 25% DS by weight waste within 14 to 16 years. The same degradation was observed in the parallel unit (Lys. 8, Table 3-1), dismantled in 1981, i.e. the majority of degradable mass was degraded under these favourable conditions (short acidic phase, diffusing oxygen) within 5 years, then the mass remained approximately constant.

Activation of exothermic degradation processes
The landfill sectional core (Lys. 7) was divided into two lysimeters with the same content; the bottom part was directly relocated without reactivation for comparison purposes and was run as an anaerobic landfill sectional core (Lys. 7, bottom) in a highly compacted state (cf. Fig. 2-1). The top lysimeter part (Lys. 7, top) was tested to see whether a substantial activation of the aerobic degradation using a simple aerobic landfill technology in accordance with the 'chimney draft method' is possible by loose stockpiling on an air-permeable base. Waste temperatures in the re-activated lysimeter (Lys. 7, top) rose only by about 10°C above the outside temperature. Aeration by means of a pressure surge (BIOPUSTER® method), proven in subsequent in-situ biochemical activation of old landfills (Chapter 11), failed to activate an efficient exothermic degradation. An extensive degradation (25% DS waste)

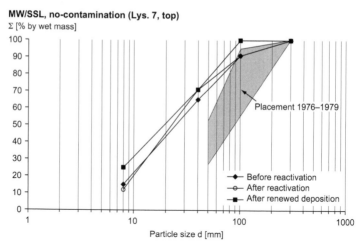

MW/SSL, no-contamination (Lys. 7, top)

Σ [% by wet mass]

Fig. 8-2 Change of the sieve curves of anaerobic municipal waste (MW) with sewage sludge lenses (SSL) (Lys. 7, top) from the first deposition to final deposition after reactivation

or poor gas exchange (silting-up the pores) can be considered as a cause. Therefore, a biochemical investigation in the activation phase is of highest importance for the assessment.

Change of sieve curves

Another proof of the degradation processes is the shift of the municipal waste sieve curve from coarse to the medium to fine fraction. The sieve curves of the reactivated landfill sectional core (Lys. 7, top) in Fig. 8-2 indicate a clear shift to the medium and fine fraction, thus a significant degradation results, which had already been reached in the parallel unit (Lys. 8) after 5 years.

Influence of reactivation on leachate contamination

The potential change of leachate contamination due to degradation of the material and the accompanied relocation of the seeping channels is essential for emission forecast. The BOD_5/COD ratio of the leachate remained fairly constant regardless of the mechanical mixing. This quotient lay between 0.013 and 0.077 in same order of magnitude in the leachate from the almost doubly compacted wastes of the final compaction. It can be proved that these typical conditions for extensive degradation are the consequences of degradation processes and not of washed-out channels. The associated COD annual transported mass was 30 000 mg/t DS waste. Other parameters also confirm that relocation combined with an interim treatment and doubling of the density did not increase the low contamination found in the long-term observation. All parameters, except for a few values, met the minimum requirements for discharge of waste water into watercourses (AbwV, 2002) even after a renewed compaction. The aerobic interim treatment, despite the moderate result, exerted an effect by reducing the contamination – in particular AOX.

Assessment of solids using deposition and landfill engineering criteria
(e.g. LAGA 1998, TASi 1993 and AbfAblV 2001)
The allocation of a solid according to the LAGA Z2 and TASi parameters indicated that, after 16 years, ignition loss, solids' TOC as well as sulphate and chloride were the only parameters that markedly exceeded the limiting values. The aerobic interim treatment had no measurable influence on these parameters.

Summary
The sum of all currently used landfill engineering criteria proved that this deposit was stable to a large extent. The fact that well-conserved, degradable material was still found, was in direct contradiction to the stability and had to be further investigated.

8.3.3   Chemical analysis of toxic residues
Jan Gunschera, Jörg Fischer, Wilhelm Lorenz and Müfit Bahadir

8.3.3.1   BTX aromatics and very volatile chlorinated hydrocarbons

BTX aromatics and very volatile chlorinated hydrocarbons in gaseous emissions
BTX aromatics were detected in the gas phase of the deposit with no industrial contamination (Lys. 7, before waste removal) within a concentration range between 5 µg/m^3 (benzene) and 420 µg/m^3 (toluene) after 16 years. Very volatile chlorinated hydrocarbons were found up to 7.8 µg/m^3 (trichloromethane). In the interim treatment phase (aeration of Lys. 7, top) increased concentrations were measured for trichloroethene (180 µg/m^3) and tetrachloroethene (6700 µg/m^3). No BTX aromatics or very volatile chlorinated hydrocarbons were found after the final compaction because of evaporation during dismantling.

BTX aromatics and very volatile chlorinated hydrocarbons in leachates
Concentrations markedly below 1 µg/l of very volatile compounds were only detected in leachates in isolated cases during the entire observation period after 16 years. The aerobic activation of biological degradation processes failed to mobilise this type of material.

239

8.3.3.2 Medium and semivolatile hydrocarbons

Medium and semivolatile chlorinated hydrocarbons in leachates

No chlorobenzenes, PCB or DDT analogues were detected in the leachate of the initial deposit, the activation deposit and the final compaction during the entire observation period before and during relocation. HCH isomers were measured in isolated cases in concentrations <1 μg/l.

The AOX concentration in the leachate was 0.3 mg/l after 16 years before and after activation. An increase to 0.5 mg/l was only observed for a short period during activation.

Medium and semivolatile chlorinated hydrocarbons in eluates

The results were not uniform due to the inhomogeneity. Thus, the maximum AOX content of the eluate from sample 8–40 mm was 2.5 mg/l and that of the sample <8 mm 0.9 mg/l. Other eluate concentrations were below 0.2 mg/l.

Only a few chlorinated hydrocarbons were detected and at low concentrations, e.g. chlorobenzenes in only one sample 8–40 mm of 2.6 mg/l, HCH (hexachlorocyclohexane) <1.2 mg/l in all samples. PCB and DDT analogues were not detected.

Medium and semivolatile chlorinated hydrocarbons in solids

Contamination by PCB, DDT analogues and HCH isomers were within a range of 0.31–0.75 mg/kg DS waste. The aerobic biochemical degradation was only able to specifically reduce the concentration of HCH isomers to values <0.03 mg/kg DS waste (Lys. 7, top).

PCB, DDT analogues and HCH isomers were detected at concentrations up to 0.13 mg/kg DS waste only in the individual relocated comparative landfill sectional core (Lys. 7, bottom) before and after the compacted re-installation. Significant differences were not observed. Increased content of HCH isomers (0.62 mg/kg DS waste) were measured before degradation in the re-activated landfill sectional core (Lys. 7, top). They decreased markedly during dismantling, while the concentrations of DDT analogues showed non-uniform high values in comparison to the bottom half (0.31 mg/kg DS waste). The PCB content was comparable only to those in the individual relocated lysimeter part. Chlorobenzenes were not detected.

8.3.3.3 Organic acids, chloro and alkyl phenols

Organic acids, chloro and alkyl phenols in leachates

The organic acid content in leachates were low outside of activation.

240

The concentration was between <3 and 190 mg/l leachate without activation and hexadecane acid exhibited the highest content (between 15 and 110 mg/l). During activation the values rose by more than one order of magnitude to a maximum of 5135 mg/l. The high value was mainly caused by benzoic acid, phenyl acetic acid and iso-butyric acid. The concentrations of benzoic acid and phenyl acetic acid were below the determination limit in the parallel test without activation (Lys. 7, bottom). These acids are characteristic of the acidic phase of a newly placed bulk field of a landfill and prove an anaerobic activation, which was not stopped by the final compaction within the observation period.

Organic acids, chloro and alkyl phenols in eluates

The eluates were not suitable for characterising the contamination of the waste using organic acids, chloro and alkyl phenols.

Organic acids, chloro and alkyl phenols in solids

The concentration of organic acids was equally low throughout the entire waste body without activation. In addition to traces of benzoic acid, the acids mainly consisted of long-chain fatty acids. Phenylacetic acid or heptanoic acid were not detected.

Due to the activation of part of lysimeter 7 (top), the extrapolated overall concentrations of organic acids increased from 9800 μg/kg wet waste to just 23 000 μg/kg, mainly by long-chain fatty acids. In connection with the strongly increased values for benzoic and phenylacetic acid, the high value for heptanoic acid, i.e. 29 000 μg/kg (fraction <8 mm) indicates (Table 8-2) the degradation of higher molecular compounds. Heptanoic acid cannot be a degradation product of aromatic compounds since this mechanism leads via benzenediol and, after ring fission, to C_6 bodies (Schlegel, 1985).

8.3.3.4 Semivolatile substances (phthalates, PAH, triazines)

Semivolatile substances in leachates, eluates and solids

The solids contained a maximum of 22 000 μg/kg PAHs. The main components were: benzo[a]pyrene, benzo[b] and benzo[k]fluoranthene. Nevertheless these materials were not found in either the leachate or in the eluate.

Simazin and the de-alkylated degradation product (desethyl deisopropyl atrazine) were detected in solids, leachates and eluates (range: from non-detectable to 600 mg/kg wet waste). The simultaneous absence of atrazine excludes the general occurrence of triazines as a

Table 8-2 Organic acids in the initial anaerobic municipal wastes with sewage sludge lenses (Lys. 7, top), <8 mm, 8–40 mm fractions and extrapolated total content

Compound[a] [μg/kg wet waste]	MW/SSL, non-contaminated (Lys. 7, top)					
	Before reactivation			After reactivation		
	<8 mm	8–40 mm	Total	<8 mm	8–40 mm	Total
Acetic acid	n.d.[b]	n.d.	n.d.	n.d.	n.d.	n.d.
Propionic acid	n.n.[c]	n.n.	n.n.	220	n.n.	26
Iso-butyric acid	n.d.	n.d.	n.d.	n.d.	n.d.	n.d.
Butyric acid	n.n.	n.n.	n.n.	200	n.n.	23
Valeric acid	n.n.	n.n.	n.n.	1000	n.n.	116
Caproic acid	1000	n.n.	150	4700	450	790
Heptanoic acid	n.n.	n.n.	n.n.	29 000	n.n.	3360
Benzoic acid	500	<500	<300	4000	400	680
Phenylacetic acid	900	<500	<400	2000	n.n.	230
Tetradecane acid	2800	1300	1000	2750	2000	1400
Hexadecane acid	31 000	5800	7400	38 000	18 500	14 500
Octadecane acid	3500	1300	1200	400	2750	1540
$\sum$ org. acids	39 700	8400	9800	82 300	24 100	22 600

(a) Nomenclature varies depending on working group, see Fischer (1996)
(b) Not determined
(c) Non-detectable

source. Simazin was not found after completing the activation of the landfill sectional core (Lys. 7, top) by aeration, only triazine was detected – a de-alkylated degradation product.

A strong increase of phthalate content from 20 mg/kg to 3100 mg/kg in the <8 mm fraction was conspicuous during activation. Obviously, a decay of coarse pieces of plastic took place. However, no significant change in the phthalate content due to biochemical activation was detected.

8.3.3.5 TOC contents

TOC content in leachates

It was conspicuous during the long-term observation that TOC values of up to 350 mg/l sometimes occurred in the leachate within the waste body, while the contamination at the base only oscillated around 50 mg/l (maximum 70 mg/l) (see Fig. 5-1(6)). During the reactivation of the upper lysimeter part (Lys. 7, top) the same peaks were mobilised again and the creation of long-chain fatty acids and phenol was identified as a cause. The relocation of the base without aeration (Lys. 7, bottom) did not lead to an increase in TOC in the leachate, although the potential of degradable materials was sufficient. Obviously, oxygen access is also necessary for a predominantly anaerobic degradation and degradation can be mobilised again after 16 years because of the existing potential.

TOC contents in solids

The extrapolated total of solids TOC decreased due to the verifiable activation from just 6% to 4%. The reduction in the fraction <8 mm from 13% to 5% was significant. Regardless of the activation of degradation processes by aeration, the relocation TOC values only increased due to relocation by 1% to 2% again. Obviously, the renewed mixing promoted the decay processes in the coarser waste fractions (>40 mm) which led to an increase in TOC in the smaller waste fractions.

8.3.3.6 Element contents

Element contents in solids

The limiting values of AbfKlärV (1992) for Cr and Hg were only just exceeded in the solid material but much more so for all other parameters.

Element contents in eluates

The eluate contents, with the exception of Ni, met the criteria of landfill class I. Ni contents with 0.011 mg/l and 0.015 mg/l were below the limiting value (<0.02 mg/l) in two samples for landfill class II, and they sat above it in the 3rd sample (0.034 mg/l).

Although the pH values were about the same in eluates (7.8–8.1) and leachates (7.6–8.1), the concentration of the elements in eluates was sometimes considerably below that found in leachates. This tendency coincides with practical experience from landfill investigations.

Element contents in leachates

A mobilisation of alkali metals (Na, K), earth alkali metals (Ca, Sr), heavy metals (Cu, Ni and Zn) and sulfate was observed in the leachates during the ventilation phase. Particularly conspicuous was the simultaneous increase of calcium and sulfate from 200 mg/l to 500–600 mg/l in the periods of biochemical activity, characterised by the increase in TOC. The process resembled the acidic phase of a landfill despite a pH value of 8.1. Since acid production was also proven, it can be concluded from this observation that a waste stored for 16 years still reacts to an oxygen influence akin to a fresh waste with acid generation and accompanied mobilisation of minerals under certain circumstances.

8.3.3.7 Summary of the chemical investigation

The residual contents of the pollutant groups were moderate and only partially mobilised by biochemical reactivation due to aeration. However, it has to be noted, for the assessment of the long-term stability, that organic acids were produced during reactivation, known from the initial phase of waste deposits. It can be concluded from this that an anaerobic reactivation potential was available, which, however, was not completely mobilised and reduced by the measures applied.

8.3.4 Biological assessment of stability
Martin Kucklick, Peter Harborth and Hans-Helmut Hanert

8.3.4.1 Ecophysiological assessment

Characteristic colouring

The organoleptic assessment showed that – depending upon oxygen access – different zones had developed in the non-contaminated municipal waste deposit with sewage sludge in 'lenses' over 16 years: a brown coloured aerobic edge zone (the PE cladding of the test body was not completely impervious to diffusing oxygen in spite of a barrier made from AL-PE composite sheet), a partially aerobic mixing zone (dark with brown marks) and markedly anaerobic zones in the pasty sludgy material.

Numbers of microorganism

In partially anaerobic zones and in the sewage sludge the number of methanogens (about 6.5×10^5/g DS waste) and sulfate reducing bacteria (1×10^7 and/or 7×10^5/g DS waste) were within the range of values found in the literature for landfills (10^5 methanogens and $10^{5.5}$ sulphate reducing bacteria per gram in 2 m depth; Sleat, 1987). Acid-forming microorganisms, which supply substrates for these groups, were also detected.

Characteristic metabolic products

Methane at 10% by volume found in the core ranges was low compared to a landfill. However, since it was generated despite the constant influence of atmospheric air, this indicates active CH_4 generation. The pH value of the waste was slightly alkaline: 7.6 in a partially anaerobic area and 7.7 in the sewage sludge. A sulphide content of 850 mg/kg in the core range indicated a marked anaerobic zone. The boundary range affected by oxygen only exhibited a sulphide content of 26 mg/kg.

Stability state
Based on these findings, i.e. moderate CH_4 generation, high sulphide content, slightly alkaline solid material, relatively high numbers of methanogens and sulphate reducing bacteria, the lysimeter can be classified as being in the methanogenic phase, possibly the late methanogenic phase due to the low CH_4 concentrations.

8.3.4.2 Ecotoxicologic assessment

The ecotoxicologic assessment indicated highly poisonous material for luminous bacteria within the core range of the pasty-sludgy material and innocuous material in the aerobic edge region. Based on luminous bacteria toxicity of the solid material, the visible zoning of the test body was clearly identified.

Very high toxicity was present with an EC_{20} reciprocal value of about 200–400 l/kg in the core range of the top lysimeter part in the sewage sludge lenses. (One would have to dilute the eluate from 1 kg waste with 200–400 litres of water to reduce the inhibition of bacterial luminosity to 20%.) Toxicity was still high in the partially anaerobic transition ranges but about an order of magnitude lower, and no luminous bacteria toxicity was found in the aerobic edge regions with intensive oxygen contact. Obviously, a complete decontamination seems to be possible through contact with atmospheric oxygen as far as this can be detected by this test.

The core ranges of the lysimeter – free of sidewall disturbances – were only considered for scaling up the results to real landfill conditions and were found to be highly toxic in this case.

The leachate did not exhibit toxicity for daphnia and luminous bacteria. This leachate was, however, ventilated when passing through the drainage gravel, thus its origin corresponded to that of the aerobic boundary range.

Gas from inside the lysimeter exhibited a medium to moderate luminous bacteria toxicity of $7–11 \, l/m^3$ and was thus about ten-times more toxic than the surrounding air. An access of outside air during waste removal reduced toxicity of the gas to insignificant values.

8.3.4.3 Biochemical reactivation

The cause of the inhibition of the aerobic degradation in the aeration phase of the re-activated landfill part (Lys. 7, top) was clarified: BOD – not detectable in the aerated mix – was approximately $140 \, mg/(kg \times h)$

in the laboratory. It was proved by adding a structural material – sand – that blocking of the fine pores impeded the passage of oxygen so preventing the aerobic degradation of the degradable organic material. Even after 16 years of storage, a deposit appearing stable according to the organic contamination of the leachate still contained a substantial amount of easily degradable organic material whose degradation was anaerobic and potentially aerobically mobilisable.

8.3.5 Comparative interpretation of the individual results for stability assessment
Peter Spillmann

The results of stability assessment according to waste management criteria are not unambiguous. Conservation was easily visible in the anaerobic zones within the sphere of influence of the sludge lenses. Nevertheless a high degree of degradation, i.e. 25% by weight of dry mass waste, was confirmed for this type of landfill, which was also reflected by the shift in the sieve curves. The reactivation based on the 'chimney draught method' failed to have any effect both concerning temperature and organic leachate contamination. The biochemical assessment indicates that the initial material was in the late methanogenic phase. Based on all these results, the deposit would have to be classified as stable despite a contrary impression.

The detailed biological investigation into reactivation – contrary to the waste management investigation – proved that there was a substantial aerobic activation potential, and the chemical investigation proved that despite a pH value >8, organic acids were produced which mobilised metals similar to the acidic phase of a fresh waste. The sensitive reaction of luminous bacteria to anaerobic conditions enabled differentiation between aerobic zones, anaerobic transition zones and anaerobic cores (ranges of sludge and standing water). The anaerobic ranges agreed with the observed re-activatable conservation zones.

A comparison of the individual results permits the conclusion that extensive stabilisation in the sense of mineralisation and humification took place only under aerobic degradation conditions, while anaerobic degradation particularly in combination with an inhibition of leachate flow led to a re-activatable conservation. The effects of these conserved zones cannot be measured in the leachate and gas composition as long as the residual substances from the anaerobic degradation are further mineralised and humified in a largely aerobic environment. The conserved anaerobic zones are activated anaerobically in the sense of an acidic initial phase of the deposit when some form of interference into the waste body, e.g. construction, take place. The generated acids can remove heavy metals despite a mainly basic leachate. The deposit was therefore reversibly conserved and not stable in the sense of an aftercare-free deposit.

8.4 Municipal solid waste without industrial contamination after extensive biological degradation

8.4.1 Selection of landfill type

The method for the intensive biological degradation of a population equivalent sewage sludge waste mix was developed as the 'Gießen model' by the working group 'Gießen University Institutes for Waste Management' (GUfA, 1970) together with the city of Gießen and introduced into waste management. In the initial version the fine material was removed after extensive decomposition by sieving and used as an auxiliary substrate in the reclamation of raw soils (Gießen version). Because there was no demand for this 'product', the entire rotted material was compacted in facilities built later (Oldenburg version). The bottom half of the landfill sectional core consisted of sieve residues of the 'Gießen model' (however with high fines adhering), the top part of total rotted and then compacted mix (similar to Oldenburg). The composition of the initial material was – within unavoidable material fluctuations – identical to the current installation technique using 'lenses' or 'cassettes'. Because of its incontestable technical advantages, the landfill type using extensive aerobic biochemical stabilisation has been a state-of-the-art model and example for methods with similar technology and objectives, e.g. 'chimney draught method' in Swäbisch Hall since 1976 (Spillmann and Collins, 1981). This type of landfill was selected as a practice-proven opposite to the usual compaction landfill (Lys. 5).

8.4.2 Landfill engineering assessment
Friederike Brammer and Hans-Jürgen Collins

Condition of materials

The appearance of the material in the landfill sectional core after 16 years revealed, without exception, brown earth-like material. Afterwards, the entire waste body remained in an aerobic condition, although the material was twice as compacted as that installed in cassettes ($\rho_{DS\,decomposition} = 1.0\,t/m^3 > \rho_{DS\,cassettes} = 0.51\,t/m^3$). From this it can

247

be concluded that the BOD had been reduced by the intensive decomposition in the entire test body to such an extent that oxygen dissolved in rain water and exchanged by diffusion was enough to supply sufficient oxygen for microorganisms.

Leachate contamination

Eluate values according to TA Siedlungsabfall (TASi, 1993), Class I, including TOC (approx. 20 mg/l) were met and only slightly exceeded by cadmium (limiting value 0.05 mg/l, eluate 0.06 mg/l). The solids TOC of Landfill Class II (TASi, 1993) was exceeded by a factor 2.1.

Mass changes

It has to be noted that degradation of materials difficult to degrade under these conditions was measured by gravimetry after the first compaction. Approximately 24 tonnes of dry substance was compacted into about $25\,m^3$ volume after extensive aerobic degradation ($\rho_{DS} \approx 1\,t/m^3$). About 21 tonnes of dry substance was removed in 1992 after 10 years. That corresponds to a mass reduction of more than 10% by weight of DS waste of the biochemically stabilised waste mix in an extremely compacted condition.

Change of sieve curves

The sieve curves (Fig. 8-3) indicate a significant change of the 'particle' composition towards fine material and thus extensive degradation.

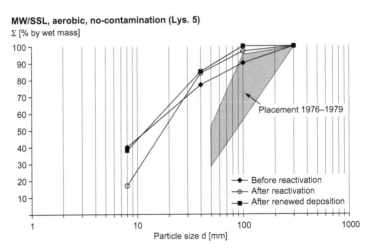

Fig. 8-3 Changes of sieve curves of the permanent aerobic sewage sludge waste mix (Lys. 5) from the first deposition to the final deposition after the ineffective reactivation test

248

Activation of exothermic degradation processes
The reactivation of biochemical degradation processes was hard to recognise from the temperatures (about 3°C higher than the average air temperature). It could only be proved by the chemical analyses of the substances contained in water.

8.4.3 Residual content of selected chemical compounds and changes imposed by reactivation of aerobic degradation
Jan Gunschera, Jörg Fischer, Wilhelm Lorenz and Müfit Bahadir

8.4.3.1 BTX aromatics and very volatile chlorinated hydrocarbons

BTX aromatics and very volatile chlorinated hydrocarbons in gaseous emissions
In a state of rest, very volatile chlorinated hydrocarbons within a concentration range of $0.1-17\,\mu g/m^3$ and BTX aromatics toluene $(270\,\mu g/m^3)$ and o-xylene $(60\,\mu g/m^3)$ were detected in the gaseous phase before opening the lysimeter. The concentrations of BTX aromatics dropped below the detection limit $(10\,\mu g/m^3)$ after aeration. The trichloromethane concentration remained fairly constant $(20\,\mu g/m^3)$ and the tetrachloroethene concentration rose slightly to $6.7\,\mu g/m^3$. Neither very volatile compounds nor BTX aromatics were detected after the final compaction.

BTX aromatics and very volatile chlorinated hydrocarbons in leachates
Very volatile compounds were only detected in the leachate in isolated cases over the entire observation period and in concentrations significantly lower than 5 µg/l.

8.4.3.2 Medium and semivolatile chlorinated hydrocarbons

Medium and semivolatile chlorinated hydrocarbons in leachates
Minor quantities of HCH isomers (<1 µg/l) but no PCB, DDT analogues or chlorobenzenes were detected in the leachate. AOX concentration was constantly between 0.1 and 0.3 mg/l over the entire period of observation.

Medium and semivolatile chlorinated hydrocarbons in eluates
The AOX values were below the detection limit (0.1 mg/l) in the eluates of the <8 mm and 8–40 mm waste fractions, so that no analysis of individual substances was performed.

249

Medium and semivolatile chlorinated hydrocarbons in solids

The concentration of HCH isomers in the solids remained constant in the range of 0.05–0.1 mg/kg DS waste over the entire sampling period. DDT analogues and PCB were found in concentrations up to 0.3 and/or 0.6 mg/kg DS waste. Chlorobenzenes were not detected.

8.4.3.3 Organic acids, chloro and alkyl phenols in leachates, solids and eluates

The content of organic acids in the eluates and leachate samples of the decomposed mix (Lys. 5) before its renewed aeration was low, between 50 and 150 µg/l, similar to the deposit without pre-treatment (Lys. 7) where acetic acid and benzoic acid dominated.

The contents of organic acids in the solid waste samples were between 2700 µg/kg before and/or 5300 µg/kg wet waste after renewed aeration and, as expected, below the residue contents of the untreated waste. A characteristic was the intensified occurrence of hexadecanoic acid. In particular, the <8 mm fraction exhibited a high content after aeration (4200 µg/kg before and 24 000 µg/kg after the interim treatment).

Similar results were found for phenol contamination in the tested samples. Traces of phenol and methyl phenol (1–3 µg/l) were detected in the leachate. The residue contents in the waste samples were equally high as in the deposit without pre-treatment and were between 80 and 320 µg/kg wet waste. Concerning the typical degradation products of organic compounds the contamination of the decomposed mix was altogether low. Only a slight increase of hexadecanoic acid indicated incipient degradation processes.

8.4.3.4 Semivolatile substances (phthalates, PAH, triazines) in leachates, solids and eluates

No significant difference was found for the group of semivolatile substances between the decomposed mix (Lys. 5) and the deposited waste without any treatment (Lys. 7).

Hardly any PAH was detected in the leachates and eluates, while the PAH content in the solid waste samples was in the range of between 500 and 3100 µg/kg DS waste. Benzo[b]fluoranthene and benzo[k]fluoranthene as well as benzo[a]pyrene contents were also the largest of PAH here. Based on the similar PAH pattern and similar concentration ranges it has to be assumed that the occurrence of PAH is in connection with their ubiquitous occurrence in municipal waste.

250

A trace of simazin (1.5 µg/l) from the group of the triazines was detected in the eluate of the <8 mm sample after interim treatment. Simazin was also found in the solid waste samples (25–280 µg/kg DS waste), although the mix had not been intentionally spiked.

Analogies apply to the phthalates as to the group of PAH and triazines. Traces were only detected in the leachates and eluates. The values found in the waste samples were between 700 and 8200 µg/kg wet waste and they correspond to the pattern found there with a dominance of DEHP (diethyl hexyl phthalate) and DBP (dibutyl phthalate).

8.4.3.5   TOC contents in leachates, solids and eluates

The long-term tendency of TOC contents in the leachate varied between 40 and 140 mg/l with an average of about 65 mg/l before reactivation. The average value only increased to about 120 mg/l for a short period in the final phase of the reactivation test. No significant change due to the intentional aeration was observed. For comparison: the COD limiting value for discharge into watercourses is COD<200 mg/l ≈ TOC<70 mg/l (AbwV, 2002).

The TOC content in the eluates of the <8 mm fraction decreased from 51 mg/l to 11 mg/l due to the aerobic activation. For comparison: landfill class I according to TASi (1993): TOC<20 mg/l.

The solids TOC decreased from 5.4 to 4.7% by weight in the <8 mm sieve pass of the samples during aerobic reactivation and again exceeded 5% by weight in the <8 mm sieve pass of the sample after compaction, i.e. the size of coarse degradable material was reduced by the compaction process.

8.4.3.6   Element contents in leachates, solids and eluates

The element content of a solid random sample exceeded the limiting values of AbfKlärV (1992) for Cd, Cu, Ni and Pb, that for Cr and Hg was just below the limiting values.

The eluate criteria for landfill class I were met with the exception of Ni. Ni, however, met the criteria of the class I after aeration and final compaction.

The change in content of the elements in the leachate provides a clear proof of re-activated degradation processes. The mobilisation of the elements aluminium, phosphorus and iron is illustrated in Fig. 8-4 as a typical process. It has to be noted for the interpretation of the values that the concentrations are low for a sewage sludge waste mix

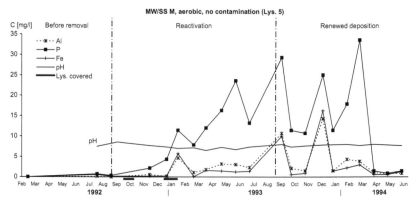

Fig. 8-4 Concentration of Al, P and Fe as well as pH value in permanently aerobic sewage sludge waste mix (Lys. 5)

and they were only detected in comparison to the very low initial contamination before aeration and to the final contamination after the final compaction. They prove that degradation processes are ongoing in this extensively stabilised material in the activation phase, which otherwise could scarcely, be proved using the common criteria such as temperature increase, BOD and methane generation. As the curves indicate, iron and aluminium are mobilised predominantly at the end of activation immediately after compaction, e.g. by the reduction of the trivalent insoluble ferric oxide to a soluble bivalent iron. The maximum value of mobilisation is therefore compared with the values after final compaction for other elements in Table 8-3. The

Table 8-3 Leachate content of selected elements at the beginning and end of the interim treatment and final phase in comparison to the current TrinkwV (2001) (Drinking Water Ordinance)

Element	MW/SS M, aerobic, no-contamination (Lys. 5)				TrinkwV (2001)
	Reactivation		Renewed deposition		
	Start [mg/l]	End [mg/l]	Start [mg/l]	End [mg/l]	
Ba	0.252	0.24	0.22	0.10	1
Ca	1367	900	910	380	400
K	275	160	160	160	12
Mg	212	144	146	91	50
Na	355	300	320	178	150
P	4.76	29	12	1.5	6.7 (PO_4^{3-})
S	610	580	630	370	240 (SO_4^{2-})
Sr	2.07	1.71	3.02	0.91	–
Ti	0.029	0.081	0.018	0.024	–

252

mobilisation can be proved without exception. The differences of other elements in the waste body between long-term contamination and mobilisation are in some cases more than two orders of magnitude smaller than those for iron, aluminium and phosphorus.

8.4.4 Biological assessment of stability
Martin Kucklick, Peter Harborth and Hans-Helmut Hanert

8.4.4.1 Ecophysiological assessment

Characteristic colouring

Two different subranges were organoleptically differentiated when dismantling the landfill sectional core: a brown-coloured top range which made up about five-sixths of the mass and a dark-brown-coloured range in the bottom part of the test body which made up about one-sixth of the mass.

Numbers of microorganisms

Regardless of the fact that methane production was extremely low in situ and non-existent in the laboratory, samples from both ranges exhibited high numbers of methanogenic bacteria before activation: 3.3×10^5 and 1.4×10^6 methanogens per gram DS waste. The number of sulphate and iron reducing bacteria was also high for an extensively aerobic waste body. Only the number of acid-producers was clearly lower than in the biologically not pre-treated material in other test bodies. No acid-producers were detected in the top part of the aerobic waste body and their number was 35 per gram DS waste in the darker, possibly partially anaerobic ranges, which was at least two orders of magnitude below other biologically non-pre-treated wastes.

The cell numbers of some anaerobic metabolic groups decreased within the aerobic interim treatment and the following final compaction. Among the anaerobic microorganisms active at low redox potential, the number of iron reducers and methanogenic bacteria was lower by one to two orders of magnitude when landfilling took place after interim treatment than before interim treatment. Also, methane generation potential from H_2 and CO_2 was no longer detected after the interim treatment and during the renewed landfilling. However, a low methane generation potential from acetate, CO_2 and H_2 was present. The number of sulphate reducing bacteria, among which there are also some spore-forming bacteria, remained constant.

253

Acid-producers, characteristic of the acetogenic fermentation phase, were no longer detected during the renewed landfilling.

The number of denitrifying bacteria increased however – a consequence of nitrate generation by nitrification during the aerobic interim treatment. The nitrification processes became apparent by the rise of nitrate concentration in the leachate to 3500 mg/l during aerobic reactivation. The activity of denitrifying bacteria reduced the nitrate content in the leachate to about 50 mg/l during the subsequent landfilling. This explains the increase of denitrifying bacteria as the only anaerobic metabolism group. The number of aerobic organotrophic microorganisms also increased, however, but only in the landfilling phase following the interim treatment.

Thus, the microbiological results indicate that further stabilisation processes still take place to a minor extent despite extensive biochemical stabilisation under aerobic conditions.

Gas composition

Gas composition before reactivation within the core part of the landfill sectional core (Lys. 5) was characterised by the presence of large quantities of O_2 (9.2% by volume), just as large quantities of CO_2 (9.3% by volume) and only very small quantities of CH_4 (0.7% by volume). A nitrogen content of 79.9% by volume, which was slightly above the nitrogen concentration of atmospheric air (78.1% by volume, Mortimer, 1987) and the simultaneous absence of N_2O, typical of denitrification, proved that no large gas volumes were produced. The large O_2 quantity of 9.2% by volume indicated a moderate biological activity. The small quantities of methane showed that methane generation took place only in a few microareas of the material.

Degradation processes during reactivation should be biologically activated in an alternating way, i.e. by the sequence of aerobic and anaerobic phases. A very high O_2 concentration of usually around 20% by volume indicated that O_2 consumption, thus respiration activity, was very low. O_2 concentration was 17–18% by volume, i.e. only somewhat lower at the beginning of the first planned anaerobic interim treatment phase and at the end of the anaerobic interim treatment in the summer. BOD was somewhat greater due to the higher outside temperature, but still very low compared with other test deposits without biological stabilisation. BOD was so low that the technically unavoidable remaining minor leakages and diffusion through the lysimeter wall were sufficient to prevent anaerobic conditions from being

254

maintained. With increased investigation time, sealing had no effect on the O_2 content at all.

No methane was detected in three out of four gas measuring probes 8 months after the final compaction. In the 4th probe 2.8% by volume methane was measured.

Oxygen consumption under laboratory conditions

The BOD of the samples from both zones of the decomposed mix without activation was very low: 8.8 mg O_2/(kg × h) (red-brown) and/ or 12.6 mg O_2/(kg × h) (dark-brown) – for comparison: waste biologically untreated: 40–140 mg O_2/(kg × h), fresh waste: 290 mg O_2/(kg × h). The BOD was roughly the same as that of a forest soil (see Fig. 8-6). In the laboratory test the BOD decreased towards the end of the aerobic interim treatment to 4.7 mg O_2/(kg × h), however, it rose again to 9.7 mg O_2/(kg × h) during the renewed landfilling activity which corresponds to the same amount as before the engineering-scale activation.

An attempt was made to increase the respiratory activity on *Pürckhauer* (*widely slotted pipe used as a driven probe*) samples of compacted initial material because of the low BOD at field and laboratory scale. Since the sample water content was 55% of the maximum water capacity (WC), and the optimum BOD in soils is usually at 60 to 65% WC, an attempt was made to increase the oxygen consumption by adding different amounts of water to the loosely filled waste packed into the containers (Fig. 8-5). The highest BOD occurred when 10 g H_2O/100 g DS waste of water was added which corresponded to a water content of 66% of the maximum water capacity. Higher moisture content in the waste once again reduced the O_2 consumption strongly. BOD only increased moderately from 13.7 mg O_2/(kg × h) to 16.5 mg O_2/(kg × h), i.e. by 20%.

In another laboratory test, a trial was made on a 2nd waste sample to try and increase the BOD by adding different quantities of a nitrogen-phosphorus fertiliser. The amount of phosphate was a quarter of the nitrogen quantity in each case. Nevertheless, no stimulation of oxygen consumption was achieved. Sufficient mineral substances required for the biological degradation were available.

Methane generation potential under laboratory conditions

Analogous to the high number of methanogenic bacteria, methane generation potential (after adding H_2 and CO_2) was also fairly high: 1.4 mg CH_4/(kg × h) for the aerobic range and 3.7 mg CH_4/(kg × h)

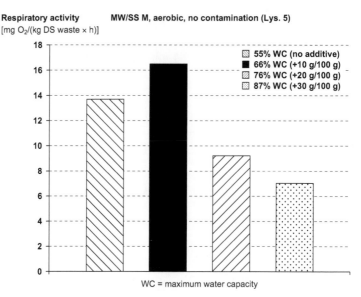

Fig. 8-5 Respiration activity of the permanent aerobic sewage sludge waste mix (Lys. 5) as a function of water content. The optimum water content for the aerobic degradation processes was approximately 65% of WC (maximum water capacity)

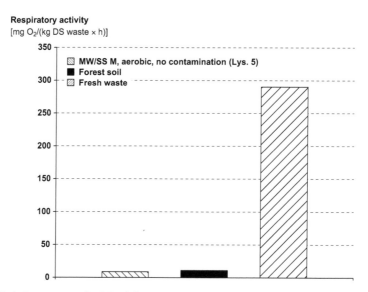

Fig. 8-6 Comparison of BOD of the permanent aerobic sewage sludge waste mix (Lys. 5) with that of fresh waste and nutrient-rich soil (mixed woodland, A horizon) – oxygen consumption of the permanent aerobic sewage sludge waste mix is about on the level of the forest soil

256

for the likely anaerobic range. No CH_4 generation took place in solid samples without any nutrient addition in either range of the biochemically stabilised sewage sludge waste mix, although an appropriate methanogenic microflora was detected based on the cell numbers and methane generation potential. It can be concluded that none or only very little organic material was available in the waste which can be metabolised under the conditions of the methanogenic phase. For comparison: the humus-rich A horizon of a forest soil exhibited moderate CH_4 generation under anaerobic conditions, while the soil under a meadow produces no methane. The substrate-dependent gas generation potential was lower than that of a forest soil.

8.4.4.2 Ecotoxicologic assessment

Leachate samples of the decomposed mix exhibited no toxicity to luminous bacteria or Daphnia before removal. The solid samples also failed to exhibit luminous bacteria toxicity. Only gas samples showed a moderate toxicity, about five- to seven-fold above the ambient air. All in all, the biologically extensively stabilised waste mix can be classified as an ecotoxicologically no-concern substance.

The luminous bacteria toxicity G_L and Daphnia toxicity G_D of leachates exhibited a value of 2 or less during the entire reactivation period. G_D only increased to 3.3 during the 2nd sealing phase of the lysimeter after 4 months of aerobic interim treatment. This can also be classified as a moderately toxic value. The limiting value of the Waste Water Ordinance (AbwV, 2002) is $G_L = G_D = 4$. While COD and nitrate markedly increased in comparison to the initial values in the spring and summer of 1993, toxicity did not increase.

No toxicity was detected after the final compaction either.

8.4.4.3 Assessment of stability state

Based on the lack of methane generation activity in the laboratory, the low methane concentration in the lysimeter, but concomitantly relatively high number of methanogenic bacteria, and sulphate and iron reducing bacteria as well as the relatively high methane generation potential indicate that the capability of degradation was still there under the conditions of the methanogenic phase. Nevertheless, hardly any conversion processes of the methanogenic phase took place. The biologically stabilised mix, therefore, was at the beginning of a postmethanogenic phase.

8.4.5 Material tests according to soil science criteria
Georg Husz

8.4.5.1 Test objective
Based on the classification of the decomposed mix as a 'forest soil' according to biochemical stability conditions, a sample obtained by wet separation from the ≤ 5 mm fine fraction was sent to the ÖKO-Datenservice GmbH, Vienna, to be tested according to soil science criteria. This institute was selected because, at its site on the 'Langes Feld' completed landfill, it successfully produces site-specific soil on an industrial-scale which can be integrated into the environment. The institute has the necessary laboratory capacity for process monitoring, final control and long-term experience in the analysis of soil-like organic-mineral mixes (Husz, 2002). The institute was asked to test a long-term decomposed sewage sludge waste mix for its conformity to a natural soil or its suitability as an initial component for the production of soils that can be integrated into the environment. This test was limited to the composition and state of elements and materials of natural origin. Positive assessments apply under the test conditions for synthetic compounds. ÖKO-Datenservice GmbH assessed whether the fine material could be integrated into the environment under these conditions.

8.4.5.2 Assessment of fine materials for integration into the environment
The results clearly indicate that the material is a decomposition product of an aerobic fermentation.

It reached a 'stability' condition which allows the material to be further treated without any special emission. In its current state, the material could be called a 'waste compost' with a potential for biological and mechanical soil amelioration. The amount of key nutrients is low, but that of trace nutrients (particularly zinc) is much higher. The material contains about 12% lime and exhibits a pH of 7.58 (H_2O) and/or 7.14 (KCl). It thus has a substantial acid neutralisation capacity.

The ion combination shows a high Na content both in soluble and in the adsorbed state.

However, calcium remains the dominating ion. The high sulphate and chloride content of the anions also provides a relatively good environment for magnesium solubility, while heavy metal cations are available although in a rather low-solubility form.

The good agreement of cation and anion equivalents, without considering the HCO_3^- ions, is conspicuous. It follows from this that hardly any dissolved bicarbonates are available, as would normally be expected in organic-mineral activated materials where CO_3^{2-} would be constantly dissolved by way of micro-biologically produced CO_2.

The outgassing of the sample or 'quantitative' removal of CO_2 by artificial aeration could cause and/or explain such a state. (Editor's comment: the extremely low respiration activity hardly provides enough CO_2).

A broad NO_3/NH_4 ratio, i.e. $(NO_3\text{-}N)/(NH_4\text{-}N) = 0.86/0.01 = 86$, substantiates the fact that oxidative conditions prevail in the sample. The mineralised nitrogen of $140\,mg/kg$ DS waste is predominantly available as NO_3 nitrogen.

If not only water-soluble but also adsorbed ammoniacal nitrogen is included in the calculation – being appropriate to the reaction,

$$(NO_3\text{-}N)/(NH_4\text{-}N) = 0.86/0.14 = 6.14$$

then a nitrogen ratio results which manifestly proves the existence of aerobic conditions.

The low solubility of metal cations thus becomes understandable (neutral to slightly alkaline reaction, trivalent oxides and/or hydroxides).

The organic substance content is about 20% and the C/N ratio is 11.7. This at least does not contradict the assumption that it is predominantly in a humified form (which would be a good indication to 'stabilisation').

However, the ratio of low-molecular to high-polymeric humic substances indicated that the organic substance is predominantly on the side of low-molecular compounds: the so-called 'humification type' can be determined using colorimetry at two different wavelengths based on production of a suitable extract (Welte) and measurement of colour extinction. The results from both measurements are shown as the quotient Q 4/6.

Q 4/6   <3: Predominantly grey humic acids – stable type
         4–5: Predominantly brown humic acids
         >5: Predominantly fulvic acids – relatively unstable

The Q 4/6 value was 6.33 in the tested sample, thus relatively on the side of the less stable 'fulvic acid'.

From the purely chemical point of view, doubts can be raised about the fulvic acid character of the organic substance, since typical fulvic acids:

(a) are rather characteristic of results of 'acidic' fermentation or conversion processes; and
(b) exhibit a C/N ratio greater than 25.

Table 8-4 Sequential humic substance explanation of the aerobic deposited material according to Husz (2002)

Fulvic acids	Hymatomelanic acids	Brown humic acids	Grey humic acids	Humins
Low	← Molecular weight →		High	Variable
Low	← Cross-linking →		High	–
High	← Solubility →			Low
High	← Reactivity by functional groups →			Low
Low	← 'Stability' →			High
C/N 80–26	15–12	16.7–12	11.6–7.75	<7.75

It has to be assumed in each case that substance groups dominate that are mainly on the left in the attribution scale in Table 8-4.

The existing cation exchange capacity is too low to fit the definition of fulvic acids either. Again, it has to be assumed that a substantial fraction of the organic substance still consists of cellulose and/or lignin-containing (relatively sturdy) substances.

The material is markedly 'humified' and so 'stable' that it can be easily processed as a component to produce soils to be integrated into the environment.

The material cannot be used as a growing medium and/or substrate without further processing with other components because of an unbalanced provision of available cations, too high electrolyte concentration and too high heavy metal content.

8.4.6 Interpretation of the results from water-management aspects
Peter Spillmann

The soil science assessment confirms waste-management, chemical-analytical and biological assessments of the material. The material is very well stabilised in terms of waste management, but cannot be integrated into the environment. The waste-related heavy metal content is very high compared to arable soils, but can only be mobilised to a minor extent. The organic material can be slowly degraded and converted under the influence of oxygen. Humification is not yet complete. Those conditions are crucial for the waste management assessment, so that the leachates can meet discharge conditions into a receiving stream without any additional treatment. It is key that the reactivation of biological activity due to interference into the waste body does not cause a critical increase in emission and the activation largely eliminates the remaining organic contamination, which is low anyway. This should prevent an uncontrollable contamination load from emerging after the guarantee of the technical barriers has expired.

Within their groundwater investigation in the same research project (cf. Section 2.1) Weis *et al.* (1995) investigated the humic-like substances in the leachate of this biologically stabilised mix characterised by the leachate COD and classified them as at the preliminary stage to natural humic substances. Filip and Smed-Hildmann (1995) identified the organic substance deposited from these preliminary stages on the particles of a porous aquifer near the surface due to microbial activity as a long-term stable humic substance which could be integrated into the environment. The same authors' measurements suggest that humic substances which can be integrated into the environment failed to be formed on particle surfaces of organic substances from leachates of poorly stabilised landfills. This provides a 1:1-scale experimental proof that an extensive biological stabilisation leads to very stable preliminary stages of soil formation. It has also been proved that the biological degradation chain of the still soluble organic wastes, e.g. in porous aquifers, yields a geologically stable form of humic substance (Filip *et al.*, 2000; Filip and Berthelin, 2001). However, the conversion process of humic-like substances to a stable humic substance does indeed induce biochemical activity. This reaction produces its own soluble organic substances which must then be degraded along their further flow path (Spillmann *et al.*, 1995; Spillmann 1995a). Therefore, despite very good stabilisation of the initial substance, the groundwater goes through an extensive change. Only a targeted transformation into soil suitable for integration into the environment solves the problem over the long term. A prerequisite for this is the efficient selective collection of industrial products in the catchment area which the industrial city of Braunschweig managed to achieve successfully (see assessments in Section 8.4.5.2 about inorganic materials and Section 8.4.3 about organic materials after activation).

8.5 Municipal solid wastes with industrial contamination and without biological pre-treatment

8.5.1 *Selection of landfilling technology*

Before the introduction of municipal waste landfills, the objective of mixed landfills with industrial contamination was to microbiologically degrade or immobilise industrial wastes with the help of organic substances in the municipal solid waste (co-disposal). This disposal method was common in Germany and is still state of the art in some other countries, as long as the disposal is generally at a low level and waste-water treatment is not fully developed. Sewage sludge deposits therefore only have a subordinate role in old mixed municipal and industrial waste landfills.

Therefore, for the reactivation studies of old landfills, municipal wastes were selected without sewage sludge and without biochemical stabilisation, but with three levels of industrial contamination to investigate (Lys. 2, bottom = no-contamination, Lys. 2, top = moderate (low + medium) contamination; Lys. 9 = high-contamination; see Fig. 2-7).

Extensive biochemical stabilisation (decomposition) with high chemical contamination (Lys. 6/10) was not investigated, because this type of landfill was not implemented industrially and therefore is not significant for landfill mining (see Fig. 2-1).

8.5.2   Landfill engineering assessment
Friederike Brammer and Hans-Jürgen Collins

8.5.2.1   Initial masses and densities
Fifty-eight tonnes of wet waste (43 tonnes of dry waste) were installed in the non-contaminated to medium-contaminated experimental deposit of sludge-free municipal waste (Lys. 2) (water content = 26% by weight (WS = wet substance)). The total volume of the deposit was $71.7\,m^3$ in 1981. This yielded then a bulk density of $0.60\,t\ DS\ waste/m^3$ with respect to the dry substance installed.

In the high-contamination experimental deposit (Lys. 9), 27.1 tonnes of wet waste (19.0 tonnes of dry substance) were installed. The volume of the lysimeter was $35.3\,m^3$ at the time of the final measurement in 1981. This yielded a bulk density of $0.54\,t\ DS\ waste/m^3$, with respect to the dry substance installed. The lower density results from the fact that the high-contaminated waste was compacted without being mixed.

8.5.2.2   Material state and material components
During removal of these wastes contaminated in 'lenses', the chemicals added during the construction at all contamination stages of the test body could be readily identified as indicated by the typical light brown, flaky and bluish, smudgy spots in the photograph (Fig. 8-7). This also enabled a targeted sampling for the chemical and biological investigations.

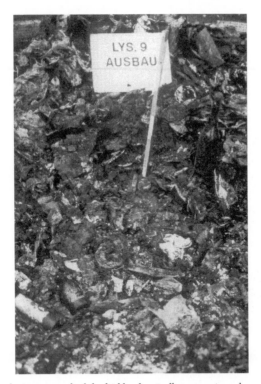

Fig. 8-7 Wastes during removal of the highly chemically contaminated anaerobic municipal waste (Lys. 9) (AUSBAU = REMOVAL)

The water content of the various material groups in the lysimeter with none to medium contamination (Lys. 2) was highly scattered and exhibited a water content in the lowest layer between 5.4% by weight (WS) for problematic wastes and 66.8% by weight (WS) for paper/cardboard. Therefore the dry substance weights are only listed in a tabular comparison (Table 8-5) of the content materials at the time of dismantling.

Extent and distribution of moisture content are important boundary parameters for the interpretation of degradation processes. Higher water contents were found for all groups of materials (except material composites) in the bottom part of municipal waste (Lys. 2, bottom, no-contamination) than in the top part of the same deposit (Lys. 2, top, low to medium contamination). The average water content of the material group fractions was 39% by weight (WS) in the top part of the municipal waste deposit and 47% by weight (WS) in the bottom part, i.e. considerably higher. Decreasing water permeability of

263

Table 8-5 *Waste composition of non- to medium-contaminated anaerobic municipal waste (MW) (Lys. 2, top and Lys. 2, bottom) on removal from the lysimeter in 1992*

Material groups	MW, moderate contamination (Lys. 2, top)			MW, non-contamination (Lys. 2, bottom)		
	Water content [% dry]	Weight		Water content [% by weight, dry]	Weight	
		[% by weight, wet]	[% by weight, dry]		[% by weight, wet]	[% by weight, dry]
Paper/cardboard	153.29	3.16	2.04	201.18	12.17	7.60
Packaging material	48.00	1.94	2.15	55.51	1.48	1.79
Fe metals	33.38	6.01	7.38	31.48	3.76	5.38
Non-Fe metals	47.11	0.44	0.49	57.89	0.10	0.12
Glass	6.28	3.29	5.06	8.00	5.27	9.18
Plastics	39.68	13.24	15.52	74.11	9.94	10.74
Textiles	90.11	1.46	1.26	120.32	1.55	1.32
Minerals	5.55	2.12	3.28	13.91	1.35	2.23
Material composites	50.86	2.45	2.65	40.46	6.31	8.45
Problematic wastes	5.66	0.28	0.43	5.71	0.19	0.34
Nappies	0.00	0.00	0.00	0.00	0.00	0.00
Wood	105.48	1.98	1.58	154.91	1.05	0.78
40–100 mm	–	36.37	41.84	–	43.17	47.93
8–40 mm	76.77	53.22	49.28	104.68	52.90	48.60
<8 mm	91.87	10.41	8.88	112.91	3.93	3.47
	$\overline{WC}$ = 63.70	$\sum$ = 100.00	$\overline{WC}$ = 100.00	$\sum$ = 88.10	$\sum$ = 100.00	$\sum$ = 100.00

the wastes with increasing surcharge is suggested as a main reason, which can be concluded from the much higher bulk density of 0.60 t DS waste/m^3 in the bottom layers (Lys. 2, bottom) as opposed to 0.51 t DS waste/m^3 for the top layers (Lys. 2, top).

The medium water content of the high-contamination municipal waste (Lys. 9) was 43% by weight (WS) at the time of dismantling.

All groups of materials, except kitchen wastes, identified during installation in 1976–1979 were still there when the lysimeter was dismantled in 1992. As expected for all deposits, the 8–40 mm unseparated sieve fraction was the largest group of materials, with approximately 33–49% by weight DS waste. The listed materials from the sortable fraction with the exception of paper and cardboard (Table 8-5), all count as difficult to degrade, and formed the major part of the sortable fraction (sortable = 100%) in the no-contamination (Lys. 2, bottom) and the moderate-contamination stages (Lys. 2, top) being about 90–95% by weight. In the high-contamination stage (Lys. 9) the difficult-to-degrade material fraction was only 60% by weight in the sortable fraction (sortable = 100%). This difference cannot be explained by the size-reducing pre-treatment mixing of the wastes before the first deposition of the non contaminated and contaminated stages alone since the sieve fraction was the same, i.e. about 49–58% by weight DS waste in all lysimeter stages. The paper and cardboard content, which is about an order of magnitude higher in the initially same waste, is therefore a proof of a retarded degradation.

Plastic represented the main group of the difficult-to-degrade material groups in all three deposits with about 10–16% by weight DS waste, while the fractions of glass (5–9% by weight DS waste, ferrous metals (approx. 4–7% by weight DS waste) and material composites (3–8% by weight DS waste) were somewhat smaller. Taking into account that the density of plastics and their material composites including textiles is around 1000 kg/m^3 while the density of minerals is at least 2500 kg/m^3 and that of iron is 7800 kg/m^3, plastic materials dominate in all three deposits. Approx. 70–80% by volume of the nonporous material of difficult-to-degrade to non-degradable wastes (sortable, but difficult-to-degrade + sortable without paper/cardboard + packaging composites = 100%) consisted of plastic. The characteristic feature in all three deposits is therefore the same 'more modern', plastic-rich municipal solid waste. The results obtained from wastes with different degrees of contamination are therefore comparable.

8.5.2.3 *Volumetric and gravimetric measurement*

The volumetric and gravimetric measurement of the deposits offers another landfill engineering assessment for the stabilisation processes. The measurement of no-contamination to medium-contamination deposit (Lys. 2) yielded a volume of about $60.5\,m^3$ at the time of removal from the lysimeter after approx. 16 years of storage. In comparison to the measurement after installing the last assembly stage in 1981, a volume reduction of about 15.6% was obtained which is almost exclusively due to settlement since the dimensions of the lysimeter did not show any significant changes. The causes of the settlement in the waste body were the collapse of cavities produced by biochemical degradation of organic substance and the loss of strength and/or softening of certain groups of materials due to an increase in moisture content of waste during storage.

With respect to the initial height of Lys. 2, i.e. 3.65 m, the measured 0.60 m change in height yields a settlement of 16.4% within about 16 years. This value is within the range of total subsidence values for landfills, i.e. about 20% of the waste height after 15–20 years (Gertloff, 1993).

The total mass of the municipal waste removed from Lys. 2 was 57.9 t wet substance (WS). 17.2 t of dry substance with a water content of 38.8% by weight (WS) was removed from the top part of the municipal waste deposit (Lys. 2, top) with low to medium contamination and 15.8 t of dry substance with a water content of 47% by weight (WS) from the no-contamination bottom part (Lys. 2, bottom). A mass loss of 10 t dry substance, i.e. 23% by weight, results from this with respect to the dry substance placed at the time of the first installation of Lys. 2 (assembly stages 1976–1979).

Since mass loss exceeded volume reduction, the bulk density of the municipal waste (Lys. 2) decreased from $0.60\,t$ DS waste/m^3 (1981) to $0.55\,t$ DS waste/m^3 (1992). This confirms the known fact that mass loss due to degradation alone by relocation of the wastes can be entirely utilised as volume gain (see, e.g. Spillmann, 1989).

Before removing the high-contamination municipal waste (Lys. 9), a volume of $32.9\,m^3$ was measured. Having an installation volume of $35.3\,m^3$ (1981), a volume reduction of 6.8% with respect to the initial volume can be calculated. This reduction is very small compared to a municipal solid waste with low contamination (Lys. 2 with 15.6%) or no-contamination municipal solid waste (Lys. 7 with 12.7%). This is another proof of a low degradation performance.

31.4 t wet substance and/or 17.8 t dry substance with a water content of 43.3% by weight (wet substance) was removed from the lysimeter containing high-contamination municipal waste (Lys. 9). This shows a mass reduction of about 6.3% by weight DS waste. This is only about 25–27% by weight of the degradation of non-biochemically pre-treated, industrially non-contaminated (Lys. 7, Lys. 2, bottom) or only low to medium contaminated (Lys. 2, top) conventionally deposited municipal solid wastes.

Only small volume reductions occurred during the 12-year period of storage in the high-contamination municipal waste deposit (Lys. 9) due to the low mass reduction. Since the mass reduction was proportional to the volume reduction between installation and removal in 1992, the bulk density did not change from 0.54 t DS waste/m^3 at the time of first installation until removal in 1992.

8.5.2.4   Temperature during aerobic activation

A key criterion for the landfill engineering assessment of stability is the temperature under aerobic conditions if gas exchange is not inhibited. Unlike wastes with no industrial contamination (Lys. 7) neither gas exchange was inhibited by sewage sludge, nor was degradation restrained by intensive biological pre-treatment. Therefore, a detailed analysis of temperature is advisable in the case of contaminated municipal solid wastes.

The undersize (<100 mm) from sieving of the waste removed from Lys. 2 was separated after homogenisation for decomposition into a no-contamination and a moderate-contamination municipal waste (the low to medium contaminated stages of Lys. 2 (see Fig. 2-7) were combined in the 'moderate-contamination' stage) and was arranged loosely in two passive bottom-ventilated lysimeters (sectional core from a flat windrow according to the 'chimney draft method').

A small part of the energy released in the aerobic microbial conversion is used for the maintenance of the microflora. The larger part of the energy, however, is conveyed as heat to the environment. The course of waste temperature during decomposition allows conclusions on the decomposition progress during aerobic degradation. An intensive warming of the waste indicates high activity of microorganisms and thus an intensive aerobic conversion of the organic substance. A lateral heat insulation of the lysimeter prevented a heat exchange between the lysimeter wall and the surrounding air other than that in a flat windrow to a large extent.

The waste temperature was measured during the aerobic activation once a week at two measuring levels ($\frac{1}{3}$h and $\frac{2}{3}$h, h = height of the lysimeter) in the no-contamination deposit (Lys. 2, bottom) and the moderate-contamination deposit (Lys. 2, top). In the high-contamination deposit the measurements were carried out at three levels ($\frac{1}{4}$h, $\frac{1}{2}$h, $\frac{3}{4}$h). Based on weekly agricultural meteorological reports, the daily average air temperatures were recorded simultaneously for comparison with the ambient air temperature. The highest temperature, 58°C, was measured in the no-contamination lysimeter, in the top level II (Fig. 8-8). The peak values of the bottom level I were 10°C less than the maxima of level II. The temperatures of the moderate-contamination lysimeter agreed in both levels (Fig. 8-9) and corresponded more to the temperatures of the bottom level I of the no-contamination lysimeter (Fig. 8-8). The temperatures in the bottom of the high-contamination lysimeter (Lys. 9) resembled those of the moderate-contamination lysimeter. However, the remaining part of the high-contamination deposit (level I + II) (Fig. 8-10) behaved like fresh waste placed for decomposition and reached start temperatures of 60°C to 75°C. A common feature of all three mixes placed for decomposition was that the temperature increased briefly and then assumed the surrounding air temperature, and, after about 1 year of decomposition, they increased again to reach 40°C. Even if cooling in the winter months sometimes retarded exothermic degradation, this unambiguously proves a very high degradation potential, particularly in the high-contamination deposit.

8.5.2.5 Change of sieve curves

The change of the sieve curve from initial material to removal from the lysimeter provides an initial estimate for degradation during storage. The change between removal from the lysimeter and the end of the aerobic interim treatment gives an indication on degradation potential still available at removal and how much has been degraded during the interim treatment.

The sieve curves of the non-contaminated deposit (Lys. 2, bottom) agreed with those of the moderate-contamination municipal waste (Lys. 2, top) to such an extent that its sieve curve suitably characterises both deposits (Fig. 8-11). To a large extent, the high-contamination deposit (Lys. 9) corresponded to the initial material (Fig. 8-12) at the time of removal from the lysimeter. After the aerobic interim treatment the sieve curve of the high-contamination waste agreed with those of

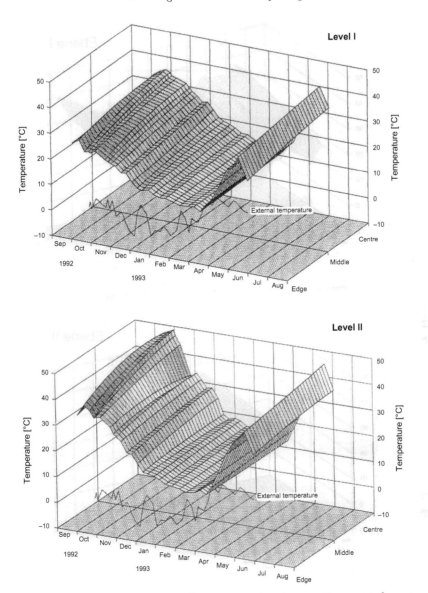

Fig. 8-8 Temperature in the chemically non-contaminated anaerobic municipal waste (Lys. 2, bottom, Levels I and II) during the aerobic interim treatment in comparison to the external temperature

no-contamination and moderate-contamination wastes. Even if one considers that the waste of the high-contamination deposit was not remixed before installation, this difference is clearly indicative of a retarded degradation, since the initial materials of the non-contaminated

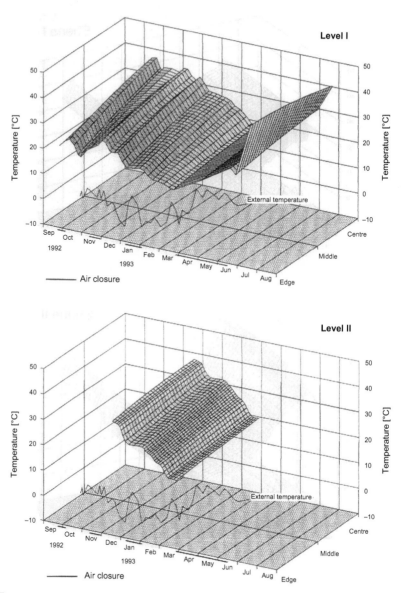

Fig. 8-9 Temperature in the chemically moderately contaminated anaerobic municipal waste (Lys. 2, top, Levels I and II) during the aerobic interim treatment in comparison to the external temperature

stage and of all three other contamination stages were collected by mixing-compacting collection vehicles, so that there was no basic difference between the mechanical pre-treatments of the materials at the time of the first installation.

270

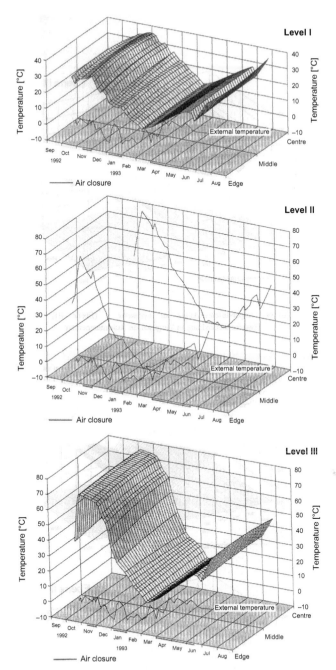

Fig. 8-10 *Temperature in the highly chemically contaminated, initially anaerobic municipal waste (Lys. 9, Levels I, II and III) during the aerobic interim treatment in comparison to the external temperature*

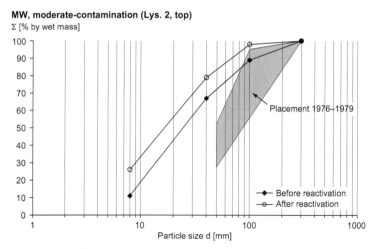

Fig. 8-11 *Changes of sieve curves of moderately chemically contaminated anaerobic municipal waste (Lys. 2, top) from the first deposition to final installation after reactivation*

8.5.2.6 Assessment of solids based on waste-management criteria
Table 8-6 displays the assessments of the solids according to material content and eluate contamination based on the LAGA allocation parameters and the TASi (1993) criteria.

As expected, the carbon-related solids criteria such as ignition loss and TOC were not met. The eluate criteria for trace elements according to LAGA allocation are determined by the initial material and only

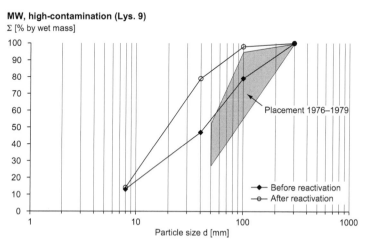

Fig. 8-12 *Changes of sieve curves of the highly chemically contaminated municipal waste (Lys. 9) from the first anaerobic deposition to final installation after reactivation*

insignificantly affected by degradation processes. Differences before and after aerobic treatment primarily indicated analysis tolerances (up to a factor 4), which had to be considered in the allocation. The limit for nickel was markedly exceeded in the high-contamination deposit (Lys. 9, galvanic sludge with extremely high nickel content) and cadmium in the moderate-contamination Lys. 2 (top) (galvanic sludge with high cadmium content).

Except for cadmium and nickel, the eluates in the extremely high contamination stage without and with aerobic interim treatment also met the criteria of landfill class II according to TASi and the stricter allocation values Z2 of LAGA for toxic trace elements. The limiting values of landfill class I according to TASi are met by most eluates. Hence the deposits tested here are within the normality of waste management despite a high increase in the chemical contamination, except for an increase of cadmium and a peak for nickel. Thus, the case tested in the high-contamination stage (Lys. 9) was by no means special.

8.5.3 Residue contents of selected chemical compounds and their change by reactivation of a predominantly aerobic degradation

Jan Gunschera, Jörg Fischer, Wilhelm Lorenz and Müfit Bahadir

8.5.3.1 BTX aromatics and very volatile chlorinated hydrocarbons

Gaseous phase

Very volatile chlorinated hydrocarbons in a concentration range from $2-10\,\mu g/m^3$ and aromatics in concentrations from $200-1300\,\mu g/m^3$ were detected in the gaseous phase of non- to medium-contaminated waste (Lys. 2). The contents of most very volatile chlorinated hydrocarbons and aromatics dropped to or below detection limits ($1\,\mu g/m^3$ for very volatile chlorinated hydrocarbons and $10\,\mu g/m^3$ for aromatics) soon after the removal for interim treatment. These substances may have evaporated during removal and sieving.

The substance classes exhibited different behaviour types in the high-contamination deposit (Lys. 9). While the concentrations of the very volatile chlorinated hydrocarbons and BTX aromatics dropped due to degassing during relocation and sieving (no BTX aromatics or very volatile chlorinated hydrocarbons were detected after the final compaction). The contents of chlorobenzenes increased from the initial state toward activation and only decreased after the final compaction below the detection limits (Table 8-7).

Table 8-6 Results of solids and eluate analyses of the initially anaerobic municipal waste with different chemical contamination (Lys. 2, top, moderate contamination, Lys. 2, bottom, initially no contamination, Lys. 9, high contamination) before and after the biological treatment in comparison with different allocation values (LC: Landfill class, MBT: Mechanical-biological treatments, LAGA: Länderarbeitsgemeinschaft Abfall = States Waste Working Group

Parameter	Unit	LAGA Z2	Allocation values AbfAblV (2001)			MW before reactivation			MW after reactivation		
			LC 1	LC 2	MBT	Lys. 2, bottom No cont.	Lys. 2, top Moderate cont.	Lys. 9 High cont.	Lys. 2, bottom No cont.	Lys. 2, top Moderate cont.	Lys. 9 High cont.
Ignition loss	% by wt.	–	3	5	–	20.8	19.6	36.4	13.9	15.2	19.9
TOC	% by wt.	–	1	3	–	7.9	6.1	6.3	5.1	6.6	8.4
TC	% by wt.	–				10.3	7.7	14.8	8.5	6.8	10.2
Arsenic	mg/kg	150	–	–	–	2.2	2.3	0.7	2.3	3.5	1.2
Lead	mg/kg	1000	–	–	–	220	310	190	487	454	348
Cadmium	mg/kg	10	–	–	–	8	43	10	5	41	6
Copper	mg/kg	600	–	–	–	150	230	210	100	178	105
Nickel	mg/kg	600	–	–	–	140	250	2350	540	198	633
Zinc	mg/kg	1500	–	–	–	550	870	700	410	770	390
Eluate											
pH	–	5.5–12	13	13	13	7.3	7.1	7.3	7.7	7.6	6.89
Conductivity	µS/cm	3000	10000	50000	50000	1436	689	790	1112	481	782
TOC[a]	mg/l	–	20	100	250	22	15	162	33	27	40
Phenol[a]	mg/l	0.1	0.2	50	50	<0.02	<0.02	<0.02	<0.02	<0.02	<0.02
Arsenic	mg/l	0.06	0.5	0.2	0.2	<0.025	<0.025	<0.025	<0.025	<0.025	<0.025
Lead	mg/l	0.2	0.2	1	1	0.1	<0.1	<0.1	<0.1	<0.1	<0.1
Cadmium	mg/l	0.01	0.05	0.1	0.1	0.05	0.05	0.06	<0.1	<0.1	<0.1
Chromium VI[a]	mg/l	0.15[b]	0.05	0.1	0.1	<0.025	<0.025	<0.025	<0.025	<0.025	<0.025
Copper	mg/l	0.3	1	5	5	0.2	0.1	<0.1	0.1	<0.1	0.43
Nickel	mg/l	0.2	0.2	1	1	0.1	0.1	0.7	0.1	0.1	0.82

Mercury[a]	mg/l	0.002	0.005	0.02	0.02	<0.001	<0.001	<0.001	<0.001	<0.001	<0.001
Zinc	mg/l	0.6	2	5	5	0.2	0.1	0.2	<0.1	<0.1	<0.1
Fluoride[a]	mg/l	—	5	25	25	<2	<2	<2	<5	<5	<5
Ammonium	mg/l	—	4	200	200	<1	<1	<1	<0.005	<0.005	0.59
Cyanide[a]	mg/l	0.1	0.1	0.5	0.5	<0.02	<0.02	<0.02	<0.02	<0.02	<0.02
AOX[a]	mg/l	—	0.3	1.5	1.5	0.42	0.21	1.2	0.107	0.073	0.108
Water-soluble	% by wt.	—	3	6	6	0.95	0.51	0.67	0.69	0.3	1.06
Chloride	mg/l	30	—	—	—	132.7	33.2	32.4	76.8	19.1	50.9
Sulfate	mg/l	150	—	—	—	287	44	57.8	270	39	82.7
Nitrate	mg/l	—	—	—	—	66.4	139	6.75	29.9	69.9	3.11

(a) Determined in subproject C (Brammer *et al.*, 1997)
(b) Total Cr

275

Table 8-7 Concentration of very volatile components in the initially anaerobic, high-contamination municipal waste (Lys. 9) in $\mu g/m^3$

Lysimeter	Unit	MW, high-contamination (Lys. 9)		
		Before removal	Reactivation	Renewed deposition
VV CHC	$\mu g/m^3$	1800	130	n.d.
Chlorobenzenes	$\mu g/m^3$	1000	2000	5.5
BTX aromatics	$\mu g/m^3$	8300	1000	n.d.

n.d.: Not detected

The main components were monochlorobenzene, 1,2- and 1,4-dichlorobenzene as well as 1,2,4-trichlorobenzene. Microbial degradation processes during the interim treatment may provide an explanation. Thus we conclude that the above chlorobenzenes are generated from the metabolism of lindane (WHO, 1991) which was intentionally spiked to Lys. 9. Altogether, a strong decrease in trace gas concentrations can be found by the activation, which results mainly from the emission into the atmosphere of the relevant substances during relocation and sieving. In the case of interference into an old deposit which is seemingly stable as a consequence of low chemical contamination, gaseous emissions have to be expected which increase proportionally to the industrial contamination both quantitatively and qualitatively. These materials will outgas over the long term.

Leachates
Very volatile compounds in the leachate of no- and moderate-contamination waste (Lys. 2) were only detected in isolated cases and in concentrations markedly below 1 mg/l during the entire observation period. BTX aromatics and very volatile chlorinated hydrocarbons were only found in the leachate of the high-contamination deposit (Lys. 9) before activation. The concentrations were 200 mg/l for very volatile chlorinated hydrocarbons and 84 mg/l for BTX aromatics in June 1992, and still 130 mg/l for very volatile chlorinated hydrocarbons and 60 mg/l for BTX aromatics in September 1992. Afterwards, no very volatile chlorinated hydrocarbons or BTX aromatics were detected because of degassing during relocation.

8.5.3.2 Medium and semivolatile chlorinated hydrocarbons in leachates, eluates and solids

Leachates
As expected, small quantities of medium and semivolatile chlorinated hydrocarbons were only detected (up to approx. 1 $\mu g/l$) in leachate

samples from non- and moderate-contamination municipal waste (Lys. 2) due to their low water solubility.

In the leachate from the high-contamination deposit (Lys. 9), no DDT analogues were found and the PCB concentration dropped below 1 µg/l. However, HCH isomers and, in addition to chlorobenzenes, γ-pentachlorocyclohexene (γ-PCCH) were found as another lindane metabolite. The base contamination of the overall concentration for HCH isomers in the leachate was low in spite of a high contamination in the waste ($\leq$1 µg/l) and mainly originated from γ-HCH (lindane). The concentration of both lindane and its metabolites increased due to the activation from $\leq$1 µg/l at the beginning of activation to about 8 µg/l for chlorobenzene (CB) and PCCH as well as to about 35 µg/l for HCH as single peaks from the base contamination. A second, markedly lower peak emerged immediately after the renewed placement in a compacted state. The concentration then decreased to the initial level. Similar mobilisation peaks occurred 3 months after the final compaction which can be seen from the AOX and CHC concentration values.

Concerning the AOX values, the leachates from the no-contamination (Lys. 2, bottom) and moderate-contamination wastes (Lys. 2, top), relocated for decomposition, show pronounced contamination. The AOX concentration in the leachate from the no-contamination deposit increased from a base level of 0.2–0.5 mg/l to 0.5–1.3 mg/l during the activation. (For comparison: the limiting value for discharge into the effluent stream is 0.5 mg/l (AbwV, 2002, Appendix 51) was exceeded by the leachate from no-contamination waste). The AOX concentration in the leachate from the waste with intentionally moderate contamination increased due to the activation from a base level of 0.2–0.3 mg/l to only 0.3–0.6 mg/l during the activation.

In comparison to the sum of the chlorinated hydrocarbons (CHC), the concentration of the sum of the adsorbable chlorinated organic substances (AOX) markedly shows in Fig. 8-13, that the activation was not limited to the technical activation phase from the example of the high-contamination waste (Lys. 9). The level of the contamination in the leachate, which was hard to detect in the initial state, rose constantly both during and after the technical activation and became easy to detect. The activation of material discharge is therefore not limited to the period regarded as the landfill-engineering activation phase.

Eluates

No DDT analogues or PCBs were detected in the eluates from the <8 mm and 8–40 mm waste fractions of the no-contamination stage

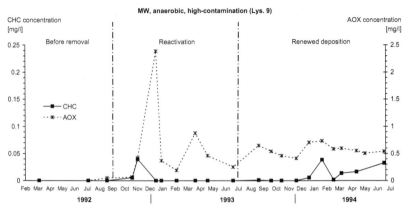

Fig. 8-13 Comparison of AOX concentration with the sum of detected chlorinated hydro-carbons in initially anaerobic municipal waste with high chemical contamination (Lys. 9)

and the moderate-contamination stage starting roughly from when the waste was first removed from the lysimeter. HCH isomers and chlorobenzenes were only contained in the eluate of the 8–40 mm waste fraction from the moderate-contamination zone (Table 8-8). Mainly γ-pentachlorocyclohexene (γ-PCCH) and 1,2,4-trichloroben-zene emerged as metabolites from the aerobic degradation of lindane. Other chlorobenzenes and α-HCH were produced as byproducts (Straube, 1991; WHO, 1991; Swannel, 1993) so that the proven substances probably originated from lindane metabolism.

Unlike leachate contamination, AOX concentration in the eluates of the moderate-contamination zone was 0.7–2.3 mg/l proportionally to

Table 8-8 Concentration of HCH isomers and chlorobenzenes in the eluate of the waste sample from the initially anaerobic municipal waste with moderate contamination (8–40 mm fraction) in mg/l (Lys. 2, top)

MW, moderate contamination (Lys. 2, top)			
Substance	Concentration in eluates [mg/l]	Substance	Concentration in eluates [mg/l]
1,2,3-trichlorobenzene	0.063	α-hexachlorocyclohexane	0.004
1,2,4-trichlorobenzene	0.26	β-hexachlorocyclohexane	n.d.
1,3,5-trichlorobenzene	0.0056	γ-hexachlorocyclohexane	3.3
1,2,3,4-tetrachlorobenzene	0.0007	δ-hexachlorocyclohexane	n.d.
Other chlorobenzenes	n.d.	γ-pentachlorocyclohexane	2.0

n.d.: Not detected

278

the intentional contamination, thus significantly higher than in the no-contamination zone (0.2–0.3 mg/l). The values of the <8 mm fraction were higher than those of the 8–40 mm fraction and AOX was no longer detected starting from the aerobic interim treatment.

A comparison with single material analysis results apparently contradicted the AOX concentration which was 2.3 mg/l in the eluate of the <8 mm sample of the moderate-contamination waste, while no individual CHC was detected. The AOX-forming substances may be components adsorbed by humic substances or those not extractable by petroleum ether due to a high polarity. In the eluate from the 8–40 mm sample of the same waste, the AOX concentration was 0.7 mg/l, while the sum of the measured single materials, converted to Cl content, was 4.0 mg/l. There is an important difference in the analysis methods: the AOX content is measured using an analysis technique that does not measure sedimented substances, while substances weakly attached to suspended matter are naturally also detected by agitated extraction. This is of importance with respect to the assessment of wastes in accordance with TASi, since, in essence, eluate criteria are used which obviously do not always correlate to the actual waste characteristics (Henseler-Ludwig, 1993).

The AOX contents of the eluates from the material of the high-contamination Lys. 9 were about 10 mg/l before the activation, i.e. two orders of magnitude greater than the leachate contamination of the same waste body (cf. Fig. 8-13). Unlike the contamination tendency of leachates, eluate contamination decreased to AOX values <0.1 mg/l during the activation and final compaction.

The eluate concentration of HCH isomers and γ-PCCH in the high-contamination deposit showed a tendency similar to that of AOX (Fig. 8-14). Chlorobenzene concentration increased during the aerobic activation but decreased to insignificant values after the final compaction. The first step of hydrogen chloride elimination to γ-PCCH within the dehydrochlorination of γ-HCH is decisive concerning reaction rates for the subsequent steps up to the formation of chlorobenzenes (Hughes *et al.*, 1953), while the main product, i.e. 1,2,4-trichlorobenzene, again represents a relatively stable stage. This explains the intermediate increase of the relevant concentrations. PCB or DDT analogues were not detected in the eluates from the high-contamination stage.

It can be concluded from the measurements that pollutant degradation, including the associated emission, was activated even if there was some reservation due to inhomogeneity.

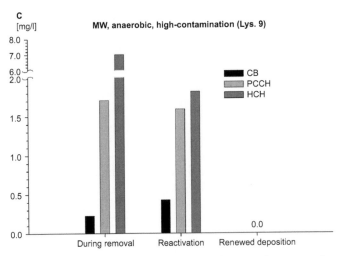

Fig. 8-14 *CHC concentration in waste eluates of initially anaerobic municipal waste with high chemical contamination (Lysimeter 9, <8 mm fraction)*

<u>Solids</u>
Both the concentrations of HCH isomers and chlorobenzenes in the moderate-contamination stage and the distinct traces of HCH in the no-contamination stage decreased from when waste was first removed from the lysimeter, over the interim treatment stage and to the final state (Fig. 8-15). Chlorobenzene concentration even dropped below the detection limit. The concentration of DDT analogues and PCB substances within the no-contamination decreased from removal to the interim treatment. Comparable concentrations to those during interim treatment were then measured in the final state.

An estimation of degradation kinetics is needed for HCH to assess the long-term effects. Since spiking was performed in 'lenses' in the deposit (cf. Fig. 2-7), the spiked chemicals were distributed extremely inconsistently, which impeded representative sampling. However, the waste was homogenised before the aerobic activation stage, therefore samples taken from activation and from the final compaction can be regarded as adequately representative, allowing the lindane degradation rate to be estimated. Degradation kinetics of the first order is assumed. The medium lindane concentration immediately after completion of deposition with moderate contamination (Lys. 2, top) was 140 mg/kg DS waste. At the time of waste removal from the lysimeter a lindane concentration of 8.0 mg/kg DS waste was detected. Considering a storage time of about 13 years, this corresponds to a half-life of about

280

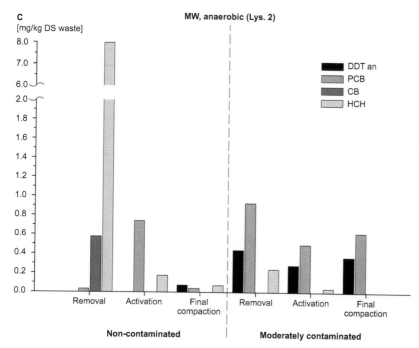

Fig. 8-15 Contents of medium- and semivolatile chlorinated hydrocarbons of the solid samples from no- to moderate-contamination anaerobic municipal waste (Lys. 2) converted to mg/kg air-dry total lysimeter waste (DDT an = analogues of DDT)

3 years. This estimation applies to the overall system including the no-contamination base, into which a substantial part of the initial contamination was relocated during the mobilisation phase (cf. Section 6.2.2.3). This contamination relocated over the medium term was no longer found in the removed waste after 13 years. Since a release cannot be fully excluded, this estimation provides the maximum degradation rate under the available degradation conditions.

The half-life of lindane in arable soils is considered to be about one year (WHO, 1991; DFG, 1982). There was possibly an inhibition of the degradation due to the relatively high concentration in comparison to pesticide application. However, comparative investigations show that degradation rates also depend on water and humic substance content (WHO, 1991), so that the difference of the half-lives may even have been caused by reduced bioavailability. The HCH concentration was 0.16 mg/kg DS waste after activation (275 days), so that the half-life was about 0.3 years during activation. At the end of the observation phase (342 days) the concentration dropped to 0.08 mg/kg DS waste.

281

In view of a half-life of about 1 year this corresponds to the half-life of lindane in arable soils indicated in the above literature.

When assessing the chemical analysis results of the degradation steps, their different degradation rates have also to be taken into account. In the chemical dehydrochlorination of lindane via γ-PCCH, the first reaction step decides the reaction rate (Hughes *et al.*, 1953), while trichlorobenzenes represent a relatively stable stage. Therefore γ-PCCH was degraded in the initial, 13-year-old deposit down to concentrations below the detection limit, while trichlorobenzenes were still detected. Since degradation of α-HCH is slower than that of γ-HCH (Straube, 1991), this substance (α-HCH) is enriched due to the microbial isomerisation of γ-HCH. However, this type of metabolism is relatively insignificant in comparison to dehydrohalogenation, so that α-HCH concentration is comparatively low. In addition, not all α-HCH degrading microorganisms are capable of isomerisation (Nagasawa *et al.*, 1993). Trichlorobenzenes and α-HCH were also detected as secondary metabolites in the initial, 13-year-old deposit similar to chlorobenzene in the gaseous phase. The HCH concentration in the interim-treatment lysimeter was then in the range of values measured in the non-contaminated deposit (Lys. 2, bottom) and thus so low that the metabolites could no longer be detected. This proof applies to low- and medium-waste contamination.

The homogenisation of the waste by mixing was not sufficient for representative solids samples to be taken because of the very high punctiform HCH release in the material placed in the highly contaminated deposit (Lys. 9), as opposed to the waste with moderate contamination (Lys. 2, top). Otherwise the observed increase of lindane concentration from activation to final compaction (from 7.4–46 mg/kg DS waste) cannot be explained. In addition, the measured lindane-concentration was 1900 mg/kg DS waste at removal from the lysimeter which was considerably greater than the average concentration of 590 mg/kg as calculated from the recorded quantities. Therefore, the ratio of lindane metabolite concentration to lindane concentration in the respective samples will be used in the following for the description of lindane degradation in the highly contaminated deposit (Table 8-9).

No lindane degradation took place in the high-contamination municipal waste (Lys. 9) before its first removal – in contrast to the moderately contaminated waste. The measured concentration of lindane-metabolites γ-PCCH, 1,2,4-trichlorobenzene and α-HCH was low in comparison to lindane concentration. A decrease of lindane content was not detected. The amounts of material mixes introduced

Table 8-9 Concentration of γ-HCH and metabolites in solid waste samples from the initially anaerobic high-contamination municipal waste (Lys. 9)

Time	γ-HCH [mg/DS kg] (A)	γ-PCCH [mg/DS kg] (B)	CB[a] [mg/DS kg] (C)	α-HCH [mg/DS kg] (D)	B/A [%]	C/A [%]	D/A [%]
MW (Lys. 9) high contamination							
Before removal	1900	4.4	0.05	0.26	0.23	0.003	0.014
Reactivation	7.4	1.5	0.62	0.23	20	8.4	3.1
Renewed deposition	46	1.3	0.44	2.58	2.8	1.0	5.6

(a) Chiefly 1,2,4-trichlorobenzene

(heavy metals, lindane, simazin, cyanide) were probably so high that no anaerobic microbial degradation could take place. After aerobic activation the concentration of the metabolites was markedly higher in comparison to the lindane concentration. Obviously, microbial activity was intensified due to the change from anaerobic to aerobic conditions. The highest concentration of the metabolite produced by isomerisation (α-HCH) was measured in the final compaction, while the contents of the degradation products resulting from dehydrochlorination (γ-PCCH and 1,2,4-trichlorobenzene) were again lower. Lindane concentration was markedly lower than before the interim treatment, but still substantially higher than the values measured in the moderately contaminated material (Lys. 2, top). The results show that a marked decrease in lindane concentration was also achieved by aerobic degradation in the extremely contaminated deposit. However, the length of activation period was not sufficient to achieve complete degradation.

8.5.3.3 Organic acids, chlorophenols and alkyl phenols in leachates, eluates and solids

Non- to medium-contaminated deposit

Leachates and eluates

The contents of organic acids in the leachate of non (Lys. 2, bottom) to moderately contaminated municipal waste (Lys. 2, top) were generally low before activation and were between 40 and 270 mg/l, often even below the determination limit. Long-chain fatty acids (C_{14} to C_{18}) dominated. Any influence on the organic acids content of the degradation processes by aerobic activation was not observed at these low concentrations in the leachate (50 and 200 mg/l). In the solid waste samples an increase of the organic acid content by activation was detected. It was primarily because of an increase in C_{14}, C_{16} from about 2200 mg/l to 8200 mg/l. Very volatile acids (C_2 to C_7) as anaerobic degradation products were hardly detected; benzoic acid and long-chain acids dominated. In the eluates of waste samples, only low acid contents of between 15 and 80 mg/l were detected.

The phenol content of non- and moderately-contaminated wastes was even lower than the acid content. The leachate and eluate contents were almost always below the detection limit, only traces of phenol and cresols were found (1–6 mg/l). The results from solid waste were similar: concentrations between 300 and 1100 mg/kg wet waste.

The contamination level of total dissolved organic carbon (TOC in leachate) increased during the activation from about 200 mg/l to approx. 1000 mg/l, i.e. about five-fold. The activation of degradation only affected the organic acids immediately after the final compaction by triggering an increase from the range of 120–290 mg/l to 3600 mg/l, i.e. exceeding an order of magnitude – a typical consequence of oxygen deficiency in activated material.

Solids

Acid content of the solids corresponded to those of the leachates. In the non-contaminated deposit (Lys. 2, bottom) the content of organic acids decreased from 26 000 µg/kg wet waste to 13 000 µg/kg due to aerobic activation. Traces of short-chain fatty acids, however, were found. The aerobic activation has obviously resulted here in a more intensive degradation of high-molecular compounds. Benzoic and phenylacetic acids are the only apparent exceptions. They are obviously the final members of a complex degradation chain with aromatic precursor substances which are finally mineralised primarily through phenol and ring fission.

Highly contaminated deposit (Lys. 9)

Leachates

The conservation of the initial material and the reactivation of the degradation processes due to aeration can be clearly proved by the organic acids. In view of the concentration (up to about 350 000 µg/l) and composition (Table 8-10), the concentration of leachates in the initial material corresponded unambiguously to the acidogenic anaerobic degradation phase – a freshly compacted waste.

The aerobic activation reduced the acid content in the extremely contaminated leachate from about 350 000 µg/l to about 1300 µg/l

Table 8-10 Organic acids in the leachate of initially anaerobic highly-contaminated municipal waste (Lys. 9) in µg/l

Compound[a] [µg/kg]	MW, anaerobic, highly-contaminated (Lys. 9)								
	Jul. 1992	Sep. 1992	Oct. 1992	Nov. 1992	Dec. 1992	May 1993	Aug. 1993	Nov. 1993	Dec. 1993
Acetic acid	15850	11200	535	240	n.n.	n.n.	n.n.	n.d.	n.d.
Propionic acid	40700	28400	36	n.n.(c)	n.n.	n.n.	21	<12.5	6340
Iso-butyric acid	12050	9300	22	n.n.	n.n.	n.n.	21	<12.5	1860
Butyric acid	72850	60000	37	23	<5	n.n.	15	<12.5	4170
Valeric acid	66000	57700	8	n.n.	<5	n.n.	11	<12.5	1500
Caproic acid	109250	100450	500	500	6	6	20	<12.5	230
Heptanoic acid	21450	19850	<250	400	n.n.	n.n.	n.n.	n.n.	n.n.
Benzoic acid	15450	22600	85	53	7	<5	20	<12.5	263
Phenylacetic acid	n.d.(b)	n.d.	35	5	<5	<5	61	<12.5	79
Tetradecanoic acid	n.d.	n.d.	8	34	37	n.n.	n.d.	28	218
Hexadecanoic acid	n.d.	n.d.	27	915	255	99	147	127	555
Octadecanoic acid	n.d.	n.d.	13	215	20	21	35	136	1288
Σ org. acids	353600	309500	1306	2385	325	126	351	291	16503

(a) Nomenclature varies depending on working group, see Fischer (1996)
(b) Not determined
(c) Non-detectable

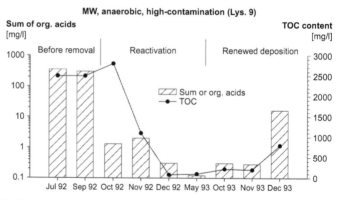

Fig. 8-16 Comparison between organic acid content (sum) and TOC values in leachate of the highly chemically-contaminated anaerobic municipal waste (Lys. 9)

and in the subsequent activation to a minimum of 120 μg/l by about three orders of magnitude (Fig. 8-16). After final compaction the acidity increased to 16 000 μg/l with a rising trend, i.e. the value of long-term contamination under anaerobic deposition conditions assumed from the trend was reduced by the aerobic decomposition processes for the period of activation by about one order of magnitude.

The contamination level of the organic compounds in the leachate of the highly contaminated waste, characterised by the sum of organic carbon (TOC), was proportional to the contamination with organic acids (Fig. 8-16). Hence it follows that the bulk of the dissolved organic carbon consisted of organic acids. This condition characterises the beginning of biochemical degradation by hydrolysis. In the leachate of the municipal solid waste with no intentional industrial contamination (e.g. non-contaminated deposit, Lys. 2, bottom) there was no connection between the clearly detected contamination by dissolved organic compounds (TOC) and the insignificant contamination by organic acids. In connection with decreasing ammonium concentration, the state of the non-contaminated waste corresponds to a highly progressed degradation towards the end of the methanogenic phase. The initial conditions were therefore basically different.

Solids

Similar clear tendencies can be seen from the results of the solid analyses, although the pattern of organic acids is basically different. Predominantly long-chain fatty acids, phenylacetic and caproic acids were detected before the aerobic activation, lighter volatile acids like those found in leachate were hardly detected. This speaks for the fact

286

that these degradation products are carried immediately into the leachate path. After the aerobic activation, only benzoic acid and long-chain acids could be detected. The very high content of hexadecanoic acid in the <8 mm fraction before activation, which dropped from 1200 mg/kg to 25 mg/kg during activation, is also conspicuous.

Altogether, the concentration of the acids dropped from 270 mg/kg to 76 mg/kg during the aerobic activation. Since the concentration also decreased simultaneously in the leachate, an intensive degradation must have taken place.

The phenol contamination – intentionally spiked behaved akin to the organic acids which was particularly well proved by the highly contaminated deposit (Lys. 9). The leachate contained 300–600 mg phenol/l before the aerobic activation and it dropped permanently by about three orders of magnitude after the aerobic activation. This was because the phenol contamination of the solids was aerobically degraded from about 100 mg/kg by two orders of magnitude to approx. 1 mg/kg. This degradation was confirmed by the phenol content of the eluate: a reduction by almost three orders of magnitude from approx. 10 mg/l to approx. 0.02 mg/l was observed in the <8 mm fraction. The high degradation rate due to the influence of oxygen was confirmed by the analysis of the moderately contaminated municipal waste (Lys. 2) before reactivation, since oxygen was able to penetrate by diffusion into the low and, in particular, the medium contaminated zone in the upper third of Lys. 2 from the waste surface during a storage time of 15 years. Despite the intentional contamination with phenol sludge, phenols were not found in the leachate at all and only in insignificant concentrations in the solids. However, in the non-contaminated zone (bottom half of Lys. 2) phenols were found in the leachate, solids and eluate. The contamination was, at the same level including an activation peak, as in the highly contaminated municipal waste (Lys. 9) during the activation phase. These measurement results suggest for the long-term assessment that phenols are stable under anaerobic conditions over the long term and are removed by the leachate, while extensive industrial contamination mixed with municipal solid waste can be biochemically degraded under aerobic conditions.

8.5.3.4 Semivolatile substances (PAH, phthalates, triazines) in leachates, eluates and solids

PAHs were hardly detected in the leachates and eluates even from the highly contaminated municipal waste (Lys. 9) despite high industrial contamination. Only the solids exhibited sporadic PAH contamination. However, matrix effects impaired the analyses very strongly, so that the detection limits in some cases were relatively high (500 µg/kg wet waste). The concentration was at all contamination stages in the same order of magnitude between 500 and 2000 µg/kg wet waste. Benzo[b]fluoranthene and benzo[k]fluoranthene were the dominating contaminants. An influence of the PAH content or shift within the waste fractions due to the aerobic activation was not detected.

Analogous to PAH, the residual content of phthalates in leachates and eluates were low throughout (from n.d. to 45 µg/l) in all contamination levels and were not influenced by the activation of the biochemical degradation. Substantial phthalate content (9700 to 49 000 µg/kg wet waste) was detected in the solids of the highly contaminated deposit which exceeded the content in low-contamination deposits by an order of magnitude. Its largest fraction consisted of DEHP (di(2-ethyl-hexyl)phthalate) and DBP (dibutyl phthalate). Only a relocation of the contamination from the >40 mm coarse material to the <40 mm fine material could be proved as an activation effect, but no degradation was detected – which might be concluded from the substantially lower contamination in the lower contamination stages.

A relocation of PAH and phthalates from the moderately contaminated zone (Lys. 2, top) to the non-contaminated zone (Lys. 2, bottom) was proved as a tendency, but not as a significant concentration relocation. The intentional spiking of triazine in both top thirds of the deposit (Lys. 2, top) resulted not in its relocation into the bottom third (Lys. 2, bottom) but triazine once dissolved was removed by the leachate.

The different levels of triazine contamination resulted in a clearly differentiated picture. Increased simazin content was expected due to the intentional spiking in the top two-thirds of Lys. 2. However, the analysis of the leachate before waste removal failed to detect any concentration of triazine. Small amounts of simazin and de-ethylsimazin (6–80 µg/l) were only found after the beginning of activation. However, the values only dropped below the determination limit after completion of the aerobic activation. This tendency was accompanied with a short-term increase of TOC content after waste removal. The analysis of waste samples provided considerably more information. 170 000 µg/kg triazine was detected before activation, this value decreased to 150 µg/kg after activation with the decrease of simazin and simaton contents dominating (Table 8-11). The low contents in the leachate exclude the possibility of a removal by leachate. If it was not due to

288

Table 8-11 Triazine content in the material of the moderately contaminated anaerobic municipal waste (Lys. 2, top), 8–40 mm fraction, and extrapolated total content in µg/kg wet waste

| | MW, anaerobic, moderate contamination (Lys. 2, top) | | | | | |
| | Before reactivation | | | After reactivation | | |
Compound	<8 mm	8–40 mm	Total	<8 mm	8–40 mm	Total
Simazin	35 500	65 000	38 279	20	150	75
Desethylsimazin	n.d.[a]	165	88	n.d.	n.d.	n.d.
Triazine	n.d.	n.d.	n.d.	n.d.	100	46
Simetryn	n.d.	20	11	n.d.	n.d.	n.d.
Simeton	372 000	171 000	129 677	110	n.n.	28
Sum	407 500	236 185	168 054	130	250	150

(a) Non-detectable.

sample inhomogeneities, so metabolic degradation must have taken place.

The eluate contamination of the individual waste samples behaved similarly to the contamination in the solids. Substantially higher residual concentrations (120 and/or 680 µg/l eluate) were detected before the activation than in the leachates, but these values dropped after the aerobic activation below the detection limit (0.5 µg/l).

In the highly contaminated deposit (Lys. 9) the effects of the aerobic activation of the biological degradation processes of natural organic substances were detected as mobilisation of some contaminants without their degradation both in the leachate and in the solids and eluate. The reverse profile of pollution emission is characteristic compared to the TOC values and the phenol and acid contents of the leachate: both TOC values and phenol and acid contents were noticeable due to high values before aerobic activation and were reduced by aerobic activation permanently to insignificant contaminations. Contrary to the contaminant profiles in the leachates of aerobically degradable organic substances, simazin was hardly detectable during the 15-year anaerobic deposition until aerobic activation: it only exhibited a concentration of about 0.005 mg/l. As a consequence of relocation and aerobic activation the contamination increased within a short period by about three orders of magnitude to 1.0–5.0 mg/l. Even after the activation peak faded away, the contamination was 0.01–0.5 mg/l, i.e. 0.5 to two orders of magnitude higher than in the leachate of the initial condition. A clear increase of the triazine concentration was also noticed in the eluates of the solid waste samples after the aerobic activation. Thus the simazin concentration increased from 590 to 5700 µg/l in the eluate of the <8 mm fraction and from 810 to 4250 µg/l in the eluate of the 8–40 mm fraction. The simazin concentration in the fine material of the solids decreased complementarily: from 30 mg/kg to 0.08 mg/kg in the <40 mm fraction. The activation of the degradation processes by relocation and aeration therefore mobilised the emission of the stored triazine in the highly contaminated deposit, but failed to contribute to its degradation. The total concentration (35 mg/kg before the activation and 29 mg/kg after the activation) remained constant within the range of inhomogeneities.

It was not possible to detect the pesticide dicrotophos spiked earlier in any of the samples.

8.5.3.5 *Comparison of various TOC concentrations in leachates, eluates and solids as stability criteria*

Conclusions can be drawn about the biochemical stability of the initial substance and the inhibition effects caused by pollutants from their relationships between eluteable organic carbon (eluate TOC according to DEV (Deutsches Einheits Verfahren = German Standard Procedure) without complex aggregates and emulsions), organic carbon dissolved by leachates (leachate TOC including complex aggregates and emulsions) and organic carbon of the solids (solids TOC), measured before, during and after aerobic activation. The TOC of the leachates reacts most reliably to conversion processes because the leachate covers the entire waste body including the finest aggregates and colloids. When interpreting the leachate TOC it has to be taken into account that this parameter primarily reflects changes in and completeness of degradation, but does not specify the extent of degradation because the bulk of degraded carbon (more than 90% by weight) is discharged from the waste body as a gas in inorganic compounds (CO_2 and CH_4) (cf. Kucklick *et al.*, 1996, paragraph 2). TOC in the leachate especially increases when incomplete degradation only to organic acids takes place. It has to be taken into account in this test that it was not possible to prevent a small amount of oxygen from diffusing both into the crushed stone of the non-contaminated bottom third of Lys. 2 and that of the highly contaminated deposit (Lys. 9). Oxygen diffusing through the waste surface mainly affected the moderately contaminated top zones of Lys. 2, top, without an earth cover. Under these conditions, TOC principally characterises the difficult-to-degrade organic materials which cause extensive contamination in ground water (cf. Spillmann

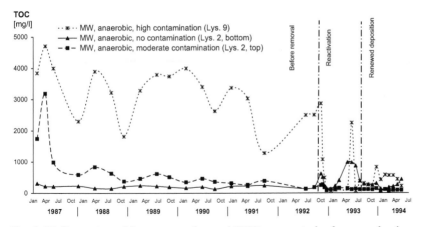

Fig. 8-17 Comparison of long-term tendency of TOC content in leachate samples from initially anaerobic municipal waste with different chemical contamination

et al., 1995). Therefore the following conclusions can be drawn from the long-term properties of the leachate TOC shown in Fig. 8-17:

- Despite favourable boundary conditions (flat deposit, air access to the base), the high industrial contamination (Lys. 9) led to a sustainable inhibition of degradation, so that carbon compounds were extensively removed not in a mineralised but in an organic form. Aerobic activation reduced the inhibition of the degradation processes considerably, but did not manage to eliminate it completely.
- The moderate industrial contamination (Lys. 2, top) did not unfavourably affect the long-term stabilisation within this comparison. Oxygen diffusion from the waste surface was decisive for stabilisation and this long-term stabilising effect kept the TOC stable against any further aerobic activation.
- Natural anaerobic conservation can be proved by the leachate TOC of the non-contaminated deposit. The anaerobic conservation reduced the leachate contamination to such an extent that oxygen diffusing into the crushed stone – virtually a trickling filter – facilitated an extensive mineralisation and humification of the carbon content of the leachate. The activation disrupted the equilibrium and the increase in TOC indicated a substantial but, to a certain extent, incomplete degradation during and after activation. Thus anaerobic 'stability' provides a level of conservation and can be aerobically activated.

291

The conclusions about biochemical activity derived from the leachate TOC were confirmed by the TOC of the solid material starting from the disruption of degradation through waste conservation to an extensive stabilisation of the organic material:

- In the biological-anaerobically conserved material of the non-contaminated deposit the solids TOC decreased due to aerobic activation particularly in the fine and medium fractions (<8 mm from about 8% to <6%; 8–40 mm from about 12–6%).
- In the extensively aerobically degraded material of the moderately contaminated deposit the coarse components disintegrated due to the activation (increase of fine fraction from 64–72%, and after final compaction to 78%). The changes of TOC stayed within the determination accuracy.
- The material of the highly contaminated deposit conserved by degradation disturbance behaved like a fresh waste after the activation: the solids TOC values increased by two- to five-fold through disintegration of the coarse material in the medium and fine fractions (Table 8-12).
- The eluate TOC without complex aggregates and emulsions confirmed the trends but provided less characteristic results.

8.5.3.6 Element contents in leachates, eluates and solids

Only elements with clearly recognisable contamination are used in the following comparison. The parameters of the results presented therefore do not include all parameters of the comparative tables contained in regulations.

Table 8-12 TOC contents of solid waste samples from the high-contamination, initially anaerobic municipal waste (Lys. 9) before and after interim treatment as well as after one year's renewed landfilling in % by weight wet waste

Sample	MW, anaerobic, high contamination (Lys. 9)		
	Before reactivation	After reactivation	After renewed landfilling
	Solids TOC [% by weight]		
Fraction <8 mm	4.7	6.9	8.7
Fraction 8–40 mm	2.5	7.5	9.4
Extrapolated total content	1.4	5.4	7.6

Comparison of the non-contaminated deposit with the moderately
contaminated one

The concentrations of elements characterised both by aqua regia dissolution of the random sample and in the sample 'Fraction <8 mm
without glass and magnetic components' as well as those of the elution
test and leachate are displayed in Table 8-13. Those elements not listed
failed to indicate any significant changes.

The elements Ba, Cd, Cr and Mn showed the same behaviour in all
four types of sample (random sample, fraction <8 mm, eluate and
leachate). The spiked zone exhibited higher concentrations. This
applies to Cu only in the aqueous samples. Sn behaved indifferently.
In the spiked zone (addition of galvanic sludge in this example) the
solid samples exhibited higher concentrations and the aqueous
(eluate and leachate) samples showed lower concentrations than in
the zone without industrial sludge addition. It is not clear why the
non-contaminated sample contained so much Pb in the sample 'fraction
<8 mm without glass and magnetic components', but it is obvious that
the Pb content is higher in the fine fraction than in the total sample.
The higher Pb concentration also appears in the aqueous samples.

The comparison of the eluate with the leachate sample shows
markedly higher concentrations for the main elements such as Al, K,
Na, Mg, Ca, Fe and S, and increased concentrations for As, Ba (only
Lys. 2, bottom), Co, Ti and Zn in the leachate. As a rule, Cd, Cu,
Mn, Ni, Pb and Sr contamination is also higher in the leachate than
in the eluate but not in all samples. Concentration differences between
eluate and leachates probably originate from different pH values and a
different solid to solvent ratio. Flushing out is unavoidably due to the
high solubility of K, Na and S. Markedly smaller values of the toxic
elements Cd, Cr, Cu, Ni and Pb were determined. The considerably
higher Ba concentration in the eluate of the material from the spiked
zone can be explained by a better percolation of the sample on the
one hand and an inhomogeneous distribution of the spiked barium
cyanide (in lenses) on the other. This also means that higher Ba
contents must be expected in the leachate in the future, since Ba
being an easily-soluble element is mobilised by micro-biological conversion processes (Fig. 8-18).

The range of pH fluctuations in the neutral to slightly alkaline
leachates was small and did not affect the heavy metals concentrations.
The activation, planned to be aerobic/anaerobic/aerobic (cf. Chapter
8.1, paragraph 3), resulted in decreasing concentrations for the
elements Co (from 0.09–0.04 mg/l), Cu (from 0.8–0.2 mg/l) and

Table 8-13 *Comparison of spiked heavy metal content and effects of the spiking of heavy metals on initially anaerobic municipal wastes between moderately (Lys. 2, top) and initially non-contaminated municipal waste (Lys. 2, bottom)*

Element	Spiked conc. [mg/kg DS waste] Moderate contam. (Lys. 2, top)	Random sample		Fraction <8 mm without glass + magn. comp.		Eluate		Leachate	
		Moderate contam. (Lys. 2, top)	No contam. (Lys. 2, bottom)	Moderate contam. (Lys. 2, top)	No contam. (Lys. 2, bottom)	Moderate contam. (Lys. 2, top)	No contam. (Lys. 2, bottom)	Moderate contam. (Lys. 2, top)	No contam. (Lys. 2, bottom)
					Concentration [mg/l]				
Ba	2070	2900–3300	400–800	2400–3300	550–850	4–12	0.05–0.1	0.8–3.1	0.1–1.4
Cd	12.4	35–60	0.6–9.0	39–68	3–10	<0.005	<0.005	<0.05–0.05	<0.005–0.02
Cr	1773	500–760	83–220	800–1200	60–470	<0.005–0.04	<0.005–0.03	0.02–0.4	0.01–0.14
Cu	111	150–370	90–160	210–320	150–200	0.05	0.05–0.18	0.03–0.2	0.1–0.6
Mn	46	900–1100	420	800–1600	370–600	0.04–0.06	0.005–0.04	0.02–0.6	0.06–1.3
Ni	439	190–650	50–570	155–900	55–580	0.11–0.35	0.02–0.05	0.05–0.31	0.14–0.57
Pb	71	250–420	130–340	460–750	2300–6800	<0.04	<0.04–0.08	0.04–0.25	<0.04–0.6
Sn	n.d.	220–270	110–150	200–265	160–900	<0.025	<0.025–0.5	<0.025–0.08	<0.025–0.15
Zn	630	460–900	300–740	900–1250	580–650	0.045	0.02–0.15	0.05–0.8	0.1–2.0

n.d. Not determined

294

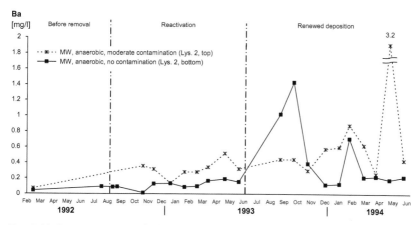

Fig. 8-18 *Barium content in the leachate from an initially anaerobic municipal waste with different chemical contamination*

Ni (from 0.7–0.3 mg/l) in the moderately contaminated stage. The sulfur concentrations decreased to one-third of the original concentrations regardless of spiking. The aerobic reactivation resulted in reductions for Ca, Mg, Mn and Na, but in concentration increases for Al, Cr, Fe and Cu (Table 8-14).

Comparison of non-contaminated deposit (Lys. 2, bottom) with that of high contamination (Lys. 9)

The element concentration of Cd, Cu, Ni and Pb in the random sample of the highly contaminated deposit (Lys. 9) exceeded the limiting values of AbfKlärV (1992). Cr and Hg were just below the limit. The main components and pollutants as well as the contamination spiked by the addition of chemicals, determined by aqua regia dissolution, are summarised in Table 8-15.

The table shows the particularly high Ba, Cr and Ni concentrations. The amount of spiked chemicals was completely recovered for Ba, one-half for Cr and up to one-third for Ni. All parameters tested, except Hg and Pb, exceeded the limiting values of AbfKlärV (1992). Increased concentrations of these metals (Ba, Cr, Ni) were also detected in the eluates. Ni and Cr exceeded the eluate criteria of TASi for landfill class II.

In comparison to the initially non-contaminated deposit (Lys. 2, bottom), the concentration of non-toxic elements was almost identical, i.e. the initial materials were identical and it was possible to compare them. The concentrations of the toxicologically relevant elements

295

Table 8-14 Element concentrations in eluate and leachate of initially anaerobic municipal waste with different chemical contamination; comparison between the moderately contaminated stage (Lys. 2, top) and the initially non-contaminated bottom layer (Lys. 2, bottom), conc. in mg/l

	Eluate concentrations				Leachate					
	Before waste removal [mg/l]		Reactivation [mg/l]		Before waste removal [mg/l]		Reactivation [mg/l]		Renewed deposition [mg/l]	
	Moderate contam. (Lys. 2, top)	No contam. (Lys. 2, bottom)	Moderate contam. (Lys. 2, top)	No contam. (Lys. 2, bottom)	Moderate contam. (Lys. 2, top)	No contam. (Lys. 2, bottom)	Moderate contam. (Lys. 2, top)	No contam. (Lys. 2, bottom)	Moderate contam. (Lys. 2, top)	No contam. (Lys. 2, bottom)
Al	0.0208	0.159	0.324	1.2	0.84	0.409	1.3–12	1.3–45	2.2–18	3.5–36
As	0.0010	0.0016	0.002	0.004	0.004	0.008	0.007–0.011	0.01–0.03	0.005–0.007	0.007–0.011
Ba	3.6	0.045	12	0.103	0.067	0.042	0.14–0.51	0.10–0.16	0.27–3.2	0.12–1.5
Ca	31	20	49	29	200	28	43–223	68–246	30–114	36–220
Cd	<0.005	<0.005	<0.005	<0.005	<0.005	<0.005	<0.005–0.013	<0.005–0.012	<0.005–0.05	<0.005–0.014
Co	0.005	<0.005	0.019	0.005	0.092	0.044	0.007–0.043	0.01–0.05	0.013–0.041	0.015–0.041
Cr	0.041	<0.005	<0.005	0.028	0.059	0.015	0.019–0.042	0.012–0.14	0.029–0.39	0.011–0.60
Cu	0.052	0.056	0.048	0.183	0.796	0.27	0.027–0.116	0.13–0.66	0.027–0.23	0.18–0.28
Fe	0.699	0.304	2.3	3.8	3.9	1.3	2.2–3.7	0.50–45	1.8–22	4.3–30
K	13.2	43	18.2	62	591	570	58–116	188–274	40–54	149–274
Mg	2.88	4.75	3.9	5.0	27	46	11–33	18–54	13–33	19–55
Mn	0.036	0.005	0.064	0.037	0.296	0.023	0.029–0.35	0.06–1.3	0.023–0.62	0.08–1.3
Na	36	99	45	134	1385	1560	311–574	890–1640	244–452	770–1340
Ni	0.114	0.023	0.349	0.046	0.728	0.306	0.087–0.10	0.16–0.26	0.078–0.31	0.14–0.57
Pb	<0.040	<0.040	<0.040	0.075	0.32	1.1	<0.040–0.048	<0.04–0.57	<0.040–0.25	0.05–0.18
S	7.18	41	10.3	81	655	138	41–91	105–699	20–69	129–360
Sr	0.137	0.040	0.292	0.590	0.227	0.100	0.15–0.64	0.18–0.70	0.15–0.66	0.19–0.90
Ti	0.009	0.010	0.008	0.025	<0.005	<0.005	<0.005–0.43	<0.005–0.41	0.01–0.24	0.03–0.4
Zn	0.043	0.018	0.049	0.151	1.7	0.122	0.07–0.63	0.20–2.0	0.05–0.79	0.17–1.5

Table 8-15 Comparison of the main components and pollutants between the high-contamination stage (Lys. 9) and the initially non-contaminated bottom layer (Lys. 2, bottom) of the initially anaerobic municipal waste and their spiked concentration

	MW, anaerobic					
Element	High contam. (Lys. 9)	No contam. (Lys. 2, bottom) [%]	Element	Spiking	High contam. (Lys. 9)	No contam. (Lys. 2, bottom) [mg/DS kg]
Ca	2.0	2.0	As	0.10	7.4	6.8
Fe	2.0	3.3	Ba	2070	3000	500
Al	1.1	0.9	Co	0.12	3.8	0.6
P	0.3	0.3	Cd	n.d.	7.4	4.9
S	0.19	0.13	Cr	2000	1066	93
Mg	0.18	0.18	Cu	269	164	87
K	0.11	0.19	Hg	0.02	1.2	0.7
Mn	0.05	0.04	Ni	6400	2100	50
Na	0.08	0.26	Pb	21	120	130
Si	12	10				

nickel (Ni) and chromium (Cr) increased by more than an order of magnitude as a consequence of high industrial contamination. Contamination by mercury (Hg), arsenic (As) and lead (Pb), however, belong to the basic contamination of municipal waste.

The percentage of element concentration of the sample 'Fraction <8 mm without glass and magnetic components' of the non-fractioned random sample was about the same percentage of the total waste fraction, i.e. the contamination of the fraction was equal to the random sample, therefore the high-contamination municipal waste (Lys. 9) has to be regarded as homogeneous (not shown in Table 8-15).

In the leachate of the high-contamination deposit the pH value (neutral) was stabilised in the interim treatment and in the final compaction at a slightly higher level than in the initial state. Mn concentration exhibited a special profile (see Fig. 8-19): the concentration reached a maximum of 12 mg/l in the initial state, dropped to a hundredth in the aerobic interim treatment and was 2 to 2.5 mg/l in the final phase. Ni concentration showed a profile similar to that of Mn, but Mn concentration was five times the Ni value before activation and three times the Ni value after final compaction. In the interim treatment, however, Ni concentration was sometimes twice as high as the Mn values.

Ba concentration showed an increasing trend which can be explained by the waste removed from the waste body and the associated improved

297

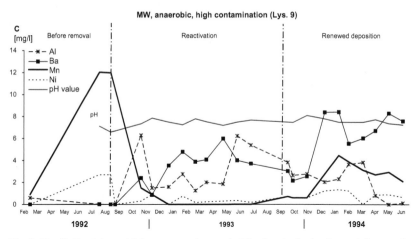

Fig. 8-19 Al, Ba, Mn and Ni concentrations and pH value in the leachate of the initially anaerobic municipal waste with high chemical contamination (Lys. 9)

homogeneous distribution of the chemicals placed in lenses (barium cyanide). This does not apply to the heavy metals spiked by galvanic sludge in greater concentrations (Cr and Ni), since, unlike barium cyanide, they were not added in a water-soluble form. Mobilisation of aluminium from the basic substance is also noticeable during aerobic activation, because mobilisation of this element is not desired either.

The removal of the highly contaminated municipal waste accompanied with aerobic interim treatment led to higher concentrations of the elements Al, Ba, Cu and P and to lower concentrations of B, Ca, Fe, K, Mg, Mn, Na, Ni, Sn and Sr in the leachate. Except for Cu, the trend for these elements changed after waste compaction following the aerobic interim treatment. Cu was again strongly absorbed in later phases so that the low level of the initial state was reached 4 months after final compaction. Fe and Sn showed markedly lower concentrations after waste removal than during the initial state, however the concentrations increased strongly during treatment and final compaction. However, the high level before waste removal was not reached again.

The difference between this and the non-contaminated deposit was that the level of the contamination was increased by spiked toxic elements at same trend.

298

8.5.4 Assessment of the influence of the chemicals on the degree of stability according to biological criteria (comparison of the three contamination stages: no contamination (Lys. 2, bottom), moderate contamination (Lys. 2, top) and high contamination (Lys. 9))

Martin Kucklick, Peter Harborth and Hans-Helmut Hanert

8.5.4.1 Ecophysiological assessment

(a) Initial state: Non- to moderately-contaminated deposit (Lys. 2)

Gas composition within the core area of the weakly contaminated
municipal waste (Lys. 2, 2nd third, at 2 m depth)

Oxygen (O_2) was only available in traces, thus an anaerobic condition prevailed. A relatively high methane concentration of 27.3% by volume proved that intensive methane generation took place despite a small waste thickness and about 15 years of storage.

Sample material from about 1.5 m depth contained a verified methane generation activity under anaerobic laboratory conditions. The number of methanogenic bacteria was 3×10^5/g DS waste which was within the range of reference values in the literature for municipal waste landfills for 2 m depth (Sleat *et al.*, 1987). The potential of this sample for methane generation at optimum H_2 and CO_2 supply was 10 mg CH_4/(kg × h) being the highest value measured before waste removal in the deposits tested.

Sulphate reduction

Dissimilatory sulphate reduction is a competitive degradation path under redox conditions of methanogenesis. The number of sulphate reducing bacteria in the slightly contaminated anaerobic municipal waste (2nd third of Lys. 2) was 5×10^6/g which was somewhat above the literature value of $10^{5.5}$/g at a depth of 2 m for landfills (Sleat *et al.*, 1987). In terms of figures, sulphate reducing bacteria were the largest confirmed physiological metabolic microorganism group in the slightly contaminated waste.

Alkalinity

The pH value of the leachate was around 8, which is within the slightly alkaline range, common for the methanogenic phase.

299

Classification
The crux of these findings proves that the stages of non-contaminated and moderately contaminated municipal waste combined in Lys. 2 were in the stable anaerobic methanogenic phase.

Highly contaminated municipal waste (Lys. 9)

Gas generation
16.8% by volume of CH_4 was measured in the highly contaminated deposit. Methane generation potential was 5 mg CH_4/(kg × h) – relatively high. On the other hand, the number of the methanogenic bacteria detected under culture conditions in the laboratory was 4.5×10^3/g DS waste which is small compared to literature values (Sleat *et al.*, 1987) and the numbers found in other deposits.

Iron and sulphate reducing bacteria
The numbers of iron and sulphate reducing bacteria, two other groups active at lower redox potentials, were very high: 9.7×10^7/g and 4.5×10^6/g.

Acidification
The number of acid-forming microorganisms was 4.4×10^4 – higher than in most deposits tested. A quantitative domination of medium- and higher-chain fatty acids confirms fermentation. Nevertheless, the number of acid-producers was too small for an acidogenic degradation phase which is dominated solely by them. The large amount of phenol of the highly contaminated deposit (1 t) has very probably contributed to the findings of a slightly acidic leachate (pH = 6.5) and the very high COD in the leachate (about 11 000 mg O_2/l). Phenol is highly soluble in water, it is a weak acid and has a conserving effect. The amount of phenol was extraordinarily high in the leachate before the beginning of waste removal.

Classification
These facts indicate that only sub-areas of the lysimeter were in the methanogenic phase, while the bulk was in the acidic initial phase. This is very uncommon for a flat municipal waste deposit without perched leachate after 12 years of storage under German climate conditions and is not due to biochemical conservation alone.

300

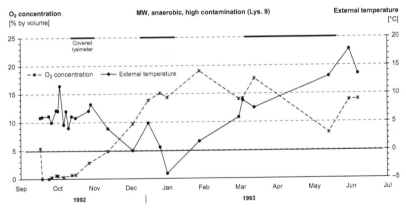

Fig. 8-20 O_2 concentration in the initially aerobic municipal waste with high chemical contamination (Lys. 9) during aerobic activation as an indirect measure of respiration activity: low O_2 concentration in the lysimeter due to high BOD at the beginning; reduction of respiration activity thus rising O_2 concentration due to cooling starting from about 50 days; slight increase in respiration activity with warming in the spring and concomitantly slight drop in O_2 concentration

(b) Reactions due to aerobic activation, exothermic reactions and oxygen consumption

The temperatures due to exothermic aerobic degradation processes have already been discussed in connection with the landfill engineering assessment (Section 8.5.2, Figs. 8-8–8-10). The typical temperature increase for a less stable waste at the beginning of the aeration phase was characteristic in the non-contaminated (Lys. 2, bottom) and the moderately contaminated deposit (Lys. 2, top). The temperature reaction of the highly contaminated deposit (Lys. 9) corresponded to a fresh waste. The similar temperature minima in the winter with a consecutive increase in the summer in all three variants can be explained by cooling.

Extent and profile of oxygen concentration in the waste of the three lysimeters (e.g. Fig. 8-20) in connection with the extent and profile of the external temperature reliably prove the effect of cooling. The characteristic oxygen concentration and, conversely, BOD, was in agreement in all 3 contamination variants, moving opposite to temperature: high BOD toward 0% by volume of oxygen at the beginning of decomposition, an increase of oxygen concentration up to almost 20% by volume in the winter and reduction of oxygen concentration to about 16 to 8% by volume in the summer.

The periodic inhibition of circulation markedly reduced the oxygen concentration in the highly contaminated stage of Lys. 9 during

301

the periods of activity at the beginning of deposition, but only in the summer months (Fig. 8-20). However, they were ineffective in the winter: the exothermic degradation came to a halt due to decreasing temperatures in the winter. Deducting this time interval of about 4 to 5 months, the mobilised exothermic degradation processes were far from being completed in any of the three variants even after 6 months of decomposition time.

It was clearly proved for the three variants that the material of the highest contamination stage corresponded to a fresh waste: consumption reduced the oxygen concentration to 0% by volume during the intensive initial phase with 60–70°C without any inhibition of circulation. Consecutively, oxygen concentration increased only slowly, at an inverse proportionality to temperature over several months – a typical process of the decomposition characteristics of a fresh material with natural draught aeration (cf. e.g. Jourdan *et al.*, 1982, Section 5.3.2). Further explanation about the cause and extent of the preceding conservation required a detailed investigation into colonisation and activity.

(c) *Change of microbial numbers and activity*

The extent of the viable cell numbers of different physiological groups directly characterises the microbial conditions of the deposit. In connection with the measurement of the potential degradation activity in terms of BOD and methane production, differentiated statements can be made about the state of stability. The increased accuracy of the differentiated laboratory tests on the material only refers to the sample tested. Tolerances result from the inhomogeneity of the deposit. An interpretation which goes beyond the determined characteristic is therefore not appropriate.

Figure 8-21 shows the viable counts of the most important groups of organisms before waste removal, during the predominantly aerobic interim treatment and after final compaction, arranged according to the extent of industrial contamination. The relevant respiration activities and potential methane generation are presented in Figs 8-22 and 8-23.

The viable counts before waste removal are similar to a large extent. Alone, the highly contaminated municipal waste (Lys. 9) contained more denitrifying bacteria by two orders of magnitude and an order of magnitude less methanogenic bacteria.

Degradation activities, measured in the laboratory, differed more explicitly: the highly contaminated municipal waste achieved an aerobic respiratory activity by about a factor of 1.7 higher and up to quadruple the methane generation rate within 2500 hours. The potential methane

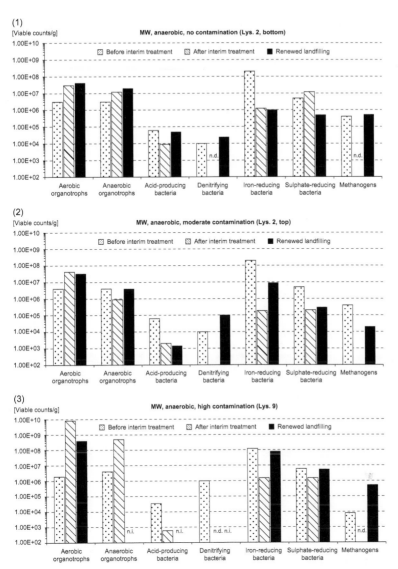

Fig. 8-21 *Influence of chemical contamination on viable cells in initially anaerobic municipal waste before, during and after aerobic activation* (n.d. = non determined, n.i. = non-interpretable). *(1) No initial contamination (Lys. 2, bottom): insufficient increase of aerobic cell numbers, but no decrease in most anaerobic cell numbers. (2) Moderate contamination (Lys. 2, top): the aerobic interim treatment resulted in a permanent increase in the number of aerobic microorganisms and a sizeable reduction of important anaerobic microorganism groups (acid-producers, iron-reducing bacteria, sulphate-reducing bacteria and methanogenic bacteria). (3) High contamination (Lys. 9): the slight dominance of anaerobic groups was replaced by a strong dominance of aerobic microorganisms. During the course of the consecutive landfilling an increase of anaerobic groups and the decrease of the number of aerobic bacteria started again*

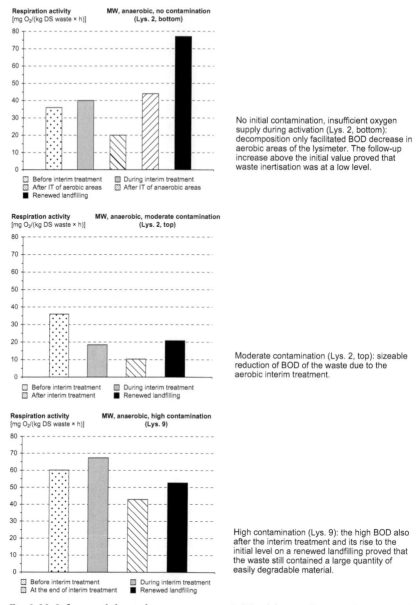

Respiration activity
[mg O$_2$/(kg DS waste × h)]

MW, anaerobic, no contamination
(Lys. 2, bottom)

No initial contamination, insufficient oxygen supply during activation (Lys. 2, bottom): decomposition only facilitated BOD decrease in aerobic areas of the lysimeter. The follow-up increase above the initial value proved that waste inertisation was at a low level.

☒ Before interim treatment   ▨ During interim treatment
◩ After IT of aerobic areas   ▧ After IT of anaerobic areas
■ Renewed landfilling

Respiration activity
[mg O$_2$/(kg DS waste × h)]

MW, anaerobic, moderate contamination
(Lys. 2, top)

Moderate contamination (Lys. 2, top): sizeable reduction of BOD of the waste due to the aerobic interim treatment.

☒ Before interim treatment   ▨ During interim treatment
◩ After interim treatment   ■ Renewed landfilling

Respiration activity
[mg O$_2$/(kg DS waste × h)]

MW, anaerobic, high contamination
(Lys. 9)

High contamination (Lys. 9): the high BOD also after the interim treatment and its rise to the initial level on a renewed landfilling proved that the waste still contained a large quantity of easily degradable material.

☒ Before interim treatment   ▨ During interim treatment
◩ At the end of interim treatment   ■ Renewed landfilling

Fig. 8-22 Influence of chemical contamination on BOD of the initially anaerobic municipal waste before, during and after aerobic activation

generation rate (= methane generation after the addition of energy and nutrients), however, decreased by almost a factor of 1/2.5 after final compaction (cf. Fig. 8-23). This allows the same conclusion which can be derived from the various cell numbers of methanogenic bacteria.

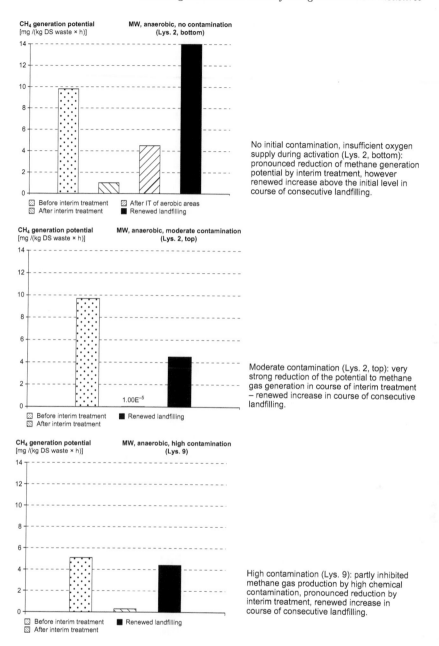

CH₄ generation potential
[mg /(kg DS waste × h)]

**MW, anaerobic, no contamination
(Lys. 2, bottom)**

Legend:
- ▣ Before interim treatment
- ▨ After interim treatment
- ▧ After IT of aerobic areas
- ■ Renewed landfilling

No initial contamination, insufficient oxygen supply during activation (Lys. 2, bottom): pronounced reduction of methane generation potential by interim treatment, however renewed increase above the initial level in course of consecutive landfilling.

CH₄ generation potential
[mg /(kg DS waste × h)]

**MW, anaerobic, moderate contamination
(Lys. 2, top)**

1.00E⁻⁵

Legend:
- ▣ Before interim treatment
- ▨ After interim treatment
- ■ Renewed landfilling

Moderate contamination (Lys. 2, top): very strong reduction of the potential to methane gas generation in course of interim treatment – renewed increase in course of consecutive landfilling.

CH₄ generation potential
[mg /(kg DS waste × h)]

**MW, anaerobic, high contamination
(Lys. 9)**

Legend:
- ▣ Before interim treatment
- ▨ After interim treatment
- ■ Renewed landfilling

High contamination (Lys. 9): partly inhibited methane gas production by high chemical contamination, pronounced reduction by interim treatment, renewed increase in course of consecutive landfilling.

Fig. 8-23 Influence of chemical contamination and oxygen supply during activation on gas generation potential of the initially anaerobic municipal waste

The differences became clearer during the activation in the interim treatment: aerobic organotrophs increased in the highly contaminated deposit by about 3.5 orders of magnitude, anaerobic organotrophs by about 2.5 orders of magnitude. In the aerobic activated material of the moderately contaminated municipal waste (Lys. 2, top) a physiological group only exhibited a maximum increase of one order of magnitude, assuming there was an increase at all. The assumed anaerobic area of the aerobic planned interim treatment of the non-contaminated deposit (Lys. 2, bottom) was proved when waste was removed for final compaction. The respiration activity was increased during the interim treatment to such an extent that anaerobic conditions emerged in the aerobic-designed deposit when a minor construction error throttled the air exchange. This confirms that the non-contaminated material had been biochemically degraded only to a certain extent and then conserved in an anaerobically (only about a 40% reduction of the potential long-term methane production was achieved in the laboratory during 2500 hours).

No stabilisation was observed in the laboratory in the highly contaminated deposit within an activation time of 10 months either by a reduction of respiration activity or by a decrease of long-term methane production despite high temperatures, good oxygen supply and a clearly proved increase of the cell numbers. Notwithstanding the initial design, an anaerobic state was only achieved for 3 months during the activation time. It is known from the long-term practice of the 'chimney draught' method, which is similar in terms of fluid mechanics, that the long-term methane production in the laboratory of a municipal waste with at least the same degradable organic waste content (e.g. Schwäbisch Hall) is reduced by about 80% after 9 months decomposition time and methane production commences in the decomposed waste but without initial delay (cf., e.g., Jourdan *et al.*, 1982, Section 6). Not only did the chemicals inhibit anaerobic degradation but also retarded or prolonged the period of the aerobic degradation. This was confirmed by an analysis of the leachates from the permanent aerobic sewage sludge waste mixes (Lys. 5 without chemicals compared with Lys. 6/10 with chemicals, Chapter 5), which found twice the residual COD in the basal discharge and about ten-fold values in the waste body due to the spiked chemicals.

The degradation of the moderately contaminated municipal waste proceeded problem-free aerobically. The substantial potential in degradable materials before the interim treatment was reduced within 10 months to such an extent that the BOD measured in the laboratory

306

before the activation was halved and methane production was reduced to the residual activity of a forest soil. The fairly large industrial contamination of Lys. 2, top did not have any unfavourable effect compared to the non-contaminated municipal waste of Lys. 2, bottom. The microbial numbers after the final compaction corresponded to the remaining activity. The numbers of the obligatory anaerobic groups (acid-producers, iron and sulphate-reducing bacteria and methanogens) were markedly below the numbers before waste removal. In addition, the dislocation of the population densities indicates an extensive aerobic degradation up to nitrification in certain sub-areas of the end-consolidated material – a process that is only possible in an extensively degraded material. N_2O was also detected for the first time under these conditions as an interim product of denitrification. The processes of nitrification and denitrification also influenced the concentrations of ammonia and nitrate in leachate. The denitrifying bacteria increased by an order of magnitude in comparison to the initial counts, while the iron- and sulphate-reducing bacteria decreased by an order of magnitude. This means that, despite compaction, oxygen surplus was sufficient to facilitate the nitrification of ammonium in certain spots and nitrate provided a favourable electron acceptor for degradation of organics in anaerobic zones.

8.5.4.2 Measurement of the ecotoxicological effect to test the initial state and its change during aerobic activation

(a) Initial conditions
Before the beginning of waste removal an ecotoxicological inventory was made for leachate, solids and landfill gas.

Non-contaminated and moderately contaminated deposit (Lys. 2, bottom and Lys. 2, top)
In the first step the toxic effect of leachate from the basal discharge for Daphnia and luminous bacteria was tested. No toxicity was found for either of the two organisms.

The investigation of the solid material from the core of the deposit with medium contamination (Lys. 2, upper third) resulted in a reciprocal value of EC_{20} of 1440 l/kg for luminous bacteria, i.e. an eluate from this material must be diluted by 1440 litres of water per kilogram waste in order to reduce the inhibition of bacterial luminosity to 20%. This means that the solids from the material of the medium contaminated zone are extremely toxic to luminous bacteria. Waschke (1994) determined a reciprocal value of EC_{20} of $106\,l/kg$ for normal fresh municipal waste using a somewhat different eluate method. The highest value in the field of contaminated land reclamation (reduction of inhibition to 50%) known to us is $1/EC_{50} = 317\,l/kg$ (Scheibel *et al.*, 1991).

$1/EC_{20} \approx 16\,l/m^3$ was determined for the gas samples from the inside this low-contamination municipal waste (second third of Lys. 2), indicating that the gas from the pores of these wastes was about 15 times more toxic than the ambient air to luminous bacteria. Nevertheless it was less toxic than the soil-air of a soil contaminated with gasoline (about $90\,l/m^3$). The gaseous phase was thus classified as medium contamination.

The large discrepancy between the extremely toxic eluate of the solid material and the non-toxic leachate is due to the technically almost unavoidable oxygen influence in the drainage system of the lysimeter. The effect of this oxygen infiltration was tested by parallel elution of a solid sample from the low-contamination municipal waste area (Lys. 2, second third) in a closed container without any gas exchange with the atmosphere (20% by volume O_2) and in another container with a nitrogen atmosphere: the eluate from the device with a N_2 atmosphere provided $1/EC_{20} = 193\,l/kg$ in the luminous bacteria toxicity test (very toxic), while the eluate of the same sample under an air influence failed to exhibit any toxicity. Obviously, a limited influence of atmospheric oxygen is already sufficient to oxidise the toxic substances during elution.

Taking into account that the leachate tests are 'disturbed' samples, it can be stated that the core area of the low-contamination municipal waste exhibited very high ecotoxicity, about a factor of ten higher than that of fresh waste. However, a short intensive aeration was capable of oxidising the toxicity-causing substances into innocuous substances. No significant Daphnia toxicity was detected.

Highly contaminated deposit (Lys. 9)

The leachate of the highly contaminated deposit (Lys. 9) proved extremely toxic. The luminous bacteria test yielded $G_L = 140$ and the Daphnia toxicity test $G_D = 770$ (see Fig. 8-24). This meant that the leachate had to be diluted $1:140$ to ensure that the luminosity of luminous bacteria was only restrained by 20% or, be diluted $1:770$ to prevent an inhibition of the swimming capability of Daphnia. For comparison, the limiting values for discharge into lakes and streams are once again: $G_L \leq 4$, $G_D \leq 4$ (AbwV, 2002). LC_{50} values are

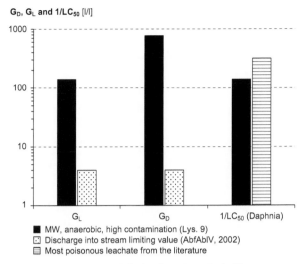

G_D, G_L and $1/LC_{50}$ [l/l]

- ■ MW, anaerobic, high contamination (Lys. 9)
- ⊡ Discharge into stream limiting value (AbfAblV, 2002)
- ⊟ Most poisonous leachate from the literature

Fig. 8-24 Comparison of ecotoxicity of the leachate from the highly contaminated deposit of the initially anaerobic municipal waste (Lys. 9) with future limiting values for discharge into streams and literature values. The leachate must be classified as extremely toxic

preferred in the literature instead of G values (dilution stages), latter ones used in DIN and administrative regulations (Fig. 8-24). For the leachate from the highly contaminated deposit (Lys. 9) the value $LC_{50} = 0.7\%$ v/v was obtained for Daphnia (the third black bar in Fig. 8-24 shows $1/LC_{50} = 143$) which is in the same order of magnitude as the minimum value obtained by Atwater *et al.* (1983) from tests on leachates of seven different landfills and 27 lysimeters (the parameter decreases with increasing toxicity). The leachate from the highly contaminated deposit (Lys. 9) was almost as poisonous as the most poisonous leachate of these investigations (see Fig. 8-24) and had to be classified, therefore, as very toxic. Since atmospheric oxygen had the same chance of influencing the leachate at the discharge of the highly contaminated deposit as that from the non to medium contaminated waste (Lys. 2), it follows from this measurement that the toxic substances from the maximum contamination were not easy to oxidise as opposed to the non- to medium-contaminated deposit and they were the only leachate pollutants in these tests which, in addition to a toxic effect on luminous bacteria, also had a substantial toxic effect on Daphnia.

The highly contaminated solid material (Lys. 9) exhibited a value of $1/EC_{20} = 850$ l/kg for luminous bacteria and had to be classified as very toxic. Toxicity of the gas within Lys. 9 for luminous bacteria was 12 l/m^3,

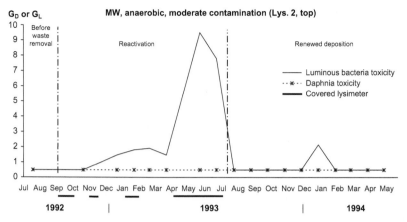

Fig. 8-25 Toxicity of the leachate from the moderately contaminated deposit of the initially anaerobic municipal waste (Lys. 2, top). At the beginning of the warmest season luminescent bacteria toxicity increased strongly in connection with a temporary encapsulation. The toxicity shows an anaerobic effect

which is moderately toxic (in comparison: the value for ambient air was $1 \, l/m^3$).

(b) Toxicological investigation of the leachates during activation and after final compaction

The toxic effect of leachates on luminous bacteria and daphnia was monitored during the interim treatment and after the final compaction.

The toxicity of the leachates both from the non-contaminated deposit and from the moderately contaminated deposit (Lys. 2) was insignificant before removal of the wastes, during the predominantly aerobic interim treatment and after final compaction (Fig. 8-25), because the unavoidable oxygen in the discharge was sufficient to oxidise the toxic substances (cf. Section 8.5.4.1). Short increases of luminous bacteria toxicity were due to anaerobic phases up to sulphide formation. The toxicity of the leachate of the highly contaminated stage (Lys. 9) was characterised by a substantial Daphnia toxicity before the removal of the wastes and during their interim treatment. Compared to these values, the toxicity of the leachate markedly decreased after the final compaction (Fig. 8-26), however, remained about an order of magnitude above the values of the low-contamination waste. The increases of toxicity occurred at the same time when the supply air to the lysimeter was closed (aerobic/anaerobic alternating treatment), but the oxygen content was only sustainably reduced in the starting phase (Fig. 8-26). Oxygen content decreased more in the moderately

310

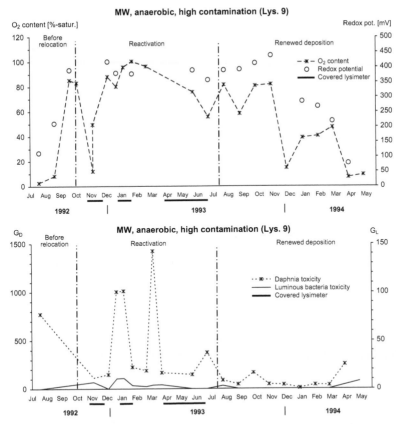

Fig. 8-26 *Comparison of* O_2 *content and redox potential as well as luminous bacteria and Daphnia toxicity of the leachate of the highly contaminated deposit of the initially anaerobic municipal waste (Lys. 9).* O_2 *content and redox potential increased while toxicity decreased drastically in course of becoming aerobic. These processes reversed under anaerobic conditions (sealing phase and final landfilling)*

contaminated material in the final phase of the interim treatment than in the highly contaminated municipal waste, without the toxicity increasing in the leachate of the moderately contaminated deposit.

The high toxicity of the leachate of the highly contaminated deposit (Lys. 9) is therefore mainly due to the substances contained within.

(c) Interpretation of toxicity as a parameter of biochemical degradation processes of non-contaminated and moderately contaminated deposits (Lys. 2, bottom and Lys. 2, top)

The luminous bacteria toxicity in the oxygen-affected leachate agreed significantly with the leachate from intensive degradation processes

311

within the aerobic area (Lys. 5). This was determined only by the oxygen influence on the degradation processes in the solids and not by the toxic effect of the industrial contamination (Fig. 8-25).

The luminous bacteria toxicity $1/EC_{20}$ was 1.400 l/kg both in the non-contaminated and in the moderately contaminated municipal waste before activation. It decreased during the aerobic degradation phase in both materials by two orders of magnitude to an insignificant level and remained insignificant in the case of a successful stabilisation of the moderately contaminated deposit even after final compaction. If the anaerobic degradation activity increased in the interim treatment – non-contaminated municipal waste – the luminous bacteria toxicity increased by about 50%. The toxicity was still about 35% of the initial value after the common final compaction of both sub-areas – a reliable indication for an anaerobic degradation activity at a lower level than in the initial material.

Highly contaminated deposit (Lys. 9)

Luminous bacteria toxicity of the highly contaminated material (Lys. 9) before the interim treatment was only about 60% of the toxicity of the non-contaminated material (Lys. 2, bottom). The direct toxic effect of the chemicals was therefore not decisive for the luminous bacteria toxicity. It can be concluded from the comparison with the moderately contaminated waste (Lys. 2, top) that luminous bacteria toxicity without Daphnia toxicity coincides with the anaerobic degradation activity (highest luminous bacteria toxicity without Daphnia toxicity in the non-contaminated deposit successfully activated anaerobically). However, Daphnia toxicity occurred in the leachate of the highly contaminated deposit in connection with reduced luminous bacteria toxicity for anaerobic degradation (Fig. 8-26). It can also be concluded from this comparison that the degradation was inhibited or prolonged under high industrial contamination. The luminous bacteria toxicity of the leachate remained low during all three phases, also in the phases of gas exchange inhibition. However, it cannot be concluded from this that the toxicity was altogether low. The leachate of the highly contaminated deposit had a more toxic effect on Daphnia by about an order of magnitude and by about two orders of magnitude in peaks than that of lower contamination stages whose effect on Daphnia was negligible. This indicates that it contained substances toxic to Daphnia which were not discharged either from the non-contaminated or from the moderately contaminated material.

The solid material exhibited $1/EC_{20} = 850$ l/kg and had also to be classified as very toxic. The toxic substances of the highly contaminated stage were not easy to oxidise.

312

The measurement confirms the requirement that toxicity of gases must be monitored and/or protection measures must be taken when working on landfills (cf. Section 11.3.2).

8.5.5 Overall assessment of the effect of industrial contamination
Peter Spillmann

8.5.5.1 Influence on stabilisation processes

Moderate industrial contamination (Lys. 2, top) affected the results of integrated but indirect acting test methods of waste management (mass loss, change of sieve curves, reactivation temperature, leachate contamination, eluate criteria) to a lesser extent than oxygen ingress from the ambient air to the top layer of the medium contamination grade (Lys. 2) in approx. 1 m depth (considerably lower activation potential in comparison to the non-contaminated base). These criteria indicate that both non-contaminated and moderately contaminated deposits went through an intensive degradation of the easy-to-degrade substances (about 23% by weight DS waste mass loss) within the first 5 years, but were then anaerobically conserved (documents and card-board packaging were preserved, the material can be re-activated by an intensive reaction both aerobically and anaerobically). Contrary to the moderately contaminated deposit (Lys. 2, top) the high- contamination deposit (Lys. 9) was conserved after a moderate degradation (approx. 7% by weight DS waste) within the acidic range.

The indirect conclusions were confirmed and specified by the detailed chemical and biological investigations. Conservation, inhibition and reactivation as well as the characteristics of the organic compounds produced were determined in a differentiated way from the ratio of organic acids to dissolved carbon (TOC), the molecular structure of organic acids and the ratio of the dissolved organic carbon (TOC) to the organic carbon in the solids (solids TOC):

- The conservation state of the organic material following an anaerobic degradation of the easy-to-degrade substances was characterised by long-chain organic acids of low concentration – untypical for municipal wastes – in the solids, eluate and leachate, whose content does not correlate with the content of the dissolved organic carbon TOC in the leachate.
- The reactivation of the extensive aerobic degradation is characterised by the increase of TOC in the leachate without an increase

313

of short-chain acids – typical for municipal wastes – because this TOC is caused by humic substance-like materials.
- The anaerobic activation causes an increase of TOC due to short-chain organic acids, characterised by a close correlation of TOC with the concentration of organic acids.

The condition of the anaerobic conservation was confirmed analytically for the non-contaminated and moderately contaminated deposits inasmuch as long-chain atypical acids were detected in the solids, eluate and leachate in low concentrations in place of the acids typical for municipal waste whose concentration did not correlate with the organic carbon material in the leachate (TOC) also dissolved at a low concentration. The stabilising effect of oxygen introduction into the moderately contaminated top layers of the waste body (Lys. 2, top) was quantified in comparison to the chemically non-contaminated base.

Moderately contaminated top layers
C14–C16 fatty acids: approx. 2000 mg/kg, spiked phenol degraded to a large extent.

Non-contaminated bottom layer
C14–C16 fatty acids: approx. 26 000 mg/kg.

The reactivation of the aerobic degradation processes was proved for the non-contaminated deposit and the moderately contaminated deposit as a fully aerobic process by a five-fold increase in the dissolved organic carbon (TOC) without any increase of organic acids. The continuation of the anaerobic degradation even after the final compaction was confirmed inasmuch as the chemical analyses detected waste-typical organic acids.

The indirect proof that degradation was inhibited in the highly contaminated deposit (Lys. 9) was confirmed in detail by the proof of organic acids typical for municipal waste in high concentration and their correlation to the dissolved organic carbon (TOC). The inhibiting effect of chemicals during the aerobic activation was characterised chemically directly by the detection of typical acids despite sufficient oxygen supply. The reduced degradation performance was quantified by the change of solids TOC: organic carbon decreased markedly in the analysed <8 mm and 8–40 mm medium to fine fractions in the solids of the non-contaminated deposit (extensive degradation) and remained constant in the moderately contaminated deposit. However, it increased by about two- to five-fold in the highly contaminated

314

material (disintegration of coarse material in the first step of the degradation processes, i.e. the first step had not yet taken place and a further degradation did not take place over the short term).

Not only do biological investigations provide further details of the influence of industrial wastes on biochemical stabilisation, but also furnish proof for the basic cause of disruption. The ecophysiological assessment confirms that the non-contaminated (Lys. 2, bottom) and the moderately contaminated deposits (Lys. 2, top) agreed to a large extent, but the highly contaminated deposit (Lys. 9) exhibited major differences. The latter contained about an order of magnitude more denitrifying bacteria before activation and an order of magnitude less methanogens despite a high long-term methane generation activity in the laboratory (2500 h) – in principle typical for the acidic phase. However, the number of acid-forming organisms was 4.4×10^4 – too small for the acidic phase dominated by acid formation. These conditions are typical for the inhibition of degradation.

Respiration activity of the highly contaminated material reached a 1.7-fold value measured in the laboratory (under ideal conditions) and at least the quadruple gasification rate in the long-term laboratory test in comparison to the non- and moderately contaminated municipal wastes. The high potential, however, was not reduced – in contrary to non- and moderately contaminated deposits – despite a good oxygen supply and a clearly proved increase of the cell numbers under aerobic, activating landfill conditions within 10 months of decomposition time (degradation without retardation is about 80% within the same time). The high content of phenol, a very soluble weak acid, might explain the inhibition by acidification in an anaerobic deposit. However, phenol can be easily degraded under aerobic conditions, thus the retardation of aerobic degradation cannot be explained by the phenol contamination.

Toxicity tests provided another proof for the negative influence of the high industrial contamination on biological processes. The very high toxic effect on luminous bacteria, observed in all anaerobic deposits and, particularly, in the leachate of the moderately contaminated stage, was no longer detected after a brief oxygen influence. Daphnia toxicity was not found in any of the phases in the non- and moderately contaminated deposit either. However, the highly contaminated deposit exhibited a considerable Daphnia toxicity, which was decreased by aerobic degradation but not reduced to insignificant values.

Summing up, the biological stabilisation processes in the weakly contaminated deposits are dominated by the deposition conditions. Therefore, the non-contaminated deposit and the moderately contaminated deposit can be treated together to assess time-related issues of the processes. The high contamination of the highly contaminated anaerobic Lys. 9 caused an extensive inhibition of the anaerobic biochemical processes, whose end was still not to be seen after 15 years despite favourable deposition conditions. A direct comparison of the permanent aerobic deposits (the non-contaminated Lys. 5 compared with the highly contaminated Lys. 6/10) proved that the aerobic stabilisation processes were also disrupted.

8.5.5.2 Degradation and release of environment-polluting substances

According to the waste management allocation criteria of TASi, AbfAblV and the LAGA leaflets, none of the three contamination stages was particularly conspicuous regarding industrial contamination. In the solids it was only the limiting values for cadmium in the medium-contamination deposit (galvanic sludge with high cadmium content) and for nickel in the highly contaminated deposit (galvanic sludge with extremely high nickel content) that were exceeded but were not extremely conspicuous. The eluates did not provide any special indication for high emissions or special emission potential either, not even for the maximum-contamination municipal waste. Not only were the criteria of landfill class II met according to AbfAblV for this high contamination, except for cadmium and nickel, but also the stricter Z2 criteria of LAGA. No serious environmental impact can be recognised from these values, not even for the highly contaminated deposit.

The detailed chemical analyses proved a differentiated contamination potential, well beyond the waste-management criteria. The toxic gases in the waste body, which were tested for very volatile chlorinated hydrocarbons and aromatics, were already mobilised and released almost completely by mechanical intrusion. This potential must be particularly considered for large waste bodies where these gases cannot diffuse out because of the long paths – unlike in the test system. Furthermore, when assessing the toxic gas potential, it must be taken into account that, due to incomplete degradation, even toxicologically less precarious materials such as perchloroethylene may give rise to highly toxic and extremely carcinogenic gases such as vinyl chloride.

Degradation and emission activation of pollutants was proved in detail by the example of the chlorinated hydrocarbon 'lindane': it was only possible to achieve the degradation rate of arable soil in the zone with oxygen diffusing inwards. Under disrupted conditions (high contamination, Lys. 9) the degradation was equal to zero. The activation of disturbed degradation processes resulted in the emission of both lindanes and degradation products such as chlorobenzene. Triazines were mobilised in the highly contaminated deposit but – contrary to the less contaminated deposit – not degraded. Phenol was continuously emitted but not anaerobically degraded, or fully degraded during the aerobic activation, unlike in less contaminated deposits.

It was also possible to markedly reduce the toxic elements by the reactivation of the degradation processes – proportionally to the initial contamination. This means that they can be remobilised from reactivatable deposits. However, it was the non-contaminated municipal waste which released the maximum lead emission. Thus the remobilisation potential of the elements from municipal waste deposits, even without industrial contamination, may not be neglected despite the very favourable-appearing results of waste-management investigations.

The high toxicity and, in particular, the lack of several species of organism cannot alone be the consequence of those substances tested and found as indicators and recognised as mobilisable materials. These findings from the biological tests complete the picture of the detailed chemical investigation: the fact that the currently valid deposit criteria were met, does not exclude the possibility that environment-polluting materials can be mobilised over the long term from old deposits and, also, from current landfills.

8.5.5.3 Comparison of the influence of operating technology with the influence of industrial waste

A population equivalent sewage sludge waste mix, intentionally contaminated with industrial waste, exhibited the greatest stability in terms of natural soil among the landfill types tested. It was stabilised biochemically predominantly aerobically before deposition up to producing almost water-insoluble humic substance-like materials (Lys. 5). The material remained permanently aerobic even after an extensive compaction and the degradation progressed measurably after compaction. The intentional reactivation about 15 years after compaction was hardly measurable, the mobilisability of the heavy metal inventory was insignificant, although it was markedly above that of a natural soil. According to biological criteria the mobilisable activity was comparable to that of a forest soil. However, soil-science investigations indicated that inertia of biochemical reactions, but not the molecular structure of the material, corresponded to that of extensively stable humic substances. But since the biochemical development of the humus soil

317

under local humid climate conditions goes from still unstable materials with low-chain molecules to stable humic substances with high-chain molecules, a constant increase in stability and a simultaneous decrease of emissions can be expected under these conditions. The fulfilment of this minimum requirement for a calculable aftercare will be used as a basis of comparison for other deposits.

In a compacted condition the initially anaerobic municipal waste deposits with and without sewage sludge appeared just as stable as the intentionally aerobic stabilised sewage sludge waste mix. Low to moderate contamination with industrial waste did not recognisably affect biological stabilisation. The reactivation tests proved, however, that this stability was based both on aerobic and anaerobic activatable conservation. If no intrusion into the waste body takes place, aerobic activatable degradation progresses by diffusion from the surface into the waste body, as soon as gas production is exhausted. Under optimum conditions (no sealing cap, no internal gas pressure, no perched water) the aerobic stabilisation penetrated into the municipal waste within about 10 years approximately 1 m deep into the waste body from the non-covered waste surface. Stabilisation was reached only within this zone in the municipal waste, which is comparable to the intentional aerobic stabilisation of the sewage sludge waste mix. Compact sewage sludge lenses were not stabilised aerobically within this time. Since activation of the degradation processes by intrusion into the waste body or by diffusion on a long-term basis are accompanied with emissions, long-lasting and sometimes increasing emissions up to biochemical transformation into stable humic substances can be expected from deposits of this type (see Chapter 10 for estimation of period of time). Similar stable conditions, as in the aerobic sewage sludge waste mix, were subsequently only achievable by an active intrusion using intentional aerobic stabilisation.

If the degradation chain is disrupted by high contamination with industrial waste which, however, is still entirely within the range of usual analytical values, the material can be conserved under very good degradation conditions (no perched water, only 1.5-m-thick waste layer, covered with permeable loamy sand) within the acidic phase so sustainably that, as in this case, neither stabilisation of the organic substance nor degradation of organic pollutants was detected within 15 years of deposition time. A deposit of this type and composition maintains its full emission potential for an incalculable time. The targeted aerobic activation also suffered some disruption, and though aerobic degradation reduced the emissions subsequently, it was not possible to convert the deposit into a state of calculable conditions.

9

Testing the material stability of soil-like substances and plastics concerning reactivation

9.1 Objectives

The tests on landfill engineering reactivation of degradation processes in waste deposits (Sections 8.1–8.5) dealt with the influence of atmospheric oxygen which will unavoidably penetrate landfills over the long term and with the effects of an interference into the waste body. It was found that in the most favourable case under these conditions the soil-like fine fraction of a biologically extensively stabilised population equivalent sewage–sludge waste mix (Lys. 5) does not react more intensively than a forest soil. The plastics contained in the waste did not exhibit recognisable changes and are even now considered as biochemically inert to a large extent. PVC was also classified as sufficiently stable and fit to be landfilled in a detailed scientific investigation at another university on behalf of the PVC manufacturers outside of this research programme. However, these positive statements about stability fail to be in agreement with the following test results and experience.

The investigation of the fine material which appeared stable using soil science methods based on Husz's certified method for the soilification of natural wastes (Section 8.4.5) indicated that the generally biologically stable organic substance is inert, reacting like a forest soil, however it most likely does not correspond to it in terms of molecular structure. PVC, at least some of its modified variants, is not completely stable in use. Emissions of materials used as additives to soften PVC, were detected in this long-term test programme in adjacent solids and in leachates. Spontaneous heating in plastic mono landfills (Section 11.3.2.3) support the assumption that the adaptation of microorganisms to hydrocarbon compounds with a high energy content for use as an energy source may be possible within a few decades due to the short generation times of microorganism.

319

The following investigation objectives for the final assessment of landfill behaviour emerge from these uncertainties:

- Characterisation of the composition of stabilised organic solids and the organic substances in the leachates for comparison with natural organic substances and for the description of the stabilisation process.
- Determination of the landfill behaviour of PVC in different modifications as a component of municipal solid waste.
- Determination of long-term biochemical stability of chemically resistant plastic wastes.

9.2 Model-compatible upscale of landfill conditions from the laboratory scale
Peter Hartmann, Günther Ballin and Peter Spillmann

9.2.1 Scales

Geometrical agreement

The conditions necessary for a suitable model to represent landfills were described in detail in Sections 2.4 and 2.8. According to those sections, full compliance with the laws of the model is only possible on a scale of 1:1. However, it was not possible to answer the remaining questions in the large-volume landfill sectional cores. They must therefore be investigated as detailed questions in the laboratory. For this purpose the special conditions of the landfill must be suitably represented in the laboratory.

The scale of 1:1 can also be geometrically maintained at the laboratory scale for the investigation of the natural organic substance. The mechanical conditions of the actual deposit, mainly the influence of sidewall leakage and subsidence processes and the accompanying pore distribution and pore form, cannot be guaranteed. It is of prime importance for the transferability of the biological stabilisation results to understand whether the mechanical deviations of the model from the large-scale equipment (here: decomposing flat windrow, see Figs 11-32 and 11-33) have a substantial effect on the stabilisation process. This possible deviation is investigated in such a way that at least one of the material variants is tested simultaneously to the laboratory test on a scale of 1:1 in the compressible large lysimeter as described in Section 2.4.

Pressure

The pressure on the solids and the pore contents may affect the reactions. The pressure on the material is produced in the laboratory by arranging the sample between two filter plates, which apply a defined pressure on the solids using clamping bars. The pore pressure is produced by placing the compressed solid sample into a gastight pressure reactor. Both compressive stresses are independently variable – as in the actual landfill.

Time scale

No time acceleration method within the model is known from their original stable plastics up to their instability point in the landfill after a minimum 10-year storage time. Therefore a verified unstable old material must be drilled from the landfill and its behaviour compared with a new material.

9.2.2   *Thermal and biological conformity with reality*

Intensive exothermic processes

The succession of microorganisms and the self-dynamics of the biological system during the exothermic aerobic stabilisation of the organic substance are only suitably thermally represented in the laboratory if the cladding temperature of the laboratory reactor is always adjusted to the interior temperature of the decomposing waste. Thermal insulation alone is not sufficient. Errors in the thermal representation, e.g. an investigation using an externally heated climate chamber, lead to major observation errors and to 'self-sterilisation' of the organisms above 60°C (clearly disproved, e.g., by Eschkötter, 2004).

Poorly reactive exothermic processes in thermally different reaction ranges

In order to represent the thermal similarity in poorly reactive deposits, it is not enough simply to adjust the cladding temperature to the interior temperature since the initial conditions for exothermic degradation processes in chemically resistant plastics are unknown and are usually higher than the ambient temperature. The thermal reactions and their causes can be established in a temperature-controlled fashion by warming the tested waste using a slow linear increase of the cladding temperature. As soon as exothermic processes set in, the cladding temperature is adjusted to the rising interior temperature. If the exothermic process ends, the temperature again slowly rises in a linear fashion similar to the starting phase, however it is shifted parallel by the amount of exothermic heating. This process can repeat itself several times within the investigated temperature range of 30°C to 90°C, so that a step-curve develops. The disadvantage of a temperature-controlled

321

operation is that the point at which the endothermic phase starts to generate easily combustible gases cannot be recognised, and these are then converted exothermally in the next step.

Change of endothermic and exothermic processes

An energetic differentiation of the processes is possible if the reactor is controlled by energy input and not by temperature. For this purpose sufficient energy is supplied through the cladding heating to the inert fill that the unavoidable energy losses at the top and bottom of the reactor are compensated for and the minimum temperature detected in the tested landfill can be maintained as a constant minimum temperature. A slowly increasing energy input is superimposed on, this constant supply which is provided so that the material temperature rises slowly and linearly in the case of an inert fill. Within this energy control, exothermic reactions are indicated as positive deviations and endothermic reactions as negative deviations in the material tempera-ture from the slowly rising straight line. Thus it is possible to distinguish between exothermic and endothermic processes.

Determination of biological and chemical processes

Biological activity can be measured simultaneous to the temperature development and samples taken from degradation products can be used to distinguish between biological and chemical components of the degradation processes. Since the needs of the determination depend substantially on the question being asked, the necessary instruc-tions are included in the description of the test being undertaken.

9.2.3 The design principle of the landfill simulation reactor using reaction-controlled cladding temperature

The requirements described in Section 9.2.2 were fulfilled by a landfill simulation reactor which suitably represented the heat accumulation of a waste body without exerting an externally controlled temperature upon the material. From the fact that the same reactions were observed in approx. 6-m-high large lysimeters (tests according to Chapters 1–8) as in high-waste bodies, it was concluded that it was generally not necessary for the pressure in the simulation reactor to achieve that of the real landfill body. Therefore, a low-pressure double-cladding reactor made from boro-silicate glass was designed (principle: Fig. 9-1: dimensions: $D_i = 300$ mm, $H = 1000$ mm without a dome, volumetric capacity $V = 86$ l), which can simulate all landfill parameters described in Section 9.2, except pressure.

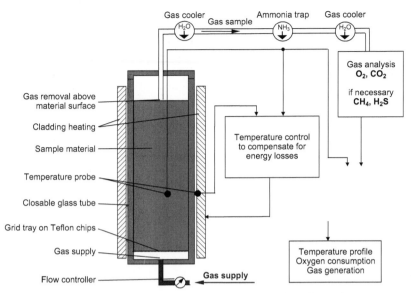

Fig. 9-1 Schematic setup of the control and measurement principle of the simulation reactor

Three sensors are arranged in the filling material (PT 100 linearised semiconductor of class 3 DIN, signal current 1 mA, operated at 500 µA in four-wire technique, stable using a high-stability reference voltage source for over 4 years), which simultaneously measure and determine the inhomogeneity of the waste. If the waste temperature rises, the cladding temperature follows suit with a temperature difference of 0.2°C and simulates the heat accumulation in the landfill. If no exothermic reactions occur in the initial temperature range, the waste is slowly warmed through the cladding at 0.2°C/week similar to an externally heated waste body. This linear rise results in a 'ramp' in the diagram. If an exothermic autogenic reaction is triggered, the material temperature rises faster than the 'ramp'. Since the cladding temperature always follows the material temperature, the curve rises step-wise in zones of exothermic processes.

The gas inflow temperature is controlled by the water in the heated bottom dome and transported in a water-saturated state (aerobic = air, anaerobic = argon), the exhaust gas flow condenses at 4°C and is passed to a continuous analysis process through two cleaning cartridges (blue gel and activated carbon): carbon dioxide and methane (2-jet infra red detector), oxygen (ion conduction cell, rare earth ceramic(s)), flow (electronic), optical absorption at 360 nm, results are reproducible without any readjustment over 4 years. Samples are taken discontinuously using activated carbon from the exhaust gas and archived by hermetically closing the tubes.

Single-wall glass reactors of smaller diameter ($D_i = 150$ mm) were built whose temperature adjustment was performed using a spiral wound hose for cladding heating to investigate simultaneous variants, based on the same principle and using the same instrumentation, but at lower costs.

A heat insulated aluminium temperature-gradient block ($400 \times 300 \times 200$ mm) was used to establish the temperature profile of bacterial multiplication, drilled to accommodate 13 reaction containers (16-mm excessive long test tubes) in each of four rows. A linear temperature gradient is formed between the freely selectable starting temperatures, and which contain all temperature levels that can occur in a test run within the reactor.

The reaction containers of the aluminium temperature-gradient blocks were operated under the following conditions: sterile plug, gas inlet 0.3 mm ID PTFE capillary, gas flow rate 2.5–3 ml/min for each container, gas inlet sterilised and moistened, aerosols precipitated. To obtain the inoculum, 2×20 g material was mixed with 80 ml of 0.85% saline solution and shaken at 55°C for 30 minutes at a frequency of 120/min. 300 µl of this extract was added to 10 ml nutrient solution (typton glucose bouillon) and left under aeration in the aluminium temperature-gradient block for 24 hours. Subsequently the cells were counted and the pH value determined.

This design was first used for the investigation of PVC (Section 9.4) in comparison to larger scales and was able to reproduce the processes in landfills with a very good accuracy. It is therefore the reference for the equipment used to investigate the reactions of natural organic wastes (Rostock University, Section 9.3; Franke, 2004; Franke *et al.*, 2006; Degener, 2005 and Weimar University, Scholwin, 2003). Furthermore it was used to investigate plastic mono deposits (Section 9.5) and mixed deposits of incineration ash and MBT wastes (Dörrie, 2000) and is the initial design for the development of the pressure reactor (Section 9.5).

9.3 Mass-spectrometric investigation of biological stabilisation of natural organic substances
Matthias Franke

9.3.1 *Objective*
A research project on the 'Use of the windrow decomposition process for extensive biological degradation of leachate substances and clarifying residues from the decomposition' was successfully completed between

July 2001 and August 2004 at the Faculty for Agriculture and Environmental Science of Rostock University, supported by the Federal Ministry for Economics and Technology (BMWi), under the project management of the Working Group Industrial Research Association 'Otto von Guericke', Berlin. Degradation and stabilisation of organic substances from leachates and solids was investigated in this research project at a molecular level (Franke, 2004) and a microbiological level (Degener, 2006). The investigations were performed comparatively on the freshly decomposing wastes and the biologically extensively stabilised material of the sewage–sludge waste mix from Lys. 5 (described in Chapter 8). The combination of two independent analytical pyrolysis methods together with waste water derived sum parameters, for the first time, enabled both qualitative and quantitative characterisation of the material composition of organic substances both for decomposition solids and leachates.

9.3.2   Test method

9.3.2.1   Test equipment

To investigate aerobic degradation processes the so-called decomposition simulation reactors (DSR) (Figs 9-2 and 9-3) have been designed and built based on the standard operating procedure 'Sampling of waste materials in landfill simulation reactors' (Heyer *et al.*, 1997) with the support by Dr Scholwin, Bauhaus University Weimar, with the financial support of the company EuRec Technology, Merkers, Germany. They consisted of six reactors, each pair working simultaneously. They offered the possibility of investigating the encapsulated intensive decomposition and subsequent composting of waste materials with a recirculation of process water under laboratory conditions (Degener *et al.*, 2004). (Editor's note: The design principle of the individual reactors, principally the constant adjustment of the temperature-controlled water jackets to the interior material temperature in the reactor, was derived from the prototype Fig. 9-1 both in the equipment of Scholwin (2003) and in that of Franke/Degener, Fig. 9-2). By excluding ambient influences a cooling of the decomposing waste due to size was avoided. Elution and conversion processes were influenced by controlling the water, air and heat balance. The reaction containers consisted of 150-cm-high glass cylinders [1] with 30 cm inside diameter. The stainless steal base plate [2] provided the air supply through a welded raised ventilation cross [3] by a compressor [4] into the reaction area as well as the removal of process water at the reactor base. The

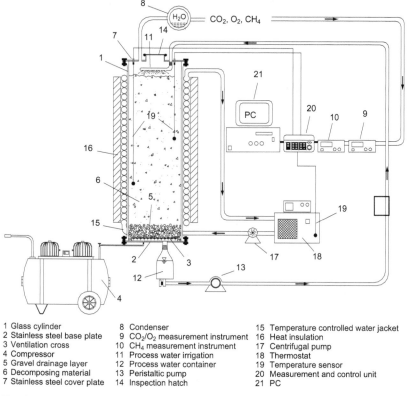

Fig. 9-2 Setup of decomposition simulation reactors (DSR)

1 Glass cylinder	8 Condenser	15 Temperature controlled water jacket
2 Stainless steel base plate	9 CO_2/O_2 measurement instrument	16 Heat insulation
3 Ventilation cross	10 CH_4 measurement instrument	17 Centrifugal pump
4 Compressor	11 Process water irrigation	18 Thermostat
5 Gravel drainage layer	12 Process water container	19 Temperature sensor
6 Decomposing material	13 Peristaltic pump	20 Measurement and control unit
7 Stainless steel cover plate	14 Inspection hatch	21 PC

raised ventilation slots prevented process water from penetrating the ventilation unit. The ventilation cross was embedded in a 15-cm-thick gravel drainage layer [5]. The air supply (20 ml/min to 1000 ml/min) ascended through this drainage layer homogeneously from the reactor base into the decomposing material [6] and prevented the accumulation of process water at the reactor base. The top opening of the glass column was also closed with a stainless steel plate [7]. The exhaust air was fed through the connecting pieces attached to the plate and via the condenser [8] to the gas analysers. Carbon dioxide and oxygen concentration was determined by CARBONOXY 10/25 gas analyser [9] (Pewatron Company, Zurich, Switzerland). Methane content of the exhaust air was measured by EGC 30 gas analyser [10] (Pewatron Company, Zurich, Switzerland). The measurement principle for the determination of O_2 concentration was based on an electro-chemical zirconium dioxide sensor. The determination of CO_2 and

CH_4 concentration was performed using IR spectrometry. Another connecting piece was used for the drip irrigation [11] of process water and condensates collected at the base in a collection container [12] and recycled using a peristaltic pump IP 8 [13] (Ismatec Company, Glattbrugg, Switzerland). Sampling and visual inspection was undertaken through an inspection hatch integrated into the cover plate [14]. The temperature-controlled water jacket [15] enabled a small section of an aerobic waste treatment to be simulated at the laboratory scale. It consisted of a PVC hose with an internal diameter of 1 cm and was spirally wound from the column base and up to the top reactor opening. The spiral was insulated on the outside by an approx. 5 cm thick layer of insulation wool [16]. A centrifugal pump [17] continuously fed the hose with water which was pre-heated by a thermostat [18] according to the actual reactor temperature. For this purpose resistance temperature probes were arranged in the thermostat and the reactor (Type PT 1000, four-wire technique) [19], whose measuring values were adjusted in the measuring and control unit 34970 A [20] (Agilent Technologies, Colorado, USA) using PC [21]. When the thermostat temperature dropped below the target temperature, which usually corresponded to the reactor temperature, heating was actuated by a relay circuit. This was maintained up to agreed measured values. To avoid a forced heating of the material, the thermostat temperature was slightly depressed.

The arrangement of aeration, exhaust air measurement and ventilation of the six DSRs is represented in Fig. 9-3. Air was supplied by a compressor [1] through a particle filter [2] and pressure reducer [3], and dosed by flow controllers [4] operated manually. Air flowed through the glass cylinder [5] and arrived at the cover plate and passed over the condenser [6] to remove the humidity taken up in the reaction space. The residual moisture content was extracted by dry cartridges [9] filled with silica gel to protect the on-line measuring technique [7, 8]. An automatically controlled valve [10] regulated the exhaust air flow of all reactors. Thus only the exhaust air of one reactor at a time flowed through the measuring gas pipe [11] to the measurement instruments, while the exhaust air of all other reactors was transported through multiple bypasses [12] to exhaust [13]. A diaphragm pump [14] sucked the exhaust air at a constant flow rate through the measurement instruments and counteracted any flow fluctuations which would have falsified the measured value. Check valves [15] in the bypass lines prevented contamination of the measuring gas with exhaust gases from the exhaust. The system for removing contaminants was flushed with

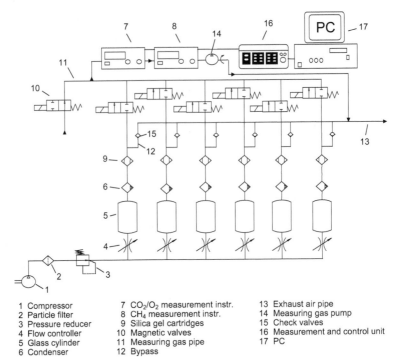

Fig. 9-3 Pattern of aeration, exhaust air measurement and ventilation of the decomposition simulation reactors

1 Compressor	7 CO_2/O_2 measurement instr.	13 Exhaust air pipe
2 Particle filter	8 CH_4 measurement instr.	14 Measuring gas pump
3 Pressure reducer	9 Silica gel cartridges	15 Check valves
4 Flow controller	10 Magnetic valves	16 Measurement and control unit
5 Glass cylinder	11 Measuring gas pipe	17 PC
6 Condenser	12 Bypass	

fresh air between exhaust air measurements of the individual reactors. The measuring and control unit 34970A [16] (Agilent Technologies, Colorado, USA) accepted the measuring signals and released the control commands. Data acquisition was performed by two multiplexer modules. A relay module was used to close or interrupt circuits for the valve and thermostat control. Furthermore, the measuring and control unit in combination with a PC [17] was used to control the magnet valves and thermostats. The PC was equipped with a program simulating decomposition, which provided the signals for locking and unlocking the relays. This program also carried out the conversion of electrical measurement signals and measuring data visualisation and archiving. Temperature and gas measuring intervals were variably adjusted. The temperature was measured using six measurements per minute so that a frequent adjustment of the set value (reactor temperature) with the actual value (thermostat temperature) reduced any undesired swinging of the thermostat. The gas measuring interval was selected according to the required flush time in order to remove the exhaust air from the pipe after a measurement procedure. An interval of 5 minutes proved to be sufficient for this purpose.

The following variants of waste were tested using a double approach:

- municipal roughly shredded fresh waste, countryside origin (fresh waste variant)
- municipal waste of Variant 1, mixed with sewage sludge (fresh waste sewage sludge variant)
- municipal waste sewage sludge mix, biologically extensively stabilised (Lys. 5, Braunschweig).

A section of an aerobic waste treatment with a natural draught was built for Variant 1 (roughly shredded fresh waste) (Franke and Degener, 2003) to check the agreement of the laboratory tests with large-scale tests in similar fashion to the lysimeter plant as shown in Chapter 2 (Figs 9-4 and 9-5). It was constructed on a circular area with a diameter of 7.5 m. The lysimeter subgrade [1] consisted of compacted crushed stone and had the form of an upturned saddle. An approx. 30-cm-high edge wall prevented the process water from escaping. The interior downslope gradient towards the drainage was 2%. A two-layer base liner [2] was placed on the subgrade. The first sealing geomembrane was placed on a 5-cm-thick compensation layer of sand. This served as a control liner and covered the entire lysimeter area. A TASi-compatible HDPE geomembrane separated by another lift of sand followed. While an approx. 30-cm-thick 16/32 mm [3] gravel layer acted as a drainage layer. A gravel-encased drainage pipe [4] with a slope of 3% was arranged on the HDPE geomembrane. A horizontal combined drainage and ventilation pipe [5] was also installed transversely to the drainage direction according to the so-called chimney effect (Spillmann and Collins, 1981; Turk, 1997; Collins *et al.*, 1998) with access the outside air at two sides of the lysimeter. The two drainage pipe ends were connected by a T-piece in the centre of the lysimeter to the vertical chimney draught pipe [6], which was connected to the chimney draught casing [7] at the lysimeter surface. Hot air, created by spontaneous heating, ascended in the vertical chimney pipe according to the thermo-dynamic principle of density reduction. The suction which developed provided an automatic fresh air supply. The shredded residual waste [9] was placed on the gravel layer and surrounded by a strong cladding reinforced geomembrane [8]. This special cladding geomembrane enabled vertical subsidence but prevented a sideways expansion of the decomposing waste. This geomembrane also prevented the outside air from accessing the decomposing material. Clamping straps were used

329

for stabilisation by supporting this cladding geomembrane at distances of approx. 25 cm. A 30-cm-thick layer of compost mulch [10] covered the decomposition body and served as an odour trap and improved the flow conditions. This layer also standardised the rain input into the decomposition body. Thirteen vertical measuring probes [11] were driven through the mulch layer into the decomposition body to measure the temperature of the material and to take samples of CO_2, O_2 and CH_4 gases. The arrangement of the probes is illustrated in the plan and in part in sections A–A and B–B of Fig. 9-4. A rain protection geomembrane

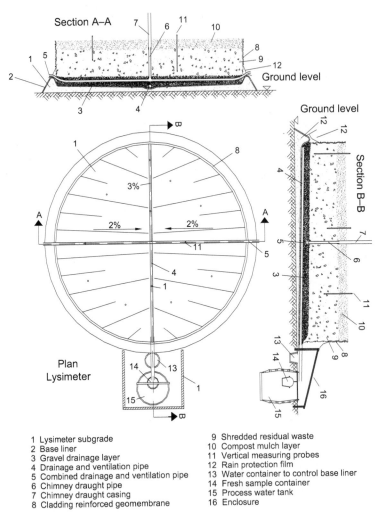

1 Lysimeter subgrade	9 Shredded residual waste
2 Base liner	10 Compost mulch layer
3 Gravel drainage layer	11 Vertical measuring probes
4 Drainage and ventilation pipe	12 Rain protection film
5 Combined drainage and ventilation pipe	13 Water container to control base liner
6 Chimney draught pipe	14 Fresh sample container
7 Chimney draught casing	15 Process water tank
8 Cladding reinforced geomembrane	16 Enclosure

Fig. 9-4 Plan and two sections of decomposition lysimeter design

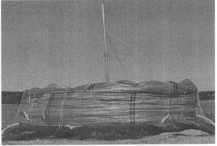

Fig. 9-5 Decomposition simulation reactors (DSR) (left) and large-scale lysimeter (LYSI) (right)

[12] prevented precipitation water from gaining direct access to the lysimeter base by diverting the external run-off water. The bottom sealing geomembrane was connected to a separate control container [13]. It would have indicated a leakage of the top HDPE geomembrane in the case of process water leakage. The regular process water discharge was removed through the drainage pipe [4] into a fresh sample container [14]. This drained into a process water tank [15] with a capacity of 500 litres. The process water containers were covered to prevent precipitation from entering the enclosure [16].

9.3.2.2 Analysis methods for characterisation of the organic substance
The combination of Py-GC/MS with the complementary Py-FIMS (independent of Py-GC/MS as a stability indicator) has proved to be an ideally suited combination of analytical methods to investigate complex mixtures of organic materials due to the ease of identification of organic substances (Py-GC/MS), the correlativity with quantitative cumulative parameters (Py-FIMS) and the characterisation of thermal behaviour (Py-GC/MS) (Franke, 2004; Franke *et al.*, 2005).

Curie-point pyrolysis-gas chromatography/mass spectrometry (Py-GC/MS)
Organic components of solid, polydisperse samples which cannot be extracted with organic solvents can be abruptly heated and evaporated under reproducible conditions in a vacuum using Curie-point pyrolysis (Leinweber *et al.*, 2002). Py-GC/MS is therefore perfectly suited for the analysis of biogenic and synthetic organic polymers (Irwin, 1982; Moldoveanu, 1998; Christy *et al.*, 1999). The assignment of pyrolysis products identified with Py-GC/MS to parent compounds is facilitated

331

by numerous investigations of polymers with well-known structures. Pyrolysis products of hydrocarbons, proteins, lipids, lignin, nucleic acids, humic substances and microorganisms can be considered here (Moldoveanu, 1998). However, the evaluation of mass spectra requires extreme care since secondary reactions due to pyrolysis such as cyclisation, recombination and cracking may occur (Saiz-Jimenez, 1994; Saiz-Jimenez *et al.*, 1994; Hatcher *et al.*, 2001) and make the assignment of pyrolysis products to their original substances difficult, particularly in the case of the analysis of undefined multi-material mixtures.

The organic residues of the freeze-dried waste water (1.2–4.1 mg) and the milled solid waste (10–13 mg) were split into small vaporisable fragments using a Curie point-pyrolyzer 1040 PSC (Fischer Company, Bonn, Germany). The Curie point of the ferromagnetic bar was about 500°C. The pyrolysis was performed by heating the bar in an electro-magnetic field in a few milliseconds and a pyrolysis time of 9.9 s. The pyrolysis products were transported by the injector in the feed gas flow (helium 5.0) into the column of the gas chromatograph Varian 3800 (Varian, Wallnut Creek, USA) for separation of the components using a 25 m capillary column BPX 5 (SGE Company, Ringwood, Australia) with an inside diameter of 0.32 mm and a stationary phase film thickness of 0.25 µm. The entire feed gas flow was fed to the column at a flow rate of 2 ml min^{-1} for 45 s. Afterwards the split ratio was 1 : 100 (45–90 s), which was reduced to 1 : 5 after 90 s. The tempera-ture programme of the GC started at 28°C (5 min) at a heating rate of 5°C min^{-1}. The final temperature was 280°C and was maintained for 40 min. The GC was coupled to a double-focusing mass spectrometer MAT 212 (Finnigan Company, Bremen, Germany). Gaschromato-graphically separated compounds were transported into the ion source where they were ionised by electron impact. The electron bombardment (70 eV) of neutral molecules produced molecular ions and fragments of various masses. They were accelerated in an electrical lens system under a 3 kV potential. Splitting the ions took place within the mass range of 48 m/z to 450 m/z by way of a magnetic and electrical sector field. A secondary electron multiplier (SEM) with a multiplier voltage of 2.2 kV was used to electrically register the split ions of various mass. The interscan time was 0.5 s at a scan rate of 1.1 scans s^{-1}. The identification of thermal degradation products took place by comparison of recorded mass spectra with library spectra (Wiley 6.0). The relative fractions of individual substances were determined by determining the ratio of identified peak areas to total peak areas of the chromatograms.

Table 9-1 Grouping of individual substances of Py-GC/MS to substance classes

Individual substances	Substance classes
Aliphatic substances	Lipids
Furans and cyclic substances	Aromatics
N-containing and N-heterocyclic substances	N-containing compounds
Substituted benzenes and aromatic substances	Aromatics
Substituted phenols	Phenols
Chlorinated substances	Chlorine-containing compounds
Sulfuric substances	Sulfur-containing compounds

Table 9-1 illustrates these substances as chemically defined substance classes based on the origin of various pyrolysis products.

Pyrolysis-field ionisation mass spectrometry (Py-FIMS)

The organic substance is temperature-resolved pyrolysed within the ion source in a high vacuum using Py-FIMS. The ionisation of the evaporated molecules is carried out in a strong electrical field (field ionisation) very gently by a modest energy transfer (0.5 eV vs. 5 eV) in comparison to electron impact ionisation, so that the resulting mass spectra are very intensive and fragmentation of the molecular ion is almost totally prevented (Lehmann and Schulten, 1976b). Loss of evaporated substances along chromatographic separating columns and dependence on the polarity of substances is excluded due to the direct inlet of molecular ions to the detector. Temperature-resolved pyrolysis enables deductions to be drawn on the thermal behaviour of evaporated substances.

Using Py-FIMS, the temperature-resolved pyrolysis of waste water (0.3–1.4 mg) and milled solid waste (1.2–2.2 mg) took place in steps of $10°C\,s^{-1}$ within a temperature range of 110°C–700°C. Ionisation of the pyrolysates was effected gently in the direct proximity of the carbon peaks of the emitter in an electrical field of $10^7\,V\,cm^{-1}$ to $10^8\,V\,cm^{-1}$. The modest energy transfer of approx. 0.5 eV (Lehman and Schulten, 1976b) produces virtually all positively charged molecular ions without fragmentation. Py-FIMS is a complementary and independent analytical method to Py-GC/MS owing to this gentle ionisation method and temperature-resolved pyrolysis. The scan range of the mass spectrometer 731 (Finnigan MAT Company, Bremen, Germany) used to detect the individual molecular ions produced was within the range of 15 m/z to 900 m/z. The samples were weighed into a quartz crucible and introduced into the ion source by way of a vacuum lock. The molecular ions were accelerated by the potential difference between the ion source (+8 kV) and the counterelectrode (−3 kV) arranged vis-à-vis, separated in the electric and magnetic sector field and detected using a SEM. The molecular masses were electronically recorded using the Maspec II software. The intensity of the nominal masses was measured in arbitrary units. Comparison to other samples was possible by normalising the total ion intensity (TII) to 1 mg sample (counts mg^{-1}). Thermograms were obtained by plotting the TII against the pyrolysis temperature, which describe the thermal behaviour of the samples. Basic information about the size of molecular fragments of pyrolysates can be obtained by calculating the medium molecular weight of the mass spectra (Leinweber *et al.*, 2002). An increase of the medium molecular weight of organic substances as a function of time indicates stabilisation (Frimmel and Weis, 1991). The averaging can be performed by weighting according to the number of molecules and their specific weight fractions. Thus two characteristic molecular weights can be obtained for organic polymers:

1. Medium molecular weight averaged according to the number of nominal mass signals

$$\overline{M}_n = \frac{\sum\limits_{i=1}^{m}(I_i \times M_i)}{\sum\limits_{i=1}^{m} I_i}$$

M_i Mass of the i-th nominal mass signal m/z
I_i Intensity of M_i (counts mg^{-1})

2. Medium molecular weight averaged according to the number of structural subunits

$$\overline{M}_w = \frac{\sum\limits_{i=1}^{m}(I_i \times M_i^2)}{\sum\limits_{i=1}^{m} I_i}$$

M_i Mass of the i-th nominal mass signal m/z
I_i Intensity of M_i (counts mg^{-1})

The polydispersity of the organic polymers of the pyrolysate concerning the molecular weight can be determined by calculating the ratio of $\overline{M}_w$ to $\overline{M}_n$. The greater the value of this ratio differs from unity, the more

Table 9-2 *Specific parameters for the assessment of waste water contamination*

Waste-water parameter	Method/standard
Chemial oxygen demand (COD)	Photometer
Biochemial oxygen demand (BOD$_5$)	DIN EN 1899-2
Total organic carbon (TOC)	DIN EN 1484
Dissolved organic carbon (DOC)	DIN EN 1484

polydisperser, i.e. the more heterogeneous the pyrolysate is concerning the individual molecular weights.

9.3.2.3 Checking the results of Py-GC/MS and Py-FIMS

The statistical calculations are based on three repetitions of the data determined using Py-GC/MS and Py-FIMS. The results are discussed based on the arithmetic averages of the repetition measurements. Significance tests of the averages of different samples were made using the t-test (bilateral question). Variation coefficients (V) were determined to assess the reproducibility of the data. Correlations between two characteristics were checked by linear and logarithmic regression. The level of significance was specified by error probabilities 5% (P $\leq$ 0.05; Symbol: *), 1% (P $\leq$ 0.01; Symbol: **) and 0.1% (P $\leq$ 0.001; Symbol: ***) (see Section 9.3.4.5 and Fig. 9-11).

9.3.2.4 Conventional sum parameters as comparative quantities

The conventional characterisation of waste water was carried out on the basis of the following sum parameters in Table 9-2.

An ISIS 9000 photometer (Dr Lange Company, Düsseldorf, Germany) was used to measure COD. BOD$_5$ was determined according to the dilution principle. Oxygen content was measured using oxygen probe OXI 330 (WTW Company, Weilheim, Germany). A Liqui-TOC analyser (Foss Heraeus Company, Hanau, Germany) was used for the measurement of TOC and DOC. After removing the inorganic carbon by phosphoric acid, the organic carbon was oxidised by sodium peroxodisulphate and the CO$_2$ generated detected using infrared spectroscopy.

9.3.2.5 Elemental analysis

Elemental analyses of C, N and S took place using a VARIO EL elemental analyser (Foss Heraeus Company, Hanau, Germany). The

335

samples were oxidatively digested at about 1150°C. The nitrogen oxides and sulphur oxides generated were reduced on a copper catalyst to N_2 and/or SO_2 and passed to a heat conduction detector after chromato-graphic separation, together with the CO_2.

9.3.3 Description of the material to be tested

Decomposition simulation reactors

The biologically extensively stabilised sewage–sludge waste mix of Lys. 5 was described prior to and after biological stabilisation in the previous Chapters 2–8.

Residual wastes from the North Western Pomerania District (Camitz landfill, Mecklenburg-Western Pomerania, Germany) were used as fresh wastes for comparison. For this purpose a sample amount of 150 kg was taken from a daily delivery. Then the total sample was analysed by sieving and sorting and the water content determined on a sub-sample. Before filling the reactors the residual waste was shredded to a maximum edge length of 10 cm and the water content was adjusted to 40% by weight. The same residual waste was used for the 2nd variant as for the 1st one. However, this batch was mixed with 25% by weight of non-treated sewage sludge (raw sludge). This had a water content of 75% by weight and was not treated with lime. The mixture was produced in a cement mixer.

Large-scale lysimeter

The residual waste for the decomposition test in the lysimeter also originated from the North Western Pomerania District (Camitz landfill, Mecklenburg-Western Pomerania). After 3 days of collection and interim storage it was shredded in a S 20.00 (EuRec Technology Company, Merkers (D)) shredder to an edge length of maximum 40 cm. Sub-sequently, the waste was transported to the lysimeter location in Rostock. A mixed sample of 120 kg was taken from the heap for sieving and sorting analysis and water content measurement. A grab excavator filled the lysimeter with 36 t of residual waste to achieve a bulk height of 160 cm. A 30-cm-thick layer of compost mulch was placed on top of the residual waste surface as an odour trap and water management layer.

Sieving and sorting analyses

A sieving and sorting analysis was carried out to determine the material composition and the particle size of the residual waste for each test. Figure 9-6 illustrates the results of these analyses.

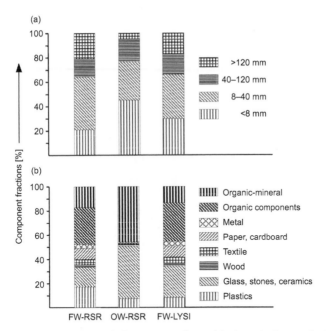

Fig. 9-6 *Results of (a) sieving and (b) sorting analyses of fresh residual waste for decomposition simulation reactors (FW-DSR) and decomposition lysimeters (FW-LYSI), and for old disposable waste material from Lys. 5 (OW-DSR)*

Figure 9-6(a) shows the results of the sieving analyses. It can be seen that the particle size distribution of the fresh residual waste for the lysimeter test (FW-LYSI) and the residual waste for the tests in the decomposition simulation reactors (FW-DSR) have very similar compositions. The non-sortable fraction <8 mm made up 20% (FW-DSR) to 25% (FW-LYSI) of the total quantity. The fraction 8 mm to 40 mm had the largest percentage: 35% (FW-LYSI) to 45% (FW-DSR). The fractions 40 mm to 120 mm and the high-calorie fraction >120 mm had approximately the same percentage: 17% (FW-LYSI) and 19% (FW-DSR). Unlike the fresh waste variants the decomposition material (OW-DSR) from Lys. 5, which was capable of being landfilled (disposable) right at the beginning of the test, exhibited a markedly different composition. The percentage of the <8 mm fraction was 45%, i.e. considerably greater than in the fresh waste variants. On the other hand the predominantly high-calorific fraction >120 mm was substantially smaller. Figure 9-6(b) shows that the fresh waste variants also exhibited very little differences concerning their material composition. About one third of the residual waste quantities could be attributed to the native organic fraction in both samples. If one considers the paper fraction and a part of the non-sortable organic-mineral fraction, about 50% of the residual waste used consisted of potentially bioavailable substances. The fraction of inert materials (glass, stones, ceramics) in FW-LYSI (26%) was different to the fraction in FW-DSR (16%). The fraction of plastics was also considerably different: 8% (FW-LYSI) and 17% (FW-DSR). A comparison with the disposable decomposition material also showed a drastic difference to the fresh waste variant. The inert fraction and organic-mineral fraction were approx. 45% each and made up almost the entire residual waste. The plastic fraction took another 8%, so that wood, paper, textile, metal and native-organic components with 0.2% to 1% were hardly represented at all.

Compared to a representative sorting analysis carried out in Germany (Kern, 2000), about the same amount, i.e. 50%, of potentially bioavailable substance was contained in the fresh waste variants of FW and FW-LYSI. The plastics fraction (8% to 17%) and the inert material fraction (16% to 26%) are however up to a factor of three greater than in the representative sorting analysis. The large fraction of inert materials and organic-mineral fraction and the absence of native-organic components in the disposable material indicated an extensive mineralisation of the organic substances. Only small amounts of the hardly bio-available wood were visually identified.

9.3.4 Test execution and results

9.3.4.1 Execution
Both the decomposition simulation reactors and the large-scale lysimeter (pilot plant scale) were operated aerobically for at least 350 days applying leachate circulation. The end of stabilisation in the laboratory scale coincided fairly exactly with the processes of the large-scale design and corresponded to the processes in the aerobic test deposits, which were investigated in the DFG (German Research Foundation) integrated programme (cf. Chapters 2–7). Therefore, the results of the decomposition process will not be described here (see details in Franke, 2004).

9.3.4.2 Stability classification based on waste water investigations of leachates
The usual waste-management stability criteria are of utmost importance for the comparison with the extensively stabilised material of Lys. 5, derived from the cumulative waste-water parameters of the leachates, which occurred during the stabilisation process (Table 9-3).

Table 9-3 Sum parameters

Test substance Months	COD [mg O$_2$ l^{-1}]	BOD$_5$ [mg O$_2$ l^{-1}]	BOD$_5$/COD	DOC [mg l^{-1}]	TOC [mg l^{-1}]
Fresh waste					
2	–	–	–	–	–
4[a]	10 890	2210	0.20	2640	3890
6	14 740	2550	0.17	4090	5340
12[a]	3550	120	0.03	1110	1230
Fresh waste/sewage sludge					
2	–	–	–	–	–
4[a]	12 410	1810	0.15	3250	4520
6	16 500	4230	0.25	4870	5950
12[a]	8600	130	0.01	2720	2890
Disposable waste material of Lys. 5					
2	700	10	0.01	160	165
4	670	10	0.01	140	150
6[a]	1033	14	0.02	230	245
12	883	10	0.01	245	260
Fresh waste lysimeter					
0	4500	2510	0.55	1380	1420
2[a]	2550	150	0.06	830	870
4	750	60	0.08	130	250
6[a]	3320	200	0.06	760	1190
12[a]	1550	50	0.03	470	490

(a) Mass-spectrometric investigation dates (Py-GC/MS and Py-FIMS)

When interpreting the analysis values it should be noted that the laboratory lysimeters were operated after an initial humidification with leachate circulation without dilution, whereas the large lysimeter also contained dilution by rain water which must be taken into account despite the circulation. The disposable sewage sludge waste mix from Lys. 5 was weathered during the stabilisation process and the subsequent observation time of nearly 10 years for weathering without leachate circulation was suspended and subjected to a closed cycle only during the laboratory tests. Considering these boundary conditions it can be concluded from the results that the reduction of organic contaminants in the leachate of fresh waste in the laboratory was only achieved by degradation. Precipitation also has a diluting effect in the large-scale lysimeter, however, the extremely slow to degrade organic substance in the leachate of the biologically very stable sewage sludge waste mix slowly enriched. In addition, a degradation inhibition by the sludge mixture can be concluded from the ratio BOD$_5$/COD = 0.01 at COD = 8600 mg O$_2$/l in the leachate of the fresh sewage–sludge waste mix (incomplete mixing or inhibitors in the sewage sludge). This inhibition of degradation has no influence on the stability investigation of the material from Lys. 5.

A comparison of the sum parameters indicates that the organic contamination of the leachate from fresh waste is degraded quickly after 1 year of uninhibited decomposition with leachate circulation, but it still has a contaminant level about half an order of magnitude higher than that of the extensively stable waste after more than 10 years of permanent aerobic landfilling. It is particularly remarkable that the organic substance of leachate from waste classified as biologically stable with a COD of 700 to 1000 mg O_2/l at a BOD_5 = 10 mg O_2/l exhibits about 30% of the degradation performance of the 1-year-old material, thus it still has a marked residual activity. Elementary analyses for carbon, nitrogen and sulphur were in line with these results (see further details in Franke, 2004).

9.3.4.3 *Curie-point pyrolysis-gas chromatography/mass spectrometry*

The reproducibility of the relative percentages of the substance classes was proven after combining the individual substances detected into chemically defined substance classes (Franke, 2004). Ratios of easy-to-degrade carbohydrates to difficult-to-degrade phenols and alkyl aromatics and ratios of carbohydrates to nitrogen compounds (similar to the C/N ratio) were calculated to improve comparability of the analysis results of different test approaches and to quickly assess the stability of the organic substance. Figure 9-7 shows the chromatograms and relative percentages of the substance classes in the process water (P-OW) and in the solids (OW) of the extensively stabilised sewage–sludge waste mix from Lys. 5. The chromatograms, relative fractions of the substance classes and the individual substances detected of the three aerobic stabilised fresh waste variants are discussed in Franke (2004). Table 9-4 displays the detected individual substances of the extensively stabilised material from Lys. 5. It should be noted that 80% of the peak areas could not be identified in the process water (P-OW) probably due to a high inorganic content (Fig. 9-7). The basepeak was assigned to benzonitrile (4), while that in the pyrolysate of humic acid extracts from residual waste compost was assigned to polypeptides (Garcia *et al.*, 1992). However, exclusively aliphatic nitriles were detected in Py-GC/MS analyses of numerous amino acids (Tsuge and Matsubara, 1985; Chiavari and Galetti, 1992; Sorge, 1995). Nonetheless, according to Saiz-Jimenez (1994) cyclisation of

340

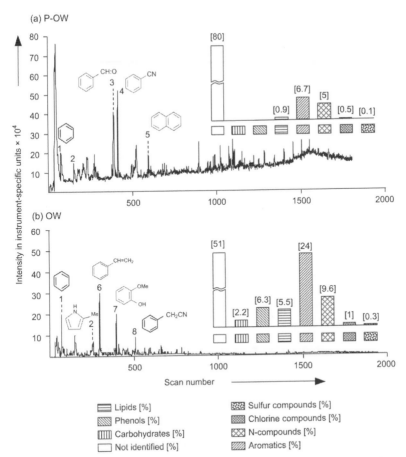

Fig. 9-7 *Chromatograms and percentage composition of the organic substance in freeze-dried process water P-OW (a) and solid OW (b) of Lys. 5 after 6 months of test period*

aliphatic amino acids in the pyrolysis may produce aromatic nitriles. Benzonitrile was traced back to residues of tire rubber (Reinhard and Goodman, 1984) in tests on leachates, though no N-containing pyrolysis products were detected in Py-GC/MS analyses of tire rubber (Cunliffe and Williams, 1998). The detection of benzonitrile in soil samples was traced back to interactions of phenolic lignin components with amino acids (Schulten *et al.*, 1997; Leinweber and Schulten, 1998), so that a biogenic origin is probable. The correspondingly intensive signal of benzaldehyde (3) was detected during pyrolysis of cellulose (Pouwels *et al.*, 1987, 1989; Scheijen *et al.*, 1989), nevertheless it was not detected in the process water of fresh waste and fresh waste/sewage sludge variant (see Franke, 2004; Franke *et al.*, 2005). Therefore

341

Table 9.4 Identified individual substances and percentages of process water (P-OW) and the solids (OW) in the extensively stabilised material of Lys. 5 as groups of substance classes (Py-GC/MS)

Substance classes	Scan. no.	Mol.	CAS	C	H	O	N	X	P-OW peak [%]	OW peak [%]
Substituted benzenes										
Benzene	72	78	71-43-2	6	6	–	–	–	2.0	2.3
Methylbenzene	149	92	108-88-3	7	8	–	–	–	0.7	3.6
1,2-dimethylbenzene	266	106	95-47-6	8	10	–	–	–	0.5	0.4
Ethynyl benzene	279	102	536-74-3	8	6	–	–	–	0.6	–
Ethenyl benzene	297	104	100-42-5	8	8	–	–	–	0.2	12.2
2-propenyl benzene	353	118	300-57-2	9	10	–	–	–	–	0.2
Propyl benzene	358	120	103-65-1	9	12	–	–	–	–	0.1
Benzaldehyde	379	106	100-52-7	7	6	1	–	–	1.8	1.0
1-methylethenyl benzene	391	118	98-83-9	9	10	–	–	–	–	1.9
1-ethenyl-3-methyl benzene	410	118	100-80-1	9	10	–	–	–	–	0.2
1,4-dimethoxy benzene	599	138	150-78-7	8	10	2	–	–	–	0.2
1,1'-biphenyl	752	154	92-52-4	12	10	–	–	–	–	0.3
4-hydroxy-3-methoxy benzaldehyde	904	152	121-33-5	8	8	3	–	–	0.3	–
1,2-benzene dicarboxylic acid dioctylester	1414	390	117-84-0	24	38	4	–	–	–	0.2
Sum in percentage of total									6.1	22.5
Substituted phenols										
Phenol	432	94	108-95-2	6	6	1	–	–	–	2.2
3-methylphenol	491	108	108-39-4	7	8	1	–	–	–	0.4
4-methylphenol	518	108	106-44-5	7	8	1	–	–	–	0.4
2-methoxyphenol	504	124	90-05-1	7	8	2	–	–	–	2.7
4-vinyl-2-methoxyphenol	702	150	0	9	10	2	–	–	–	0.4
1-phenylethanon	486	120	98-86-2	8	8	1	–	–	–	0.3
Sum in percentage of total									0	6.3

Substituted furans

			CAS							%
2,5-dimethylfuran	94	96	625-86-5	6	8	1	—	—	—	0.2
2-furan carboxaldehyde	240	96	98-01-1	5	4	2	—	—	—	0.5
5-methyl-2-furan carboxaldehyde	384	110	620-02-0	6	6	2	—	—	—	0.4
Sum in percentage of total									0.0	1.1

Aliphatic compounds

			CAS							%
Decane	404	142	124-18-5	10	22	—	—	—	—	0.1
1-undecene	494	154	821-95-4	11	22	—	—	—	—	0.2
Undecane	500	156	1120-21-4	11	24	—	—	—	—	0.1
1-dodecene	584	168	112-41-4	12	24	—	—	—	—	0.3
Dodecane	590	170	112-40-3	12	26	—	—	—	—	0.2
1-tridecene	466	182	2437-56-1	13	26	—	—	—	—	0.4
Tridecane	673	184	629-50-5	13	28	—	—	—	—	0.2
1-tetradecene	748	196	1120-36-1	14	28	—	—	—	—	1.0
Tetradecane	752	198	629-59-4	14	30	—	—	—	—	0.3
1-pentadecene	822	210	13360-61-7	15	30	—	—	—	—	0.2
Pentadecane	825	212	629-62-9	15	32	—	—	—	—	0.1
1-hexadecene	891	224	629-73-2	16	32	—	—	—	—	0.3
Hexadecane	894	226	544-76-3	16	34	—	—	—	—	0.2
1-heptadecene	956	238	629-78-7	17	34	—	—	—	0.1	0.1
Heptadecane	980	240	55044-98-9	17	36	—	—	—	0.1	0.1
1-octadecene	1020	252	112-88-9	18	36	—	—	—	—	0.2
Octadecane	1023	254	593-45-3	18	38	—	—	—	0.4	0.1
1-nonadecene	1070	266	18435-45-5	19	38	—	—	—	0.3	0.1
Nonadecane	1083	268	629-92-5	19	40	—	—	—	—	0.1
1-eicosene	1136	280	3452-07-1	20	40	—	—	—	—	0.1
Eicosane	1139	282	112-95-8	20	42	—	—	—	—	0.1
Heneicosane	1193	296	629-94-7	21	44	2	—	—	—	0.1
Heptadecane acid methylester	1213	298	2790-25-7	19	38	—	—	—	—	0.1
Docosane	1245	310	629-97-0	22	46	—	—	—	—	0.1
Tricosane	1296	324	638-67-5	23	48	—	—	—	—	0.1

Table 9-4 (continued)

Substance classes	Scan. no.	Mol.	CAS	C	H	O	N	X	P-OW peak [%]	OW peak [%]
Tetracosane	1343	338	646-31-1	24	50	–	–	–	–	0.1
Pentacosane	1389	352	629-99-2	25	52	–	–	–	–	0.1
Hexacosane	1433	366	630-01-3	26	54	–	–	–	–	0.1
Heptacosane	1476	380	593-49-7	27	56	–	–	–	–	0.1
Octacosane	1517	394	630-02-4	28	58	–	–	–	–	0.1
Nonacosane	1557	408	630-03-5	29	60	–	–	–	–	0.1
Sum in percentage of total									0.9	5.5
Cyclic compounds										
1,4-cyclohexadiene	63	80	628-41-1	6	8	–	–	–	–	0.6
2-methyl-2-cyclopentene-1-on	319	96	1120-73-6	6	8	1	–	–	–	0.5
Sum in percentage of total									0.0	1.1
N-heterocycles										
1-methylpyrrole	124	81	96-54-8	5	7	–	1	–	–	0.7
Pyridine	139	79	110-86-1	5	5	–	1	–	–	1.0
4-methylpyridine	232	93	108-89-4	6	7	–	1	–	–	0.6
2-methylpyrrole	251	81	636-41-9	5	7	–	1	–	–	1.8
3-methylpyrrole	257	81	636-41-9	5	7	–	1	–	–	1.4
3-methylpyridine	277	93	108-99-6	6	7	–	1	–	–	0.3
1-ethylpyrrole	346	95	617-92-5	6	9	–	1	–	–	0.2
3,4-Dimethylpyridine	359	107	583-58-4	7	9	–	1	–	–	0.0
2-Ethylpyrrole	361	95	617-92-5	6	9	–	1	–	–	0.1
2-Pyridine carbonitrile	492	104	100-70-9	6	4	–	2	–	–	0.2
Sum in percentage of total									0.0	6.4
Nitrogen-containing compounds										
2-Methylene butane nitrile	168	81	1647-11-6	5	7	–	1	–	–	0.3

				C	H	O	N		(%)	(%)
Benzonitrile	406	103	100-47-0	7	5	–	1	–	3.5	0.8
4-methyl benzonitrile	492	117	104-85-8	8	7	–	1	–	0.3	–
4-methyl-1-isocyanobenzene	530	117	7175-47-5	8	7	–	1	–	0.2	–
2-nitrophenol	550	139	88-75-5	6	5	3	1	–	0.3	–
Benzyl cyanide	560	117	140-29-4	8	7	–	1	–	–	1.1
Benzene propane nitrile	645	131	645-59-0	9	9	–	1	–	–	0.8
1,2-benzene dicarbonitrile	687	128	91-22-5	8	4	–	2	–	0.2	–
1,3-benzene dicarbonitrile	692	128	626-17-5	8	4	–	2	–	0.2	–
Alphamethylene benzyl cyanide	695	129	495-10-3	9	7	–	1	–	–	0.1
Heptadecane nitrile	1092	251	5399-02-0	17	33	–	1	–	–	0.1
Sum in percentage of total									5.0	3.3
Aromatic compounds										
Indene	456	116	95-13-6	9	8	–	–	–	–	0.6
3-methyl indene	554	130	767-60-2	10	10	–	–	–	–	0.3
Azulene	565	128	275-51-4	10	8	–	–	–	–	0.1
Naphthalene	591	128	91-20-3	10	8	–	–	–	0.6	0.3
2,3-dihydroindene-1-on	679	132	83-33-0	9	8	1	–	–	–	0.1
Indole	696	117	120-72-9	8	7	–	1	–	0.1	0.2
2-methyl naphthalene	700	142	91-57-6	11	10	–	–	–	–	–
Sum in percentage of total									0.7	1.5
Chlorine-containing compounds										
Chloromethane	31	50	74-87-3	1	3	–	–	Cl	–	0.2
Chloro acetic acid	32	94	79-11-8	2	3	2	–	Cl	–	0.8
1,1-dichloro ethane	67	98	75-34-3	2	4	–	–	Cl2	0.4	–
Sum in percentage of total									0.5	1.0
Sulfur-containing compounds										
Isothiocyanomethane		73	556-61-6	2	3	–	1	S	0.1	–
Benzothiophene	698	134	95-15-8	8	6	–	–	S	–	0.3
Sum in percentage of total									0.1	0.3

to assign benzaldehyde to cellulose, of which up to 13% is contained in fresh residual wastes, is not plausible. The origin of polystyrene (Alajberg *et al.*, 1980) or lignin components (Faix *et al.*, 1987) is more probable, they enrich as biologically difficult-to-degrade substances in the process water and are in part available as lignocellulose complexes (Lott-Fischer *et al.*, 2001). Naphthalene (5) can be attributed to synthetic sources such as rubber (Cunliffe and Williams, 1998), polystyrene (Audisio and Bertini, 1992), polyvinyl chloride (Alajberg, 1980) and pesticides, paints and solvents (Öman and Hynning, 1993) with a high probability. Concerning the percentages of substance classes, an enrichment of aromatic compounds opposite to nitrogenous compounds (6.7% vs. 5%) was observed compared to the process water of fresh waste and fresh waste/sewage sludge variant, which indicates a further stabilisation of the remaining organic substance. Furthermore, no signal could be attributed to the class of carbohydrates which indicates the stabilisation of organic substance by the complete loss of easily available carbon. However, it is conspicuous that no phenol compound was detected in this process water which could have indicated lignin as a refractory organic component, in addition to anthropogenic substances.

In the solids of the disposable residual waste (Lys. 5), in which 49% of the peak areas were identified, the basepeak could be assigned to ethenyl benzene (styrene) as in the fresh-waste variants. However, the signal of 2-cyclopentene-1-on, originating from carbohydrates, and other cyclic compounds, could not be detected which indicates a higher degree of stabilisation of the organic substance in the solids in comparison to the solids of the fresh waste and sewage sludge fresh waste variant (Franke, 2004; Franke *et al.*, 2005). In contrast to the process water (P-OW) intensive signals of lignin-bound phenol compounds such as 2-methoxyphenol were detected in the solids (OW) which shows that the absence of phenol compounds in the process water (P-OW) cannot be attributed to a complete microbial degradation of lignin compounds. A polymerisation of the organic substance to high-molecular compounds (Liang *et al.*, 1996) of low water solubility during stabilisation is probably the cause. Lipids were detected as a homologous series of alkanes and alkenes in the range of C15 to C29 and confirm a higher stability of the material by enrichment of long-chain lipids (Keeling *et al.*, 1994; Dinel *et al.*, 1996) compared to the fresh waste solids (C7 to C29). The enrichment of lipids in the solids is confirmed due to the nonpolar character of the alkanes compared to the aliphatic signals of process water (P-OW).

Also, compared to the solids of the fresh waste and sewage sludge fresh waste variant (cf. Franke, 2004) the high percentage of nitrogenous compounds (9.6%) can be explained by the higher content of hetero-cyclic nitrogen compounds such as pyridins and alkyl pyrroles, which were identified as refractory organic components in investigations of leachates (Kettern, 1990). The enrichment of hetero-cyclic N-compound is due to the high microbial availability of the amide-bound nitrogen, which decreases during stabilisation (Derenne and Largeau, 2001).

Due to the complete loss of carbohydrates in the process water (P-OW), the ratios of carbohydrates to aromatics and phenols and of carbohydrates to nitrogen compounds are equal to zero. The ratio of carbohydrates to aromatics and phenols was significantly lower in the solids OW (0.07) than in the solid materials of fresh waste (0.12) and sewage sludge fresh waste variant (0.12) (P $\leq$ 0.05; n = 3) – a sign of advanced biological stabilisation. This higher stability is confirmed by a significantly smaller ratio of carbohydrates to N-compound in the solids OW (0.23) (P $\leq$ 0.001; n = 3).

In addition to the detectable materials it is conspicuous that the fraction of non-detectable material almost doubles regardless of the progress of stabilisation due to sewage sludge addition.

9.3.4.4 *Pyrolysis field ionisation mass spectrometry*

The reproducibility of the total ion intensities of substance classes was proved in detail by Franke (2004). Mass spectra and thermograms of fresh waste and sewage sludge fresh waste variant are also presented in Franke (2004). Figure 9-8 shows the cumulative mass spectra and thermograms of the process water (P-OW) (Fig. 9-8a) and the solids (OW) (Fig. 9-8b) of Lys. 5 after 6 months of subsequent composting in the DSR. The P-OW spectrum indicates an agreement with the spectra of the fresh waste and sewage sludge fresh waste process water (Franke, 2004; Franke *et al.*, 2005), the highest signal intensities in the range of 15 m/z to 350 m/z and the basepeak at 59 m/z. The thermogram exhibits a markedly lower intensity (0.03) compared to the fresh waste process water (0.15–1) and so indicates a low organic contamination of this process water. The maximum ion intensity was at 400°C between the maxima of the fresh waste process water (350°C–440°C). A further enrichment of thermally stable organic compounds by the 24 months of biological treatment therefore cannot be proved. However, in comparison to the thermograms of the fresh waste process water, the thermogram indicates the presence of homogeneous molecular ions and

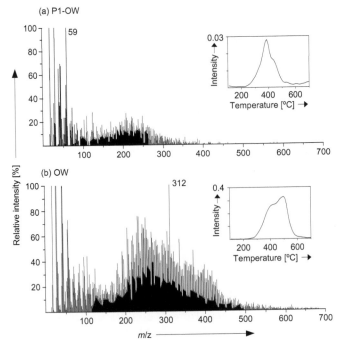

Fig. 9-8 Pyrolysis field ionisation mass spectra and thermograms of freeze-dried process water in Lys. 5, (a) P-OW after 6 months, and (b) of the solids OW in Lys. 5 after 6 months of testing

similar thermal behaviour in the narrow temperature range of the intensive ion release of 300–500°C, which can be attributed to the stabilisation of the organic substance (Sorge, 1995; Liang *et al.*, 1996).

High signal intensities were found in the solids (OW) within the range of 15 m/z to 600 m/z. The basepeak 312 m/z could be attributed to the fresh waste variant similar to the solids and shows that persistence of synthetic organic compounds such as polystyrene (Jungbauer, 1994; Otake *et al.*, 1995; Pantke, 1996) causes high signal intensities of these compounds in the residual waste even after 24 months of biological treatment (Franke, 2004). The thermogram of the solid OW exhibited a shelf at 390°C and maximum ion intensity at 510°C and is comparable to the solid samples of the fresh waste and sewage sludge fresh waste variant.

The TII in counts mg^{-1} (Fig. 9-9a) show a drastic difference between the process water P-OW (0.38×10^6 counts mg^{-1}) and the solids OW (5.8×10^6 counts mg^{-1}) ($P \leq 0.01$; n = 3), which is to due to the low organic contaminant content of the process water (cf. Table 9-3). In

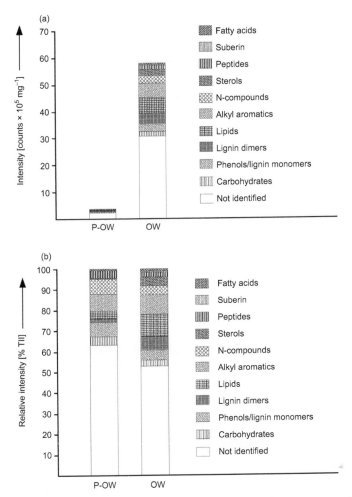

Fig. 9-9 Fractions of ten classes of organic substances of freeze-dried process water P-OW in Lys. 5 after 6 months and the solids OW in Lys. 5 after 6 months: (a) absolute intensities; (b) relative intensities

contrast to the Py-GC/MS analysis of the process water (Fig. 9-7a), in which no carbohydrates and phenols were detected, Py-FIMS was able to detect 4% of carbohydrates and 7% of phenols and lignin monomers. This difference is due to the greater sensitivity of Py-FIMS, which results from the direct transfer of the ions produced to the mass spectrometer without chromatographic separation (Lehmann and Schulten, 1976a, 1976b). The detectability of the pyrolysates is therefore greater in Py-FIMS than in Py-GC/MS (Abbt-Braun, 1987).

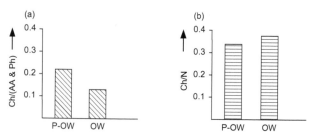

Fig. 9-10 *Ratios of (a) carbohydrates (Ch) to alkyl aromatics (AA), phenols, lignin mono-mers and dimers (Ph) and those of b) carbohydrates to N-compounds and peptides (N) of process water P-OW and solids OW in Lys. 5*

The intensities of carbohydrates in the solids are significantly higher than in the process water in agreement with the results of Py-GC/MS and they confirm the stability of the organic process water components by the absence of biologically easily available carbohydrates. The carbo-hydrates in the solids are probably components of high-molecular, polymerised organic compounds of low water solubility. The ratio of carbohydrates to alkyl aromatics, phenols, lignin monomers and dimers in the process water (P-OW) (Fig. 9-10a) hardly differs from the ratio of fresh waste/sewage sludge-process water (Franke, 2004) which, due to increased intensities of aromatic compounds, is signifi-cantly lower than in the variant without sewage sludge (Franke, 2004; Franke *et al.*, 2005).

The biological stabilisation of process water of the disposable decomposition material from Lys. 5 can thus be attributed to a relative enrichment of aromatic and phenolic lignin compounds and an exten-sive degradation of carbohydrates. Enrichment of aromatic compounds with a simultaneous degradation of biologically easily available carbo-hydrates was also detected in eluates from composted residual wastes (Chefetz *et al.*, 1998a, 1998b). The relationship of carbohydrates to alkyl aromatics, phenols, lignin monomers and dimers in the solids (OW) is significantly lower than in the process water (P-OW) ($P \leq 0.05$; n = 3) due to a significantly higher intensity of lignin dimers and alkyl aromatics ($P \leq 0.001$; n = 3). The intensity of the lignin dimers in the solids (OW) is significantly higher than in the solids of the fresh waste (FW) and the fresh waste sewage sludge variant ($P \leq 0.001$; n = 3) and can be assessed as a further indication of stabilisation of organic substance due to the polymerisation of lignin monomers (Liang *et al.*, 1996). The ratio of carbohydrates to N-containing compounds in the process water (P-OW) (Fig. 9-10b) does not differ from the ratios of fresh waste and fresh waste/sewage sludge process water (Franke, 2004; Franke *et al.*, 2005). In the solids the ratio of carbohydrates to N-compounds corresponds to the ratio determined in the fresh waste/sewage waste variant.

9.3.4.5 Correlation of the total ion intensities with sum parameters
A relationship was found between the reduction in TII (count mg^{-1}) and the reduction in sum parameters C, COD and TOC in the process water of the fresh waste variant which was confirmed for all tested process waters (Franke, 2004). The regression analysis between TII and C, COD and TOC parameters (Fig. 9-11) of the analysed process water (n = 8) indicated significant linear relationships between (a) total C (y) and TII (x) with $y = 7.33x + 70.59$ and $r^2 = 0.928^{***}$, (b) COD (y) and TII (x) with $y = 595.63x + 1688.2$ and $r^2 = 0.967^{***}$ as well as (c) TOC (y) and TII (x) with $y = 216.62x + 539.26$ and $r^2 = 0.956^{***}$. Another significant correlation was found by Py-FIMS analysis of soil samples between the parameters C_{org} and TII (Sorge, 1995). The significant relationship between COD and/or TOC parameters and TII (Fig. 9-11b and c) indicates a primarily organic character of the oxidisable process water substances. These significant correlations indicate that the Py-FIMS technique, which was used for the first time for the investigation of process water from the aerobic mechanical-biological treatment of residual waste, is a suitable analysis method for the characterisation of organic process water components.

9.3.5   Waste-management conclusions from the analyses
For the first time, the combination of the mass spectrometric analysis methods Py-GC/MS and Py-FIMS with waste water-specific sum parameters proved that easily available organic substances are degraded under aerobic ambient conditions and microbially difficult-to-degrade substances are enriched. If this aerobic process is maintained in a permanent aerobic landfill, no further pollutant emissions, caused by the retarded degradation of biologically easily available substances are expected. The emissions from a landfill of this type remain foreseeable and manageable for coming generations. However, since the area available for landfills is limited, but the demand will be unlimited over time, the landfill technology investigated cannot provide a permanent solution for waste disposal.

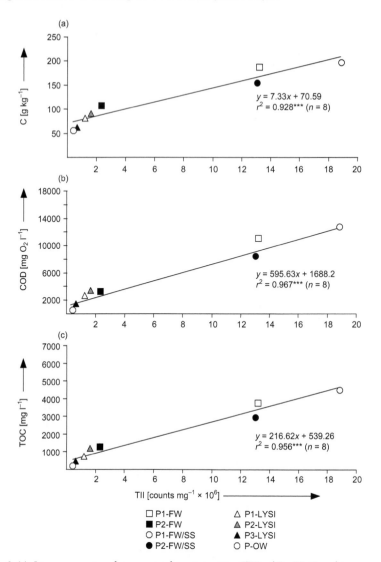

Fig. 9-11 Linear regressions between total ion intensities (TII) of Py-FIMS and parameters (a) C, (b) COD and (c) TOC of the analysed process water (n = 8) FW: fresh waste, SS: sewage sludge, OW: old waste

As far as the long-term integration of the treatment product into the environment is concerned, the exact analysis justifies Husz's doubts based on soil science considerations (Section 8.4.5) as, although the product behaves similarly to a soil, it cannot be classified as a soil in terms of its material composition, therefore, further process steps are necessary for its transformation into a product that can be integrated into the environment.

9.4 Landfill behaviour of PVC as a representative of temporarily stable plastics
Günther Ballin, Peter Hartmann and Frank Scholwin

9.4.1 Fundamentals

The ARGUS Engineering Company, Berlin, Germany, together with three other scientific working groups, was commissioned by the European Commission, DGXI.E.3, to investigate the behaviour of PVC in landfills. The work contains a detailed literature study on the material and its additives, the amount of production, and the masses deposited, the known landfill processes and the results of the experimental investigation on the behaviour of PVC in different landfills. It was shown that the manufactured and disposed masses cannot be ignored, particularly in view of the potential effect of additives being released. It was experimentally proved that biologically active additives can be released, especially in large landfills and during a thermal pre-treatment. In addition, an impact on the polymer itself under thermophilic aerobic conditions was also proved. The experimental results were checked and confirmed on behalf of the industry by an independent institute. The full report (100 pages) was published by the European Commission in the internet and is available to the reader (European Commission DG XI.E.3, 2000). Therefore a brief excerpt is only included here, which, in the opinion of the authors, contains the essential information for the assessment of the long-term behaviour.

9.4.2 Methods and extent of investigation

The task, to assess the landfill behaviour of a substance group produced in many variations, requires the following conditions to be met:

- The gap in the geometrical scale between landfill and the laboratory must be bridged.
- The time scale must be suitably accelerated.
- The variety of the products must be restricted to materials with product-typical reactions.

The problem was solved by first covering the range of variation to a large extent.

Geometrical model scale

The influence of the geometrical scale was examined starting from the landfill through the technically controllable large container ($45\,m^3$) and large-volume simulation reactors in the laboratory scale to chemical and biological analysis.

- Time scale and temperature effect:
  Time and temperature influences were estimated by the differentiation of degradation intensity within the thermophylic range (spontaneous thermal control of the system):
  - aerobic at approx. 80°C,
  - aerobic/anaerobic approx. 60–80°C changing
  - anaerobic 60°C.
- Influence of PVC on the degradation processes:
  Six laboratory reactors were operated simultaneously in pairs following the three above degradation paths, loaded with and without a PVC additive.
- Variety of the composition and influence of the surface to mass ratio:
  16 different PVC variants were tested, divided into
  - seven industrial products:
    - five soft PVC: packing film, passenger car ceiling panel, used and new; floor covering, used and new
    - two hard PVC: window frame, used and new
  - nine laboratory products with a small cross-section:
    - seven identical to the samples of the industrial products, but with substantially smaller cross sections for the exact determination of the impact effects on surfaces
    - two with clearly defined chemical composition:
      - soft PVC from PVC powder and a defined softener
      - hard PVC from PVC powder without any additive
- Proof of clear effects of background influences of the reacting municipal solid wastes:
  Investigation of the effect according to seven different, independent criteria, one to four without potential background influence:
  1. Clearly visible and perceptible changes.
  2. Impacts on the surfaces (scanning electron microscopy).
  3. Change of mechanical properties.
  4. Change of molecular weight distribution.
  5. Chemical changes:
     softeners

stabilisers

gaseous emissions

elution processes.

6. Microbial impact.

7. Carcinogenic effects of leachates and condensates.

The most important process step in increasing the investigation accuracy and simultaneously reducing the extent of work was the restriction of the geometric scale. It was proved that the landfill simulation reactor with reaction-controlled external temperature modification (see Section 9.2.3) designed for this test fairly accurately represents the processes in a landfill and/or a biological waste treatment plant, so that the results obtained therein are transferable up to the large-scale design. The informative value of the temperature variants changed due to the new findings. Temperatures of 80°C not only occur briefly in biological waste treatment plants, but also over the long-term in mono waste landfills (cf. Section 9.5). Since the cause for concern principle applies to the assessment of landfill behaviour according to the Federal Water Act, the temperature range with the most intensive reactions, i.e. the top temperature, is relevant for the assessment of PVC.

The industrial samples from among the material samples proved to be characteristic of the bulk of the PVC produced. A substantial increase of the surface in the laboratory samples failed to provide any additional information, so that this part of the investigation is not included in this excerpt of the final report.

Unambiguous effects could be identified based on the criteria independent of the background, while the chemical analysis indicated a large bandwidth of measuring tolerances even in the analysis of the initial material tested. Clear-cut results will only be discussed here. The same applies to biological investigations.

9.4.3 Test procedure and results of different investigation methods

9.4.3.1 Extent of illustration

The test equipment and the operating technique were described in Section 9.2. The detailed description of the test procedure can be studied in the report of the European Commission published on the internet. Test procedures will only be described here if they are necessary to understand a result.

9.4.3.2 Proof of agreement between landfill simulation reactors and large-scale facilities

Before statements can be made about the behaviour of PVC, the agreement of the degradation processes and the degradation performance between laboratory and large-scale facility and the influence of PVC addition on the degradation process has to be tested.

Concerning waste management criteria the laboratory data agreed with the large-scale facility and will not be described here. The agreement of the degradation performances is important. The degradation, related to the waste dry substance (Table 9-5), depended on the operational technique and, within the measurement accuracy, was independent of adding PVC samples. As expected, a characteristic performance difference was observed between the purely anaerobic degradation of approx. fifteen percent by weight DS within about 120 days and the degradation under the influence of oxygen (aerobic or aerobic/anaerobic) of approx. 20–25% by weight DS and/or approx. 25–30% by weight DS within about 120 and/or 100 days. The degradation performance under anaerobic operation was fast in comparison to degradation processes in a landfill (cf. Chapter 3) and proceeded to an extent that corresponded to a targeted anaerobic operation without an acidic initial phase. The degradation under the influence of oxygen (purely aerobic or aerobic/anaerobic operation) corresponds to a controlled biological pre-treatment (cf. Chapter 3) as far as its extent and time requirement are concerned. The results obtained in the landfill simulation reactors are thus transferable to the large-scale facility.

A higher degradation performance of the alternating operation had been expected, however it was not possible to confirm it in other tests. It is important for the investigation of PVC in the purely aerobic operation that the extensive impacts on PVC are most likely due to higher temperatures and not to a higher biological aggressiveness in comparison to the alternating operation.

Table 9-5 Degradation performance in landfill simulation reactors

Lysimeter: operational technology	Test time: days	Degradation performance: % DS by weight
1 aerobic, no PVC	124	25
2 aerobic, with PVC	124	22
3 aerobic/anaerobic, no PVC	105	28
4 aerobic/anaerobic, with PVC	105	27
5 anaerobic, no PVC	118	16

Fig. 9-12 Visible changes of soft PVC packing film (from left to right): (1) initial material fresh from the factory, (2) after aerobic ageing (approx. 80°C), (3) after aerobic/anaerobic ageing (approx. 60°C–80°C alternating), (4) after anaerobic ageing (60°C), (5) ageing in a real biological waste treatment in aerobic operation (50°C)

9.4.3.3 Changes of PVC under landfill conditions

Visually recognisable changes

Obvious changes of the soft PVC packing film fresh from the factory were recognised optically and by feeling (Fig. 9-12). The higher the temperature, the more intensive the contraction and hardening of the film while the remaining samples of soft PVC exhibited changes after ageing at 80°C. As single journey packaging material this mass product is expected to get to landfills in large quantities.

Surface impact

In the scanning electron microscope the clearest traces of surface effects were found in the soft PVC packing film after ageing under aerobic degradation conditions (Figs 9-13 and 9-14). Other samples of soft PVC also exhibited marked traces of surface impact under aerobic conditions. Soft PVC was impacted to a lesser degree under anaerobic conditions. No impact was observed under aerobic/anaerobic conditions and on the surface of the hard PVC samples within the available period of observation.

Change of mechanical properties

The stress–strain behaviour of soft PVC during the degradation process under aerobic thermophilic conditions changed characteristically: tensile strength almost doubled and the elongation limit was reduced almost to half (Fig. 9-15). This confirms the optical impression of an

357

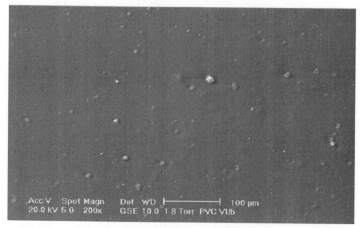

Fig. 9-13 Surface of a soft PVC packing film fresh from the factory; image by scanning electron microscope: very homogeneous surface, bright spots are dust particles

extensive embrittlement, i.e. large loss of softeners. The changes under the other deposition conditions can be recognised in their incipient state.

The stress–strain behaviour of the hard PVC of the window frame designed for long-term use also changed characteristically. It must be taken into account when evaluating the result that this material must be able to tolerate temperatures up to 60°C over the long term without losing its mechanical properties under exposure to direct sunlight. The maximum tensile strength decreased by 10% under all three deposition

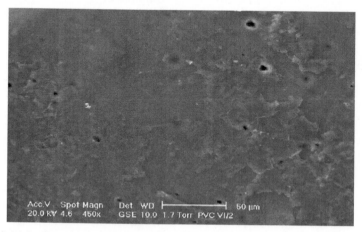

Fig. 9-14 Surface of soft PVC packing film after the end of the aerobic stabilisation process; image by scanning electron microscope: dark spots show erosion craters, the scale-like coverings consist of firmly attached colonies of adapted microorganisms

358

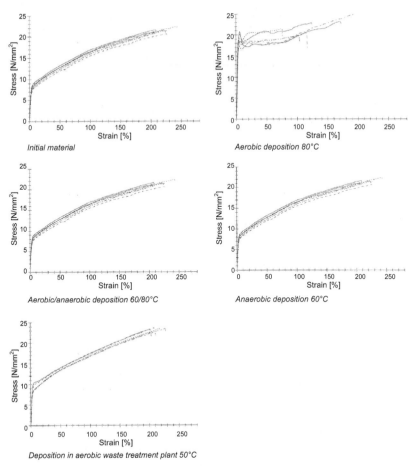

Fig. 9-15 Change of the stress–strain behaviour of soft PVC packing film under different deposition conditions

conditions in about 120 days. In addition, the maximum strain decreased by half, i.e. to the same extent as the soft PVC packing film under aerobic thermophilic conditions. This proves that the structure of hard PVC has also changed under landfill conditions.

Change of molecular weight distribution

The molecular weight distribution of a polymer is directly proportional to its chain length. If changes of the distribution are proven, this indicates an impact on the polymer structure. The investigations were performed on the same materials as the stress–strain investigations: soft PVC packing film fresh from the factory and hard PVC for the production of window frames fresh from the factory.

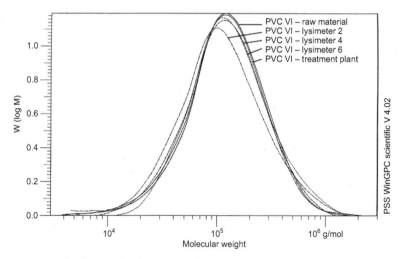

PSS WinGPC scientific V 4.02

Fig. 9-16 Molecular weight distribution of a soft PVC packing film before and after deposition under different landfill conditions; PVC VI = soft PVC packing film; raw material = initial material fresh from the factory; Lys. 2 = aerobic thermophilic degradation (80°C); Lys. 4 = aerobic/anaerobic degradation (80°C/60°C); Lys. 6 = anaerobic degradation (60°C); treatment plant = biological waste treatment plant (max. 50°C)

The measurable changes of hard PVC remained within the range of measurement accuracy. Significant changes were found in the molecular weight distribution of soft PVC by the effect of thermophilic aerobic degradation processes (Fig. 9-16). The shift of the distribution in the direction of low molecular weight indicates an impact on the basic substance of PVC. For the aerobic/anaerobic degradation and even for the purely anaerobic degradation at 60°C the attack can be recognised. Since the basic PVC substance is the same in soft and hard PVC, the impact on the window material was not measurable due to a substantially grater material cross-section and better stabilisation.

Chemical influences

Softener
The loss of softeners depended mainly on the chemical composition of the softener and secondarily on the landfill conditions. A loss of DIDP, found in samples of new floor covering, was not observed under any of the tested deposition conditions. 90% of DEHP, used as a softener in a new car liner, was expelled under thermophilic aerobic conditions (80°C) within 120 days. At temperatures below 60°C the loss was still markedly recognisable within the same time interval.

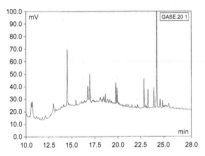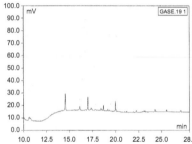

Fig. 9-17 Influence of PVC addition on the trace gas contamination from the final phase of the thermophilic aerobic degradation by finger print chromatograms – left: with PVC; right: without PVC

Stabilisers

No significant loss of the heavy metals used as stabilisers was observed within these investigations.

Organic trace gases

The available finances only enabled a qualitative investigation of the out-gassing trace gases between the units operated simultaneously without and with PVC addition. For this purpose the trace materials of the exhaust gases were adsorbed on activated carbon and submitted to a finger-print analysis and then the chromatograms of the same operating techniques without and with PVC addition were compared. In the thermophilic aerobic degradation the chromatograms themselves differed markedly even in the final phase of the degradation processes (Fig. 9-17), although the easily expelled gases were ejected rapidly, often within the first 20 days. An increase of the trace gas portions was also observed in the case of PVC addition under the aerobic/anaerobic and purely anaerobic conditions.

Chemical contamination of leachates

The influence of PVC on the contamination could not be proved by analysis due to the high background values of the leachate from deposits without targeted PVC addition.

Microbial impact

The changes in PVC samples proved so far may have been caused by a direct microbial impact or a chemical or physical impact at temperatures above 60°C. Therefore it was specifically investigated whether and to what extent a biological attack was present. For this purpose specified 40-µm-thick films were manufactured from PVC powder in the

361

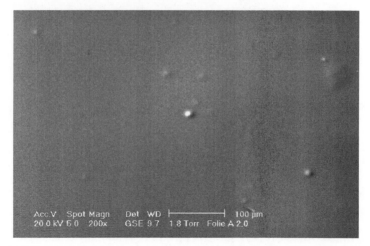

Fig. 9-18 Scanning electron microscopic image of a film of pure PVC after production. Bright spots = dust. The material exhibits a homogeneous surface

laboratory adding well-graded softener, on which the following tests were performed:

- Testing the surface changes using a scanning electron microscope.
- Growth of bacterial cultures which only use PVC or its additives as a carbon source (counting cell numbers).

Microorganisms were primarily cultivated from landfill material and the film samples were exposed to this culture with a mineral addition but without any carbon source at 60°C (= permanent temperature of high landfills). 0.5% glucose as an easily available carbon source was added simultaneously to the same tests in pairs.

The most important result is that when no glucose was added, all PVC samples were impacted (cf. Fig. 9-18 with Fig. 9-19).

PVC was not impacted when glucose was added. The bacterium-free control with mineral solution without glucose was not impaired either. Thus it is proved that the bacterial impact is a result of it being used as a carbon source. It is only impacted – according to the universal minimum principle – when no easily accessible sources are available. The use of PVC as a carbon source can only be expected in the landfill after decades at the earliest – after a slow degradation of the natural organic substance. The molecular weight distribution did not change at 60°C – as in the preceding tests. The impact on the molecular structure therefore occurs only over the short term within the thermophilic to hyperthermophilic range of the degradation processes.

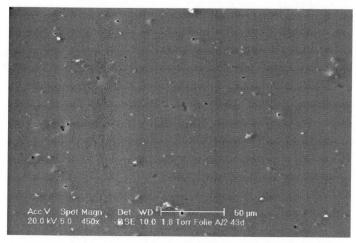

Fig. 9-19 Scanning electron microscopic image of a film of pure PVC after 43 days of incubation in a mineral salt solution without glucose at 60°C. Dark irregular spots are erosion holes, i.e. parts of the material are degraded. Bright spots = dust

Ten strains determining the degradation were isolated from the bacterial impact after microscopic characterisation (Table 9-6). Without exception they belong to the genus 'Bacillus'. It can be concluded from the differentiated growth behaviour that the strains 'Bacillus brevis', 'Bacillus sphaericus' and 'Bacillus stearothermophilus' determine the impact on PVC under aerobic thermophile conditions.

Table 9-6 Identification of bacteria strains involved in the degradation based on API 20 E and API 50 CH: Probability of identification >75% = safe identification

Strain code	Identification (% probability)	Similarity to
D 2	Bacillus sphaericus to 85.8%	Bacillus firmus
		Bacillus brevis
D 5	Bacillus sphaericus to 96.0%	Bacillus stearothermophilus
		Bacillus laterosporus
D 12	Bacillus stearothermophilus to 99.9%	–
D 16	Bacillus stearothermophilus to 99.6%	Bacillus pumilus
D 17	Bacillus brevis to 99.6%	Bacillus sphaericus
		Bacillus stearothermophilus
D 20	Bacillus spec. (possibly Bacillus circulans)	–
D 23	Bacillus brevis to 68.7%	Bacillus sphaericus
D 25	Bacillus brevis to 98.1%	–
D 26	Bacillus sphaericus to 68.0%	Bacillus brevis
		Bacillus laterosporus
D 28	Bacillus brevis to 94.0%	Bacillus firmus
		Bacillus stearothermophilus

Testing for carcinogenic effects

Leachate and condensate were tested by Professor Schiffmann, Rostock University (see method in Section 11.2.1.2). He found that the deposited PVC samples did not significantly increase carcinogenity of the leachates and the condensates. Softeners such as DEHP, whose carcinogenic effect has already been proved (Schiffmann *et al.*, 1993), are contained in the waste without intentional PVC addition therefore providing a background value.

9.4.4   Assessment of long-term stability

The assessment of the long-term stability of PVC and the assessment of undefined plastic mixtures is summarised in Section 9.5.4.

9.5   Long-term instability of chemically highly stable plastics in undefined material mixtures with high plastic content or in plastic mono landfills

Günther Ballin and Peter Hartmann

9.5.1   Objective of the investigation and selection of the tested material

It has been known for more than 20 years from temperature measurements in large landfills that temperatures as high as 70°C may occur even under anaerobic conditions (Section 11.2.3.2). Currently, these temperatures are only attributed to the degradation of the natural organic substance due to heat accumulation within the landfills. New measurements in concentrated plastic deposits indicate the same and higher temperatures are also possible in plastic deposits without natural organic substances. Local gas fires may develop in the bottom of such old landfills, whose emergence has so far been attributed to external influences (e.g. pieces of broken glass, lightning, campfire). These plastics are chemically highly stable in normal use and biochemically inert before deposition. Only mixed plastics from mono landfills have been used in order to be able to reliably test whether ubiquitous microorganisms can also cause intensive exothermic reactions in plastic wastes. The mixing ratios of the plastics and the deposition conditions differed (e.g. with and without intermediate earth layers), however, the samples were always drilled from high temperature zones of a greater than 10-year-old deposit. The reactions were compared with plastic mixtures from recycling (plastic residual fraction = PRF). The results are therefore valid for the waste spectrum specified in the sample descriptions.

364

9.5.2 Structure of the reactor

The test reactor described in Section 9.2.3 was also used for the investigation of plastic mixtures. However, since the influence of pressure on reactions in plastics is not known, a stainless steel reactor was constructed based on the design principle of the glass reactor, which, in addition to the measurements and controls of the glass reactor, can also be adapted to provide the material stresses and internal pore pressures (up to 20 bar) of a sample similar to real deposition conditions.

The temperature stability of the reactor was increased to 120°C. Because of the possibly low intensity of the thermal processes in the plastic, the resolution of the temperature measurement signals was increased to 0.01°C and the measurement detection to 0.1°C. The constancy of this measurement technique is ensured for at least 1 year.

9.5.3 Test execution and investigation results

9.5.3.1 Concept

The investigations on undefined plastic mixtures were developed step by step, so that a subsequent test was based on the results and interpretation of the preceding test. Test and result must therefore be described together for each step. The following steps have been performed so far:

A   Testing the biological and chemical reactivity of plastic mixtures from recycling (PRF)

    A.1 Without inocculation

    A.2 With inocculation by eluates from drilled landfill material from a highly active 80°C zone

B   Activation of degradation processes in drilled plastic mixtures from landfill zones with high temperatures (approx. 80°C)

    B.1 Activation by eluates from identical waste with and without nutrient enrichment

    B.2 Activation by eluates from activated soil

    B.3 Influence of water content on exothermic reactions

9.5.3.2 Testing the biological and chemical reactivity of plastic mixtures from recycling (PRF), test section A

A.1 Inoculation testing

The wastes are currently produced by recycling plastic casings and their composition is therefore not completely identical to that of the deposits. Only the order of magnitude of the fractions is important for the

Table 9-7 Composition of PRF in % DS. Water content at 80°C: approx. 40%. 45–50% of the water is contained in cellulose, PP and synthetic fabric. pH value of the dripping water: 4.5

Product	Description	Fraction in % by weight, related to DM[a]
Polypropylene	PP (pan-crushed material, strips)	50
Polyethylene	PE (foils, compact too)	2
Polymer fabric	Main component nylon, others not definable	10
Synthetic rubber	Ebonite, rubber and rubber-like substances	1
Cellulose	Cellulose acetate, paper, paperboard	4
Non-reactive minerals	Earth, demolition waste, loam	8
Polyvinyl chloride	PVC (compact material such as screw connections, etc.)	25
Total		100

(a) Dry mass

investigation. Therefore only rounded values of the single material portions will be indicated (Table 9-7).

The potential colonisation of the initial material with bacteria was investigated by testing two samples of 20 kg each for colony forming units (CFU) and by a direct count (phase-contrast microscope). Neither of the two methods was able to detect any colonisation (all counting results = zero). This result had been expected because of the low pH value.

A.2 Testing using inocculation by eluates from drilled landfill material from a highly active 80°C zone

In the first step the microbial colonisation was characterised on two drill cores 20 kg each from two equally active zones from different landfill sectional cores. Both samples resulted in the same colonisation per 1 g original sample (Table 9-8).

Table 9-8 Colony forming units (CFU) as well as vegetative cells and spores (direct counting) in biologically highly active plastic deposits per g original substance at 55°C incubation temperature

CFU in 4 days	CFU in 7 days	CFU in 21 days
1×10^6/g	1×10^6/g	2×10^6/g

Total cell number	Vegetative cells	Spores
63×10^7/g	1×10^7/g	62×10^7/g

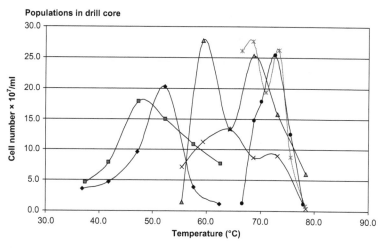

Fig. 9-20 Temperature-dependent multiplication of different bacteria from the drilled active landfill material (80°C)

The concentration by a factor 10 of the extracts resulted in the same results. The temperature requirements of the organisms were tested in the temperature-gradient block (Fig. 9-20). Multiplication of different species was proved over the entire temperature range of real landfills between 35°C and almost 80°C. Furthermore, it was found that the pH value dropped to or below 5.0 in a nutrient solution with limited multiplication in all multiplication processes, if it was not stabilised by a control at an approximately neutral range.

In the second step, the reactor was filled with sterile plastic residues and, analogous to landfill environment, the conditions for biological degradation were created. First, the degradation inhibition was eliminated and acidic pore water produced, analogous to the effect of rain water, was flushed through by high-purity water, until the pH value increased to between 4.5 and 6.5. The colonisation by all bacteria populations was reached without any selectively influenced cultivation by means of covering the sterile filling by a grid tray and distributing the original waste extracted by drilling from the active landfill zone on top of the grid tray. Leachate circulation transported the microorganisms from the active material gradually into the sterile material. 1000 ml of easily metabolised TD nutrient solution (10 g of trypton, 5 g of glucose, pH value 7.0) was applied to start the acceleration of the degradation activity. The landfill analogy is established by the fact that microorganisms and nutrients infiltrate from overlaying soil layers, such as earth from carrot processing. The liquid volume was topped up after

Table 9-9 Test series with PRF

Alternative	Condition	Start temp. (°C)	Temp. ramp	Length (d)	Nutrients
1	aerobic	45	0.4°C/d	130	1000 ml TD
2	anaerobic	35	0.3°C/d	240	0
3	aerobic	67	0.4°C/d	28	0
4	aerobic	30	0.4°C/d	140	Mineral (N, P, K)
5	aerobic	65	0.4°C/d	14	2 × 250 ml TD

4 hours in the lower dome with deionised water to 2000 ml. This filling was invariably used in all test runs. At the end of the investigation it was removed layer-wise and tested for the effects of degradation processes. Five test alternatives were investigated (Table 9-9).

The three characteristic parameters of the activity (temperature, carbon dioxide production and cell number) are shown for the long-term observations (Alternatives 1, 2 and 4) in Figs 9-21–9-23.

In the first degradation step (Alternative 1, Fig. 9-21) the initial acceleration caused by easily metabolised nutrients was terminated after 60 days at the latest (minimum of cell numbers and CO_2 production). The maxima arising thereafter coincided with the investigation in the temperature-gradient block (Fig. 9-20) and are connected with exothermic reactions, which substantially exceed the planned increase of the 'temperature ramp'.

The rate of increase due to biological activity decreased with increasing temperature (the difference to the 'ramp' becomes smaller),

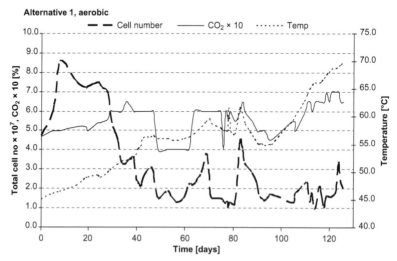

Fig. 9-21 Test process of Alternative 1

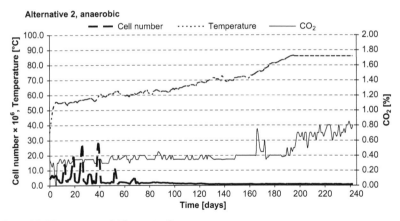

Fig. 9-22 Test process of Alternative 2

because microorganisms active in these temperature ranges only proliferate very slowly, e.g. 80 h generation time at 76°C. The markedly detectable microbial activity after consumption of the starting nutrients proves that the initially sterile production wastes contain biologically degradable components, which are degraded by a heat surplus up to a temperature of 70°C.

In the second step (Alternative 2, Fig. 9-22), the activities are tested in the same temperature range under anaerobic conditions. No

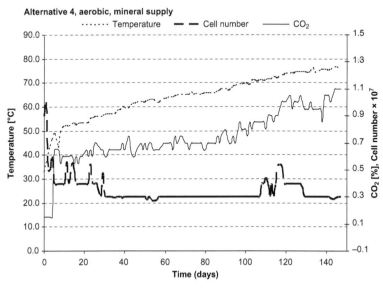

Fig. 9-23 Test process of Alternative 4

exothermic reactions, which would clearly exceed the externally forced 'ramp increase', were detected.

Temporary cell number multiplications only occurred up to a temperature increase of 60°C and then dropped to almost zero as the temperature rose further. However, CO_2 production increased constantly as the temperature rose above 70°C, even if the maximum temperature was maintained at a constant 85°C for 60 days. Above 70°C continually increasing degradation processes occurred due to chemical reactions. The irregularities in the temperature and the associated peaks of the cell numbers and CO_2 generation are reliable high-resolution measurement values. They indicate a very short-term change in degradation processes with very different heat balance. This observation was investigated subsequent to Alternative 5.

Alternative 3 is a supplement to Alternative 2. It was checked whether a biological degradation could be activated after the conversion of Alternative 2 to aerobic degradation conditions within the bio-logically inactive anaerobic temperature range of approx. 65–75°C analogous to the initial aerobic phase (Alternative 1). The biological activation did not emerge in the now impoverished material. Tempera-ture and CO_2 generation ran analogous to the anaerobic condition. Hence, it follows that, after a biological start of exothermic aerobic degradation processes, chemical degradation processes set in above 60°C when a heat source provides the energy. The impoverished material need not be biodegradable.

In the waste body, both supplementary minerals needed for biological degradation and easily degradable organic substances can be supplied by the leachate, which activate a further biological degradation in the impoverished material. The provision of minerals (Alternative 4) is the normal case in a landfill. 0.05 mol of N as NH_4NO_3 and 0.025 mol of P as KH_2PO_4 were added daily in an aqueous solution along with the circulating water at a pH value of 7.0. Nevertheless, the exothermic activity was not activated. The temperature increase almost completely follows the externally forced 'ramp'. Although colonisation remains at a higher level in comparison to the anaerobic condition and grows slightly in the range around 70°C, it has no signif-icant relationship with the constant increase of CO_2 production clearly recognisable above 70°C. The chemical degradation reactions of Alternatives 2 and 3 are thus confirmed. The biological degradation was not inhibited by absent mineral nutrients (Fig. 9-23).

Organic nutrients can also be supplied by the leachate from activated soil layers. In addition to Alternative 4, it was therefore checked to see whether reactivation is also possible under extreme conditions such as 'hygienisation' of landfills. For this purpose, after completion of Alternative 4, the material was heated up and maintained at 92°C for 7 days. Afterwards, the material was cooled down to the 65°C starting temperature, and after extracting the pore water from Alternative 4, 250 ml of TD solution each was distributed over the waste surface in 2×24 h (cf. Alternative 1) and an additional 250 ml of de-ionised water each within 2×24 h. Despite intermediate temperatures of 92°C an intensive exothermic biological degradation started immediately after a nutrient addition, which heated the material to 71°C within 6 hours and completely exhausted the nutrients. Afterwards the reactions again resembled those of the final phase of Alternative 4. The process was reproducible. It follows therefore that high landfill temperatures do not lead to a sterilisation of the material. The available microorganisms degrade easily degradable organic substances exothermally under aerobic conditions at high intensity and produce temperature increases which trigger chemical degradation processes in the biologically impoverished plastic waste.

The short-term fluctuations in temperature and CO_2 were investigated after completion of the high-resolution test series. In particular, the fluctuations exhibited a very clear agreement in time where the degradation processes above 70°C were wholly chemical. The characteristic of such brief temperature increases with a synchronous reaction of carbon dioxide generation corresponds to the spontaneous chemical conversion of gases with a limited available volume. Reactive gases may be available in real landfills in such large quantities that the reaction does stop. Since chemical degradation processes proportional to temperature increase were detected above 70°C in Alternatives 2 to 4, such exothermic processes may cause an intensive chemical degradation of plastics – possibly including fires. The fact that CO_2 was released in Alternatives 2 to 4 above 70°C under endothermic conditions, suggests that energy-rich degradation products were developed. Twenty-one energy-rich chemical compounds were detected by a GC/MS screening of the materials adsorbed by activated carbon from the exhaust gas and the materials contained in the discharge (Table 9-10) – without a guarantee for completeness.

Thermal stress and possibly chemical impacts must be measurable as a change in the characteristics of the plastics. Micro differential thermal analysis (DTA) offers a very sensitive measurement technique of the changes, performed by Professor Schick, Institute for Experimental Physics of Rostock University. For this purpose, when removing the

Table 9-10 Compounds detected using GC/MS

Compound	Condensate 02w0727.04	Nutr. sol. 02w0727.02	Anaerobic 02w0727.03	Aerobic 02w0727.01
Pinene $C_{10}H_{16}$			1	1
Alkane $C_{12}H_{26}$			1	1
$C_{12}H_{24}$		1	1	1
Terpinene $C_{10}H_{16}$			1	1
Terpinene $C_{10}H_{16}$ carene/terpinene			1	1
Dimethyl disulfide			1	
Dodecane			1	
Limonene $C_{10}H_{16}$			1	
Trimethyl dodecane/tetradecane			1	
Branched and straight-chain alkanes/ alkenes			1	
Alkane C_9H_{20}				1
Alkene C_8H_{16}				1
Aromat C_8H_{10}				1
Dithiocarbonic acid – O,S-methylester		1		1
Dithiocarbonic acid $C_3H_6OS_2$		1		1
Phthalate	1			1
Alpha-pinene $C_{10}H_{16}$				1
Undecane $C_{11}H_{24}$				1
Compounds with fracture pattern such as dithiocarbonic acid		1		1
5-methyl-2-hexanone	1			
Ketone	1			

tested material, polypropylene samples (50% mass proportion, generally accepted as 'inert') were taken at distances of 10 cm. Even a visual observation revealed different stresses. A visibly changed material without a polymer structure was recovered from the zone of proven exothermic temperature increase (Fig. 9-24).

Minor or no visible changes were detected in other areas. The specific thermal capacity of the original material differs unmistakably in all samples and in all temperature ranges from the material at the end of the test:

- In the melt range substantial displacements and changes of the maximum amounts of the maxima occur (original samples Fig. 9-25, reactor samples Fig. 9-26).
- In the post melt range (Fig. 9-27), the thermal capacities have somewhat similar values and agree approximately in 25% of the measured range (154–160°C), but in the range of the glass transition (Fig. 9-28) agreements are hardly recognisable.

Fig. 9-24 *Area of changed structure in the reactor filling (black area)*

Thus the 'inert' polypropylene, the most chemically durable compound of the deposited main components, was extensively changed in the investigated temperature range up to 90°C within only 1 year.

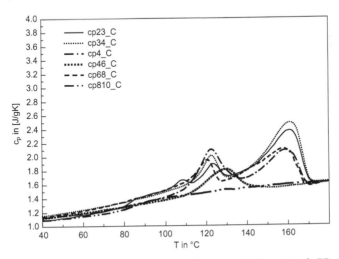

Fig. 9-25 *Specific thermal capacity in the melt range of the original PP samples.* $\Delta T = 1K$, $t_{iso} = 30\,s$

373

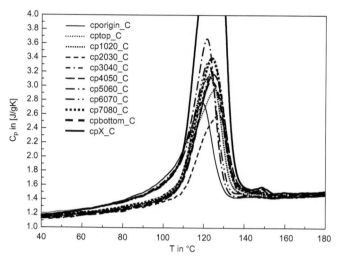

Fig. 9-26 Specific thermal capacity in the melt range of PP samples at end of test. $\Delta T = 1K$, $t_{iso} = 30\,s$

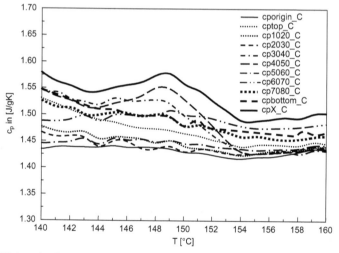

Fig. 9-27 Specific thermal capacity in the post-melt range of PP samples at end of test. $\Delta T = 1K$, $t_{iso} = 30\,s$

9.5.3.3 Activation of the degradation processes in drilled plastic mixtures from landfill zones with high temperatures (c. 80°C). Test phase B

B.1 Activation using eluates from identical waste and slow heating

Material and test conditions

In the 2nd main test phase, plastic wastes from landfill material extracted by drilling were used (approx. depth of extraction 4 m) from

374

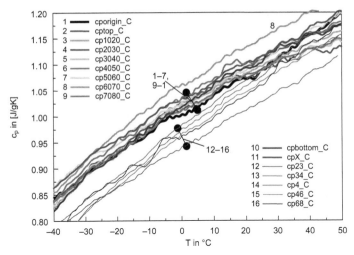

Fig. 9-28 Specific thermal capacity in the range of glass transition of PP samples. $\Delta T = 1K$, $t_{iso} = 30\,s$

a hot landfill zone of about 80°C as filling material. Since this material remains under pressure in the landfill, the stainless steel pressure reactor was used instead of the glass reactor and the analysis was supplemented in the second measurement series by a transmitted light measurement at 360 nm. Two samples were implemented (15 kg and 10 kg DS) and two temperature series performed up to 90°C for each sample. When interpreting the results it must be noted that the composition of the samples vary. Similar, but not identical mixes were tested. In order to be able to measure the presumably maximum emission values within the potential bandwidth, a sample material with more than 45% DS of undefined mixed waste was selected as the 2nd sample, since the emission of combustible gases was expected from this fraction.

Test programme
Table 9-11 illustrates the test programme together with a summary of the results. The main difference between the test conditions of the two samples was the pressure. Assuming a medium density of 1.0–1.5 t/m³ for the plastics and the intermediate earth layers, the normal stress corresponds to an overburden pressure of 55–80 m of landfill height. A gas pressure of 4 bar (corresponding to 40 m water column) can develop under this overburden pressure when degassing is poor. Greatest differences were expected from this test in comparison to the nearly pressure-free test conditions of Phase A. Compression of

375

Table 9-11 *Test programme and summary of the results*

Series	1	2	3	4	5
Filling	Landfill original material, drilled	Landfill material from Series 1	Landfill original material, drilled	Landfill material from Series 3	Landfill original material, newly drilled (4 m)
Quantity DM[(a)]	15.16	15.16	10.4	10.4	20 kg
% water cont.	20.6	20.6	47.8	47.8	5.42
Compaction	82.5 g/cm²	82.5 g/cm²	24 h 43.5 g/cm²	24 h 43.5 g/cm²	24 h 50 kg
H_2O addition (ml)	3000		3000		3000
Cell number at start	8.2×10^7 ml⁻¹	1.1×10^7 ml⁻¹	1.4×10^7 ml⁻¹	1.6×10^7 ml⁻¹	37.9×10^7 ml⁻¹
Physical conditions					
Gas pressure (bar)	4	4	1	1	0.5
Temperature schedule	Gradual 5°C 5 days each 40–90°C	Gradual 5°C 5 days each 40–90°C	Continuous 40–90°C 1.3°C/d, 90°C constant for 8 days	Gradual 5°C 5 days each 40–90°C	Gradual 5°C 5 days each 1.2°C/day 40–90°C
Aeration (ml/h⁻¹)	Air 100	Air 100, and no aeration at 90°C	Air 100	Air 100	Air 100
Results					
CO_2 generation measured in exhaust gas	Temperature-dependent, steep from 60°C to 6.5% at 90°C	Temperature-dependent, steep from 60°C to 6.5% at 90°C	Continuous increase 0.2–6.5% CO_2, pronounced from 60°C	Moderately temperature-dependent below currently 60°C. (0.1–0.4%)	0.15–0.3% below 70°C
pH					
Condensate from exhaust gas	6.3–4.7	5.6–5.0	5.3–5.9	5.6–5.3	
Discharge	5.5–3.0	4.6–4.0	5.7–5.5	5.4–5.7	4.8–5.1

Cell numbers in discharge ($\times 10^7$/ml^{-1})	6.3–4.7	1.1–1.3 very small rods ($<1\mu$)	1.4–2.1 very small rods ($<1\mu$)	1.6–1.8 very small rods ($<\mu$)	3.1, max. at 50°C and 65°C
Transmitted light at 360 nm	No measurement technology	No measurement technology	Value correlates with spontaneous temp. increase	Moderate changes from 60°C	No change
Temperature behaviour	Spontaneous temperature increases	Spontaneous temperature increases	Spontaneous temperature increases	No spontaneous temperature increase below 70°C	Increase in range of bacterium growth
Special features	Spontaneous temp. increases during heating, CO_2 to max. 6.5%	Spontaneous temp. increases during heating below 12.5% CO_2. No increase above 12.5% CO_2	Marked temperature irregularities from c. 70°C. CO_2 changes as temperature	No temperature irregularities to 60°C	
Sensors	Condensed water from gas cooler: marked odour of crude oil components	Condensed water from gas cooler: marked odour of crude oil components	Condensed water from gas cooler: marked odour of crude oil components	Condensed water from gas cooler: marked odour of crude oil components	No 'oil odour' detectable below 70°C

(a) Dry mass

377

the second sample corresponds to an overburden pressure of approx. 30–40 m landfill height. A gas pressure up to 1 bar (corresponding to 10 m water column) is expected in landfills without active degassing also in the case of pathways enabling gas migration in the plastic layers inside the waste body.

A higher initial water content of the 2nd sample in comparison to the 1st one with the same requirement for circulating water (3000 ml) indicated a higher fraction of water-storing solids in the 2nd sample (probably cellulose basis) and corresponded to the intended increase of biologically degradable components within the composition bandwidth possible in this waste.

The order of magnitude is most important for the assessment of cell numbers at the beginning of the test. It was the same (10^7) in all series (see Table 9-11).

The temperature of the material in the reactor was started at 40°C and increased on average by 1°C per day in the cladding. If the enforced temperature increase triggers exothermic reactions, these can be recognised by a steeper temperature increase. The final temperature of the enforced heating was limited at 90°C, since water was used for heat transmission in the cladding. Since the highest emissions were measured at this temperature level, the emission process of the highest temperature level was measured over 45 days.

The slow heating from 40–90°C was performed in the 1st, 2nd and 4th series in exactly the same steps of 5°C each and over 5 days per step, i.e. on average at 1°C/d. In the 3rd series a continuous increase of 1.3°C/d was used to test the influence of the increase on the reactions (no measurable influence at almost the same average value of increase).

The same moderate air flow rate of 100 ml/h was chosen in all series. In real landfills, where there is no perched water, hardly any oxygen consumption and good gas pathways prevail as found in the plastic layers, the atmospheric oxygen penetrates by diffusion to a comparable extent, so the minimum influence of the oxygen was presumably simulated. In addition, the air supply was temporarily turned off at the end (90°C) of the 2nd series, so that measurements with an extremely low oxygen supply and high carbon dioxide percentage were also performed.

Results

Overview

Table 9-11 displays an overview of the most important results. The similarity of all results is conspicuous in all four columns. Neither the

378

differences in the initial material nor the different pressures have resulted in any characteristic differences in the results. In the same way there are no clear differences between the 1st cycle (40–90°C) and the repetition in the 2nd cycle with the same temperature steps in the same filling.

Carbon dioxide generation was moderate in all four series up to a temperature of 60°C and then increased substantially up to the maximum at 90°C. The pH values are within the acidic range mainly between pH 5.9 and pH 5.3, in extreme cases between pH 6.3 and pH 3.0 (according to sulphuric acid) both in the exhaust gas and in the discharge. The percolate was always neutralised to pH 7.0 before recirculation.

The cell numbers remained in the same order of magnitude as in the initial phase, although the material was heated up in two warming cycles to 90°C. The form (very small rods) suggests heatproof continuous forms. Spontaneous exothermic reactions of limited duration occurred in all four series particularly above a material temperature of 70°C when the carbon dioxide content did not exceed 12%.

The spontaneous temperature increases are in agreement with the fluctuations (short decrease in concentrations) of carbon dioxide with regard to time.

The condensed water of all four series contained materials with a marked odour of crude oil components.

Absorptions of UV light, applied to the investigation of the 2nd sample (series 3 and 4), indicated emissions simultaneous to the spontaneous temperature increases.

Differentiated measurements evaluation

The characteristic reactions had nearly the same history in all four series of measurements. Therefore, examples of continuous measurements of temperature, carbon dioxide and oxygen (Fig. 9-29), temperature-dependent microbial colonisation (Fig. 9-30) and the relationship between spontaneous increases in temperature and fluctuations of carbon dioxide generation (Fig. 9-31) are illustrated for detailed evaluation. (Measurement of material temperature is independent of the measurement of carbon dioxide concerning measurement technique, therefore simultaneous fluctuations are no coincidence.)

Figure 9-29 does not show the cladding temperature. It was slightly higher in all four measurement series than the material temperature. Nor is it not shown that carbon dioxide was also produced under oxygen exclusion at 90°C until a level of 11% by volume was reached.

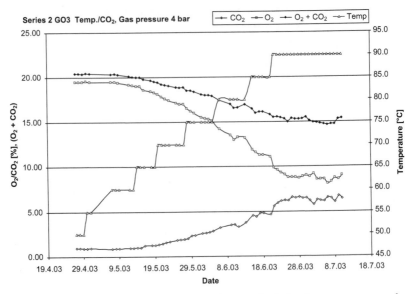

Fig. 9-29 Comparison of prime parameters (overview), Series 2: temperature, carbon dioxide, oxygen content, oxidation balance of carbon

The 2nd sample produced less carbon dioxide than the 1st sample, although it contained more undefined mixed waste than the first one, i.e. a higher production was expected. Furthermore, the information that the crude oil odour emerged above 70°C and intensified with increasing temperature, is important for the interpretation of the analyses.

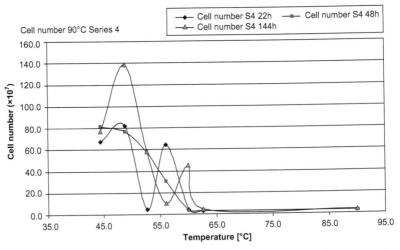

Fig. 9-30 Cell numbers as a function of temperature and length of their effects, Series 4

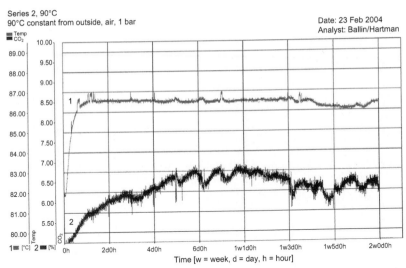

Fig. 9-31 Temperature (top graph) and carbon dioxide concentration (bottom graph). Series 2, cladding temperature 90°C, 1st part aerobic

The following conclusions can be drawn from the overview in Fig. 9-29 and the information about cladding temperature:

- Despite carbon dioxide production, which increased in all four series with increasing temperature, no heat was released.
- Up to the limit of biological activity measured here (approx. 60°C) the sum of carbon dioxide and oxygen resulted in 21% in all four series. This proves a complete degradation of natural organic substances such as starch and cellulose.
- The sum in all four series remained under 21% above 60°C and up to 6% by volume was missing in all four series at the complete oxidation of natural organic substance above 80°C. The difference to the sum of 21% (oxygen content) is proportional to the production of carbon dioxide and the increase of the oil-like smell.

The following findings have been obtained in connection with the microbial colonisation (Fig. 9-30):

- The biological starting activity in the material recovered by drilling was markedly lower than in the inoculated PRF material, test series A.2. In addition, the species were impoverished and sporulated above 70°C.
- They stopped their activity above 60–70°C, survived a heating to 90°C as spores and became active again after a cooling down below 60°C.

381

- It is common to all samples that the characteristic high carbon dioxide concentrations were produced above 80°C without any microbial activity.

The following relationships can be established from the synchronous records of temperatures and carbon dioxide production:

- Short-term temperature increase with a simultaneous reduction in carbon dioxide concentration repeatedly measured in the preceding experimental phase was verified.
- The shape of the peaks corresponds to the heat loss of an exothermic chemical reaction broken off due to the lack of mass.
- Reaction products of the 2nd series adsorbed on activated carbon were submitted to a screening. The result in Table 9-11 proves that a substantial amount of hydrocarbon compounds are released, particularly within the temperature range around 90°C. This is confirmed by the increase of the smell of crude oil-like substances.

B.2 Activation by nutrients

Objectives

It had been found in phase A that the colonisation of microorganisms was impoverished and the bioavailable fractions decreased in the material from heated landfill zones (80°C) both in the laboratory simulation of new wastes (A.2) and in landfill material recovered by drilling (B.1). However, the degradation potential of microorganisms itself was possible to prove in the material fresh from the factory after a year's degradation in the laboratory simulation, even after a short heating period of the wastes to 90°C in order to degrade easily available nutrients. It was not possible to re-activate biological degradation above 60°C, therefore test phase B.2 was aimed at checking whether the activity in the material recovered by drilling could be activated by nutrient enrichment in the eluate of the same sample (circulating water) (Test B.1) or by an eluate from activated soil after discharge of the original percolate (Test B.2).

Material and investigation steps

The reactor material consisted of the same drilling sample (4 m depth, from a zone of 80°C) as the reactor filling in Test B.1. The investigation took place in three different test runs.

Reactor filling
Material: 20 kg original material

382

6.5% water content

18.7 kg dry mass

Placement: Pre-compacting of the filling material by 50 kg surcharge in the reactor over 24 h, followed by removal of surcharge load and performing of the tests.

The reaction behaviour was investigated in two main steps, B.2.1 and B.2.2, which were sub-divided into two runs:

B.2.1   Inoculation with 3 litres of a suspension from the same material as the filling material and addition of a nutrient solution (TD solution, cf. A.2). This suspension was incubated for 24 h each at temperatures of 50°C, 60°C and 70°C before inoculation.

B.2.1.1:   Sample not water-saturated

B.2.1.2:   Water saturation of reactor content (approx. 50%), water supply over the water vapour phase at 90°C in the final phase of the test run B.2.1.1; then cooled to 35°C, testing the degradation processes in the inoculated material after water saturation at the same temperature levels as in run B.2.1.1.

B.2.2   Cooling of the material after the end of the test programme B.2.1.2, new inoculation of the water-saturated material after extracting the percolates: a compost-earth mixture was extracted using a 0.85-% sodium chloride solution and the reactor content inoculated three times with the extract analogous to the start of B.2.1.

B.2.2.1:   Test series with leachate circulation in the water-saturated material.

B.2.2.2:   Continuation of the tests according to B.2.2.1 with water drainage and successive desiccation up to the initial water content.

The process simulated in test phase B occurs in practice, e.g. when the intensive degradation processes in the landfill zone, from where the material was extracted by drilling, were interrupted by mechanical measures such as the injection of liquid nitrogen through a bore hole (11.2.3.2) and then ubiquitous microorganisms and associated soluble soil components flowed through the plastic wastes with the leachate from soil layers of intermediate layers and the remediation layer. Desiccation occurs in practice when the landfill is carefully covered and sealed with the aim of preventing spontaneous heating.

To achieve desiccation, approx. 300 ml of air was blown through the reactor from top to bottom in the flow direction of leachate. The air supply temperature was first adjusted to 54°C and increased gradually to 63°C and 69°C, with increasing desiccation. Following the investigation of the desiccation phase the material was removed in individual lifts. A differentiated test of the material was supposed to decide whether the conclusions drawn from the temperature measurements were correct, according to which the intensive exothermic reactions were concentrated to defined zones and possibly due to special material combinations.

Results

Test runs B.2.1.1 and B.2.1.2

After an interruption of the exothermic degradation processes in the waste by removing it from the landfill, no intensive exothermic degradation processes in the material extracted by drilling could be induced by the inoculation with a suspension from the percolate of the waste sample, enriched in the normal laboratory, easily metabolised TD nutrient solution (cf. A.2), which would go beyond the degradation of easily available nutrients. Principally the carbon dioxide production did not increase above 70°C. The endothermic reactions corresponded to those in PRF during the aerobic reaction tests in the 4th test run, i.e. the new material adjusted its degradation behaviour to the material extracted by drilling after 1 year of degradation in the reactor. The figures of these results (A.2) therefore also apply to the measurements series B.2.1.

Test series B.2.2: Inoculation of the water-saturated waste of the series B.2.1 with a compost – soil eluate after drainage of the original percolate
B.2.2.1 Reaction measurements in the water-saturated operation

As already observed in the previous test series A.2 and B.2.1.2, the material temperature increased to the cladding temperature by spontaneous heating after adding the nutrient-containing suspension and ran to approximately 85°C simultaneous to the cladding temperature without conspicuous spontaneous temperature peaks. The carbon dioxide content first increased to 9% due to the initial biological activity and then dropped to 4% as the controlled heating rose towards 85°C during the course of the test. Apart from the initial range, heat production was low, although cell multiplication was measured until heating reached 80°C. It can be concluded from this that the active microorganisms exhibited a metabolic behaviour different to that of compost or municipal solid waste due to the special substrate. Decarboxylisation

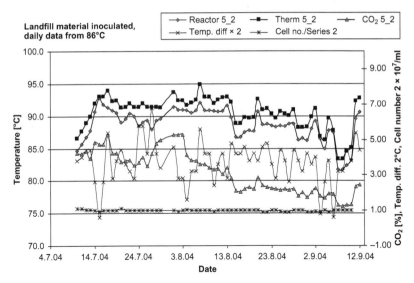

Fig. 9-32 Test process on landfill material inoculated with high-activity bacteria suspension starting from 86°C

under heat release did not seem to occur. These results are not illustrated in figures because of the good agreement with the results of the measurement series A.2.

Alternating spontaneous temperature increases up to 7°C occurred starting from the 90°C cladding temperature. The heat released was sufficient to heat the thermostat water and to switch off the temperature control of the reactor by overheating (Fig. 9-32). Since the cladding heating of the reactor did not simulate the heat accumulation of the landfill above a temperature of 95°C to protect the equipment, but extracted heat from the system by evaporating water, only the beginning of the exothermic processes was measured. The high-resolution illustration of the process (Figs 9-33 and 9-34) indicates the synchronous run of the spontaneous temperature peaks and the carbon dioxides generation.

The initial temperature of exothermic processes was determined in such a way that by switching off the heat accumulation simulation the reactor content was cooled to a temperature below spontaneous heating and then by a slow heating by means of a 'temperature ramp' the spontaneous heating was triggered. The process was repeated several times (cf. Figs 9-33 and 9-34). The initial temperature dropped from an initial 87°C to 65°C–70°C. Heating following each initiation always had to be broken off to protect the reactor.

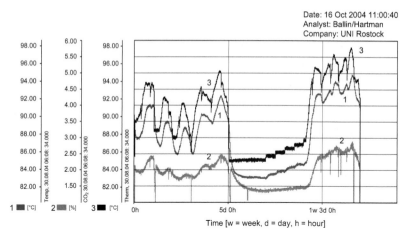

Fig. 9-33 *Determination of the initial temperature for spontaneous heating*

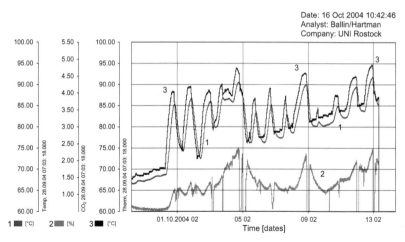

Fig. 9-34 *Temperature profile and carbon dioxide generation, connection to Fig. 9-33*

B.3   Influence of water content on exothermic reactions

Test programme

Objective

This series of investigations was aimed at determining the influence of water contained in the landfill material on heat development, microbial activity and chemical reactions in the material recovered by drilling from the original state.

386

Test equipment
The same test equipment and measuring techniques were used as in phase B.1.

Material and test steps
- The material provided for phase B.1 taken from about 4 m depth and was used. The temperature at the bottom of the drilling was 80°C. The material composition corresponds to the sample provided for phase III.
- 20 kg of fill was used in the reactor with an initial water content of 16% corresponding to 16.8 kg dry mass. The water content was adjusted to 45% for the test. That corresponds to the maximum water retaining capacity of the fill material.
- The test was started at a material temperature of 30°C and 300 ml/h constant aeration (with air) and was carried on continuously with a temperature rise of 0.5°C/day up to 60°C. Desiccation took place at a constant temperature of 60°C. The control temperature of 60°C was maintained for 60 days and the parameters of interest determined daily. This test temperature was selected because it supported the best microorganism colonisation with only one dominating bacteria population having a multiplication optimum around 60°C.
- The subsequent desiccation was performed by injecting pre-dried air into the reactor. Control of the water content in the filling material took place by measuring the volume of the condensate and separating it in a gas cooler.
- The relevant test data were measured and recorded at 24-h intervals:
  - cladding and interior temperature in the reactor
  - water content (calculated from the water content of the input material and condensate)
  - cell numbers
  - carbon dioxide portion in the reactor exhaust air.
- The start temperature of the test series was 30°C, the water content at 45%. The desiccation process took 270 days. The water content was reduced to 2.3%.

Results

Microbial colonisation and temperature-dependent multiplication
Before starting the test, the microbial status was determined. This included the determination of cell numbers in the original material and the determination of the temperature-dependent multiplication

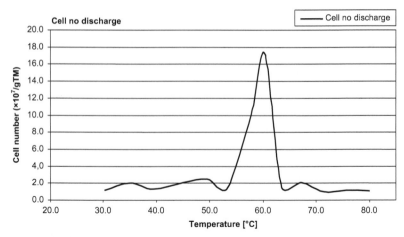

Fig. 9-35 *Determination of multiplication optima of the populations in the placed material*

of the existing populations. The material proved very deprived of bacteria with 0.4×10^7 cells per gram of dry mass. With only one population detected at considerable concentration with a multiplication maximum around 60°C (see Fig. 9-35), selection took place during the storage of the landfill material over a period of 2 years, which coincides with earlier findings on freshly drilled material from a greater depth. Other populations were only present at very low concentrations. However, they showed a strong multiplication in the start-up phase.

Results of the start-up phase up to 60°C and 45% moisture content in the material

The actual desiccation process was preceded by a 60-day start-up phase with a slow temperature increase from 30°C to 60°C. The objective was to compare the spontaneous heating processes in the landfill material after the 2-year storage time with the processes in phase B.1 in the material freshly recovered by drilling. Microbially induced temperature increases also occurred at the same time intervals and at different test temperatures. These showed good agreement with the multiplication ranges of the bacteria. Despite the long storage time, a reactivation of bacteria population occurred due to aeration and at the water content being adjusted to 45%. Exothermic reactions were conspicuous at lower temperatures, the maximum was found at 60°C. The ranges of the spontaneous temperature increases show good agreement with the multiplication optima of the individual populations (see Fig. 9-35). Thus it is guaranteed that bacteria populations existing even after long storage

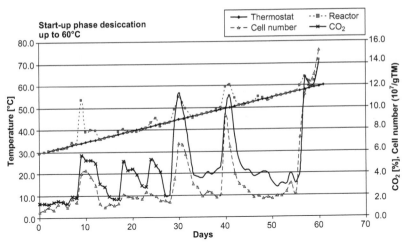

Fig. 9-36 *Start-up phase with spontaneous heating before the beginning of the desiccation process – temperature range: 30–60°C; water content: 45%; length of test: 60 days*

times are activated when suitable multiplication conditions such as water content and oxygen are established, and continue the degradation of organic substrates in the deposits with exothermic reactions.

The landfill material contains still greater fractions of biologically usable components. These substrates can be metabolised by the existing bacteria with development of heat.

A temperature increase of 0.5°C per day was chosen because of the low bacteria colonisation. Thus it was possible to consider different generation times. The measurements indicated a bacteria multiplication with heat development and CO_2 generation at different temperature levels. As previous tests had indicated, the contribution of individual populations to heat development was fairly different (see Fig. 9-36).

Test results at 60°C constant temperature during the desiccation of the reactor fill-material

By blowing pre-dried air into the filled reactor the water content of the filling was reduced. The test was carried out at a regulated temperature of 60°C. The time limited increases in temperature in the thermostats and in the reactor are due to heat development in the reactor.

The desiccation process is illustrated in Fig. 9-37 over a period of 270 days. Starting from 45%, the water content first dropped within 130 days to approx. 27% by almost linearly eliminating the capillary water according to the maximum water retaining capacity of the fill. The adsorbed water content then was reduced over about 70 days to

389

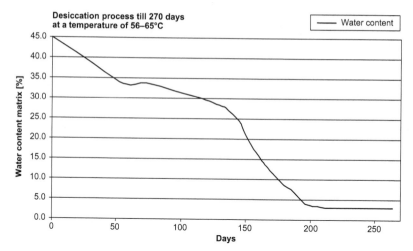

Fig. 9-37 Desiccation process at a control temperature of 60°C – initial water content: 45%; reduction to: 2.3%; length of test: 270 days

approx. 3%. After this phase the capillary water content remained at a steady 2.3% until the end of the test. The method used was not able to reduce the water content any further.

Metabolism and multiplication of bacteria are bound to a minimum water content of 20–25%. Special osmophile bacteria can tolerate a lower water content only as an exception. Microbial activity corresponding to a water content of 25% should therefore reproduce during the test of up to maximum 150 days.

The reduction of the water content to approx. 25% within a period of 150 days did not significantly affect the multiplication of the bacteria, the CO_2 generation and the temperature increases which occurred over intervals of about 20 days. With a few exceptions, a clear correlation between the multiplication of the existing bacteria, the CO_2 generation and temperature increases can be established (see Fig. 9-38a). Thus these effects are doubtlessly bound to bacterial activities.

As indicated in preceding tests, chemical reactions also occur at 60°C. In this test exothermic chemical reactions at 60°C are assumed, which are not accompanied with CO_2 generation.

In 135 days the water content dropped to <25%. Temperature increases beyond the adjusted control temperature did not occur after this time. The bacterial multiplications and CO_2 generation remained at a very low constant level. Exothermic reactions did not occur either at a water content of 2.3%.

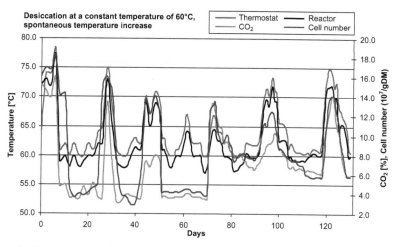

Fig. 9-38a *Spontaneous heating at a material moisture content of 45 to 25%. Length of test: 135 days*

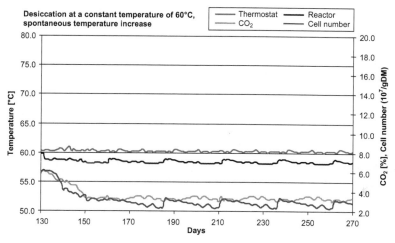

Fig. 9-38b *Test process at a material moisture content of 5 to 2.5%. Length of test: 135 days. Connection to Fig. 38a*

Results of the test at a constant control temperature of 60°C after desiccation and re-wetting of the reactor fill

Following the 270-day desiccation process the material was rewetted up to the original water content of 45%. All other parameters were kept unchanged. After a 2-day adaptation time bacterial multiplication and CO_2 generation restarted accompanied with heat development (Fig. 9-39). However, unlike the results according to Fig. 9-38a, spontaneous temperature increases occurred at shorter time intervals

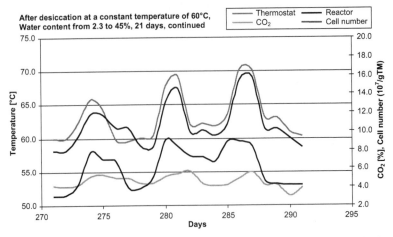

Fig. 9-39 *Temperature after increasing the water content from 2.3 to 45%. Length of test: 21 days. Connection to Fig. 9-38b*

of 6–7 days. Again, clear correlations between the measured parameters were established. The long desiccation phase (Figs 9-38a and 9-38b) had therefore no influence on the spontaneous heating processes after reactivation by rewetting.

Summary of the results and conclusions

Water content was continuously reduced from 45% to 2.3% in a long-term test in the reactor at a regulated cladding temperature of 60°C by blowing dry air into the material which had been recovered from the landfill stored over 2 years. Measurement data were recorded during this process as in preceding tests (Figs 9-38a and 9-38b).

The predominantly bacterial reactions were not impaired by the longer storage of the recovered landfill material. Several bacteria populations were detected in the material with one dominating population at 60°C (Fig. 9-35).

Spontaneous heating in the phase of the reduced water content to 25% occurred periodically in correlation with bacterial multiplication and CO_2 generation. The time difference between 2 exothermic reactions of approx. 21 days was conspicuous. The components needed for spontaneous heating were probably produced by microbial metabolic processes at the necessary concentration within this period. Simultaneously, chemical reactions without CO_2 generation were likely to occur (Fig. 9-38a). The exothermic reactions did not take place (Fig. 9-38b) starting from a water content of the matrix between 25% and 2.3%.

392

The process shown in Fig. 9-38b indicates further minor maxima, they can be interpreted as residual activities.

After the rewetting of the material, a reactivation of the microbial activity took place causing CO_2 generation and temperature increase (Fig. 9-39).

According to these results, desiccation prevents the spontaneous heating processes for the time interval of water deficit. On rewetting, the spontaneous heating processes recommence. It has to be assumed that metabolic products are produced by microbial activities which initiate the spontaneous heating process when a limiting concentration is reached. SiH_4 generation is postulated as an ignition source in this relationship. The generation of CS_2, which was also detected, may provide an explanation for the reaction mechanism. The investigations undertaken are insufficient to explain the overall relationships between the chemical reactions.

9.5.4 Assessment of long-term stability of landfills containing plastics

The tests on PVC confirm the expectations that softeners migrate away. This process has already started during normal use. Hence it follows that the use of health-endangering softeners may contaminate the drinking water resources in each case through leachate escape after the deposition of PVC over the long term. The proof of a biological attack, in particular, on soft PVC is new, which is expected mainly over the long term. Hence it follows that PVC cannot be considered as a stable long-term deposited material. The findings discussed in particular (Section 9.5.3.3, B1) suggest that even pure hard PVC changes under high temperatures and therefore may contribute to the development of extreme exothermic reactions in concentrated plastic deposits. The deposition of PVC is therefore only acceptable without harmful softeners in small, non-sortable residues in biologically extensively stabilised waste (permanently aerobic landfills, cf. Section 9.3).

It can be concluded from the investigation of concentrated, chemically highly stable and biologically inert plastic/mixed wastes that ubiquitous microorganisms can supply the launch energy over the long term, even in primarily stable mono deposits, which enables the splitting of inflammable gases from the plastics starting from 60°C. If this process is introduced just the once, it remains effective even after the desiccation of the landfills, if a start temperature of about 60–70°C arises from another source. A nutrient material source supplied by the leachate is sufficient, since the microorganisms in their spore state remain re-activatable and also if the temperature temporarily exceeds 90°C. The deposition of plastics in concentrated form therefore has to be classified as unstable over the long term. The arrangement of controlled intermediate storage facilities requires detailed preliminary investigation and the experience storage of straw and hay bales is useful since these materials have similar inclinations to spontaneous ignition.

For the assessment of high-density polyethylene (HDPE) geomembranes on top of the mineral layer in composite liners it can be concluded from these findings that even changes in polypropylene were detected in mono landfills so that this arrangement is sensitive to impacts over the long term. Placement of the geomembrane in the middle of the mineral barrier, as the inventor Professor August, Federal Institute for Materials Research and Testing Berlin, suggested (cf. Section 10.4 and Section 11.4.1, Fig. 11-31), is therefore absolutely necessary for the reliable long-term operation of the leachate barrier at the base of a landfill – as our current results indicate.

10

Establishing the long-term effects using the relationship of the test results

Peter Spillmann

10.1   Criteria for establishing the relationship

10.1.1 *Agreement of the physical and biochemical processes in the landfill sectional cores with those in the relevant real landfill (Excerpt from Spillmann, 1986, Chapters 1 and 2 and Summary)*

The investigated deposits were designed as cylindrical sectional cores from drained landfills provided with a basal liner and without any perched water and differing in their operating technology through the alternative deposition of sewage sludge, the influence of recycling and the spiking by environmentally polluting industrial wastes. The construction speed of about 2 m per year and the rather low final height of approx. 5–6 m correspond to a flat rural landfill, so that very favourable boundary conditions existed for a fast stabilisation of each of the operating alternatives. It was previously proved during the first 5 years of investigation that the landfill sectional cores tested accurately represented the processes in real landfills (Spillmann, 1986).

Mechanical agreement
- The wastes were deposited identically to the investigated landfill with or without a pre-treatment at the same density as in an operating landfill.
- Subsidence ran vertically without straining the cylinders as in the interior of real landfills.

Hydraulic control
- It was proved that hydraulic boundary disturbances were avoided.

- The sealing effect of initially permeable intermediate sandy covers and local stagnation effects of sludge lenses could be produced as in real landfills and their minimum effect was quantified.
- A reproducible relationship was established between climatic water balance, storage process, degradation of organic substance and leachate discharge.

Conformity of biological processes

- All degradation conditions occurring in standard compacted land-fills starting from the strictly anaerobic acidic phase through the strictly anaerobic methanogenic phase, from the consecutive partially aerobic transitional phase to the beginning of the aerobic final phase were properly simulated in the model.
- It was proved that the targeted aerobic pre-treatment managed to achieve an extensive biological stabilisation and permanently aerobic, highly compacted final deposition.

Unlike common operating landfills, air access at the leachate discharge of anaerobic sectional cores was not completely avoidable, so that well degraded constituents that were still detected within the waste body, were not available in the discharge. Therefore, to scale the results up to operating landfills, the same contaminants from within the waste body must be added to the leachate discharges which contaminate the groundwater. The leachate quality in the discharge of the crushed stone corresponds to the discharge of the biological cleaning stage of a sewage sludge treatment plant or a groundwater flow along a length of approx. 100 m (cf. Spillmann *et al.*, 1995).

The long-term investigations over 15 years accurately simulated long-term processes, despite the small model size:

- The long-term anaerobic conservation due to industrial contaminants in the acidic phase was also proved in flat (only about 1.5 m thick) deposits free of perched water and sludge (cf. Chapter 8).
- The transition of the stable methanogenic phase (Section 8.3.4.1) through the beginning of the post methanogenic phase (Section 8.4.4.1) to the atmospheric phase (Section 8.4.4.1) was successfully modelled. A model law was established for the scaling-up of the results to deep deposits (Section 2.8).
- The influence of an extensive recycling (e.g. Sections 2.2.2 and 3.1.1.2) was proved.
- The long-term tendencies of the activity in permanently aerobic deposits ('atmospheric phase') (Section 8.4) were proved.

Overall assessment of the equipment

All stages of all major landfill types occurring or possible in practice were properly simulated both physically and biochemically (acidic phase, methanogenic phase, air phase) as cylindrical sectional cores of flat landfills free of perched water. The results obtained by this equipment can be directly transferred to flat rural landfills and represent the most favourable storage conditions for the respective operating type concerning degradation and stabilisation processes. The same degradation processes will occur at the same or a lower speed in comparable operating landfills. Incomplete degradation, delays and conservation processes and the resulting emissions of landfill sectional cores should therefore be regarded as minimum values for operating landfills. The changes with regard to time can be calculated for deep landfills using the model laws (Section 2.8).

10.1.2 Conformity of the laboratory stability test with the processes in the waste body of real landfills

The laboratory determination of the causes of the increasing chemical and biological instability of plastic wastes within landfills, for the most part stable over the long term, puts the greatest challenge on the model technology. For this purpose, all known landfill conditions were transferred to large-scale laboratory equipment:

- original material, tested in different age groups
- leachates from real soils for the precise agreement of biological colonisation
- exact simulation of heat accumulation within a landfill
- the same solids stress and gas pressure as found inside a landfill
- exact energetic determinations of the initial conditions and energetic difference between exothermic and endothermic reaction phases
- simultaneous measurement of biological activity
- differentiation of microorganisms according to temperature ranges of their vitality
- differentiation of biological and chemical reactions.

The results obtained on primarily inert mixed plastics (Section 9.5) agree with the processes in aged mono landfills. Thus it has been proved that landfill processes, otherwise difficult to simulate at a laboratory scale, can actually be fully and properly simulated. It was proved by simultaneous tests in investigations on plastics that the solids stress and

Long-term hazard to drinking water resources from landfills

pore gas pressure do not affect the increasing instability of plastic materials. Of prime importance was the agreement with real landfills in terms of materials and thermal conditions. The results determined under normal pressure (Section 9.4 and Section 9.5 first phase) can therefore be transferred to landfills.

10.1.3 Limiting the conclusions

There are two principally different approaches for solid wastes to achieve a geologically stable integration into the environment:

- conversion into rock using complete oxidation of carbon
- production of a sediment suited to the location by converting of the organic substances to stable humic substances.

The 1st approach (production of rock-like minerals) was intensely and consistently pursued in Switzerland (obligation of incineration) concerning the long-term effect. Investigations of residues on large-volume deposits were performed and published by the EAWAG/ETH Zurich (Baccini and Gamper, 1994). They have created the basis of current regulations for waste deposit in Switzerland which are used for the concept of future ground water protection by material conversion of wastes (Section 11.4.2).

The hygienically harmless integration of natural wastes into the top soil has been solved by the invention of high-temperature decomposition and has therefore not been included in this investigation programme.

The 2nd approach aims at producing sediment which can be directly integrated into the environment because the strongest rock, available as crushed stone, represents the preliminary stage of soil which is produced by weathering. Its constituents, e.g. heavy metals, are released by means of soil production and are biologically available (cf. details in Baccini and Gamper, 1994). They can only be immobilised again in the sedimentation process. The DFG integrated programme was therefore aimed at looking into the issue of to what extent and when current practice waste deposits can be considered as sediments integrable into the environment.

The following questions had to be mainly answered for the deposits ('landfills'):

- To what extent and within which period of time can stable waste bodies be produced aerobically?

398

- To what extent are industrial contaminants really degraded or immobilised in 'reactor landfills'?
- Can emissions be controlled on the flow path in time at a reasonable cost?
- What advantages does an extensive biological stabilisation offer before final compaction?
- How stable are apparently inert materials (plastics and earth-like fractions)?

The results of reactivation experiments (Chapter 8) changed this first priority. Time scale which, according to these new results depends crucially on the pre-treatment, proved to be of utmost importance for the governing of controllable landfills during stabilisation. The extent and type of chemical contaminants proved to be major additional negative influences concerning controllability in terms of time and material. Stability will therefore be investigated as a prime criterion for the assessment of long-time effects, which influence the emissions of chemical contaminants in various ways.

10.2 Assessment of waste deposits after biological stabilisation up to the production of soil-like substances

10.2.1 Municipal solid waste with sewage sludge as a mix without direct addition of industrial residues

10.2.1.1 Classification of biological stabilisation possibilities of the alternative as tested

The biological pre-treatment of the wastes to produce a geologically stable sediment was mainly developed and implemented by the working group Giessen University Institutes for Waste Management ('Giessen model', Götze *et al.*, 1969; GUfA, 1970). The criteria to reach extensive stabilisation were not adhered to by landfill operators in practice and the success of the cost/benefit ratio was questioned both by the representatives of standard landfills (deposition without pre-treatment) and the representatives of undifferentiated waste incineration.

More than 50 alternatives of biological waste stabilisation were tested at different scales up to large-scale equipment as preparation for and extension to the DFG integrated programme. The results obtained have been published (Chammah *et al.*, 1987; Spillmann and Collins, 1978, 1979, 1981; Jourdan *et al.*, 1982; Spillmann, 1989, 1993a). The most favourable alternative was implemented for the DFG integrated programme both for tests on landfill sectional cores (large-scale lysimeters) and for the test landfill to produce leachate for groundwater investigations (cf. Spillmann *et al.*, 1995). The concentrations of the leachate components from the compacted material met the criteria for discharge into the receiving river according to current provisions. It was not possible to achieve improved results concerning leachate contaminants using currently advertised and technically complex biological treatments applying frequent relocations. These findings therefore also apply to all current biological treatment methods (status as of end of 2007).

10.2.1.2 Stability assessment

The permanently aerobic deposit tested has to be classified according to landfill engineering criteria as very stable since it fulfils the following criteria:

- The leachate components already fulfilled nearly all the criteria for discharge into the receiving river inside the compacted deposit (Section 5.1).
- Nitrogen (N_2) reached the same concentration in landfill gas as in the outside air, oxygen content (O_2) was 50% of outside air and methane was present only in traces despite an extremely high density (Section 8.4.4.1).
- All eluates, including TOC (20 mg/l), corresponded to landfill class I according to TASi (1993) without an extension for organic wastes (Section 8.4.2). (The TASi (1993) contained stricter limiting values than the legal AbfAblV (2001)).
- BOD did not correspond to that of a forest soil, nor could it be substantially increased under optimised laboratory conditions including nitrogen-phosphorus fertilisation (Section 8.4.4.1).

The differentiated physical and biological investigations also indicate an extensively stabilised material:

- There were significant differences between the isotope composition in the leachate of aerobically stabilised material and that of other deposits. Changes in isotope composition due to evaporation were detected up to the base discharge without any influence by degradation processes in the aerobically stabilised material alone. In contrast, influences from degradation processes were detected in all other deposits. However, differences in isotope composition between anaerobic stabilisation in the methanogenic phase and a long-lasting anaerobic conservation in the initial acidic phase could not be detected (Section 4.4.2). Therefore, the crucial step for stability is the permanent aerobic situation.

- The biological parameters of the medium-term investigations resulted in no less than a five-times faster stabilisation by a targeted aerobic degradation in comparison to the most favourable anaerobic degradation in the waste body. Homogeneous humification processes were detected as early as within 2 years (Sections 7.4.1.3 and 7.7.2.1).

- The long-term biological assessment (colonisation, ecotoxicity, gas composition, O_2 consumption and CH_4 generation in the laboratory, Section 8.4.4) indicated a high microbial degradation potential for the consistently aerobic degradation on a stable substrate with the gas generation potential of a forest soil.

In view of the soil-like characteristics it can be assumed that waste management objectives of integration into the environment have been reached and a stable sediment has been produced in terms of soil science. The comparison of the fine material according to soil-science criteria (Section 8.4.5) indicated that although the biochemical inertness of the organic substance resembled an extensively stable humic substance, the molecular structure agreed only with the small-molecular initial phase of humic substance generation and the mineral composition, principally the high trace element content did not match that of a natural sediment.

The analytical stabilisation test of the natural organic substance by the simultaneous use of pyrolysis GC/MS (robust analysis but only indirect conclusions from fragments) and pyrolysis-field ionisation/MS (sensitive analysis of the initial substances) compared with cumulative parameters (Section 9.3) indicated that, despite a large number of non-determinable compounds, the biological waste stabilisation can be characterised unambiguously, particularly by the degradation of soluble carbohydrates. The leachate of the permanently aerobic sewage–sludge waste mix (Lys. 5) contained only traces of carbohydrates. However, the overall picture of the determinable organic substances unequivocally substantiates Husz's assessment (Section 8.4.5) that the substance, although appearing soil-like, cannot be classified as a natural soil.

The analysis of organic chemical contaminants (Section 8.4.3) proved that neither very volatile compounds nor BTX aromatics were detected after the renewed biological activation, which can be compared to the activation for soil generation (Section 8.4.3.1). The HCH isomers from the group of medium and semivolatile chlorinated hydrocarbons remained within a range of 0.05 and 0.10 μg/kg waste DS regardless of the activation in the solids. DDT analogues and PCBs reached 0.3 and/or 0.6 μg/kg waste DS in the solids, chlorobenzenes were not detected (Section 8.4.3.2). Mobilisation was only detected as AOX traces in the leachate. It could not be proved that the semivolatile substances (phthalates, PAHs, triazine) were influenced by the aerobic biochemical degradation (Section 8.4.3.4) but they proved nearly immobile. PAH content values between 500 μg/kg waste DS (ubiquitous background value) and 3100 μg/kg waste DS, 25–280 μg/kg waste DS of simazin and 700–8200 μg/kg waste DS of phthalates (DEHP and DBP dominating) were detected in the solids. It has been proved that substances of this group can be biologically degraded on an industrial scale (Hanert *et al.*, 1992). However, the conditions of both aerobic and anaerobic waste stabilisation proved insufficient in the form implemented so far. Further development is needed to achieve the objective of degrading these materials.

The analysis of the elements (Section 8.4.3.6) confirmed that the content of non-ferrous heavy metals was too high from a soil science point of view. A change in mobility indicated a low level of reactions during reactivation and confirmed the soil-science assessment (Section 8.4.3) as to which this substrate cannot be considered an environmentally compatible soil.

If one reduces the requirements necessary for a complete integration into the environment from the geological point of view to the general current requirement of minimum-aftercare landfilling, the extensive biological stabilisation including nutrient compensation up to a permanently aerobic final deposition fulfils the following requirements:

- The leachates fulfil current discharge conditions into surface water (Waste Water Regulation 2002) including soluble organic substances with a decreasing trend for contamination (Chapter 5 and Section 8.4.3.5) over the long term.
- Non-ferrous heavy metal content of the leachate fulfils the limiting values for drinking water even when the non-ferrous heavy metal contents in the solids substantially exceed those of a natural arable top soil (Section 8.4.3.6).
- Volatile substances including BTX aromatics – if not degradable – are expelled in the high temperature phase (Section 8.4.3.1) and can be captured in adsorption filters. Semivolatile organic compounds

with a solids content typical for wastes are retained to a large extent (Sections 8.4.3.2 and 8.4.3.4).

- The emissions due to conservation phenomena of limited time periods are not expected to increase unpredictably over the long term. Soluble substances – as far as available – are preferentially mobilised by squeezing out the consolidation water at the end of the intensive decomposition (Sections 4.3 and 4.4).
- The biological degradation and change of the organic substances into geologically stable organic humic substances starts before at the end of intensive decomposition (Sections 7.4.1.3 and 7.4.1.4) and continues after compaction without interruption (Section 8.4.2: 10% mass loss within approx. 10 years in an extremely compacted material; Section 8.4.5: biological inertness akin to a stable humic substance by about 15 years after the beginning of decomposition).

It can be concluded from these findings that extensive biological stabilisation leading to an aerobic final state can help capture the volatile substances and those emitted from the flow path including gas and water management in the total landfill system. A sufficiently long aftercare period can also prevent them from polluting the environment. If the biological stabilisation process is not disturbed (no sealing cap, avoidance of stagnating leachate by systematic drainage of the entire waste body, permanent aerobic conditions maintained by gas exchange in the entire waste body by means of natural draught), contaminant peaks due to a biological reactivation are not expected to emerge at a later date after completion of aftercare. The extent of aftercare is moderate for municipal solid wastes with little industrial contamination, if the gaseous emissions are emitted and captured during the intensive decomposition and the leachates already meet the discharge criteria during operation to a large extent, i.e. lined ponds are sufficient to provide long-term biological control. The current view of the length of aftercare being about 30 years is however unrealistic even for this extensively stabilised material. Given the fact that pre-stages of natural humic substances were detected after 2 years of decomposition and inert but low-molecular, thus still unstable humic substances were detected after 15 years, the production of geologically stable, high-molecular humic substances cannot be expected under any circumstances within at least another 15 years. Since, in addition, the content of the waste body does not conform with natural sediments even after the generation of stable humic substances either, the production of sediments that can be integrated into the environment is required as a long-term solution over the geological time scale (Section 11.3.2).

10.2.2 Municipal residual waste without sewage sludge addition and without direct industrial influence

In the supplementary alternative of the investigation programme, deposits of municipal solid wastes with all valuables removed and organically enriched were compared to waste deposits which had undergone an intensive recycling of all useful wastes (recycling ratio: 65% waste DS, no visible compostable wastes) (Section 2.3.2).

No influence from the recycling on aerobic stabilisation could be detected within waste-typical ranges of the measurement results based on landfill engineering criteria. Only the extremely low initial water content of the residual waste needed to be adjusted to a targeted biological degradation (Section 3.2). In view of the temperatures and leachate contaminants (Section 5.3) an extensively separated collection of compostable wastes does not reduce the intensity of the biochemical degradation reactions, in particular heat development. Mass reduction decreased – as expected – proportionally to the reduction of degradable organic substance from approx. 20–10% by weight waste DS. If heat emission is the same, it follows that degradable materials responsible for energy delivery remain largely in the residual waste despite an extensive waste separation (it is known from composting that the addition of paper may increase heat generation during aerobic degradation processes).

Detailed biological investigations (Section 7.6) confirm the agreements. Regardless of the fraction of easily degradable organic wastes, aerobic microorganisms reached the highest population density during the intensive aerobic degradation (Section 7.6.2). The activity characterised by oxygen consumption and gas generation (Section 7.6.3) decreased in both waste compositions to the same extent, with the aerobic degradation being faster than anaerobic degradation. Both the dehydrogenase activity, alkaline and acidic phosphatase, and leachate parameters that characterise stability (Section 7.6.5) failed to provide any proof about the influence of recycling. It was only a faster reduction in autogenous heating capability in the residual waste (Section 7.6.3.3) that indicated a smaller fraction of organic substances was present.

The influence of recycling was clearly proved by the analysis of isotopes (Section 4.4): leachates from deposits without compostable waste contained no water based isotopes from compostable substances, i.e. fresh organic cell material.

An unambiguous indication of an equivalent biological activity between total waste and residual waste and the clearly different origin of the isotopes proves that though the dry organic residual substance of the residual waste differs markedly from the total waste, it is biodegradable to the same extent and activity as the total waste. Therefore the findings compiled in Section 10.2.1 also apply to residual wastes with the restriction that the nutrients have an unbalanced composition in the municipal solid waste without any sewage sludge. This restriction is without any major importance if, in a second step, the material is used as one of several components for the production of an environmentally compatible soil on a concrete stand and based on detailed chemical analysis. If it is deposited, it has to be taken into account that leachates from municipal solid wastes, that were biologically stabilised and then compacted without any nutrient adjustment, have so far failed to meet discharge conditions in terms of dissolved organic carbon (probably peatification).

10.2.3 *Influence of targeted deposits of typical industrial residue on aerobic stabilisation*

10.2.3.1 *Test conditions*

The possibility of degradation of typical, potentially degradable industrial wastes and persistent products, and the immobilisation of heavy metals was tested in three contamination stages which were built on a non-contaminated base (Section 2.6; hardening salt (cyanide), phenol sludge, galvanic sludge, lindane (chlorinated hydrocarbon ring), simazin (atrazine ring)).

It was barely possible to detect the first two contamination stages over the short and medium term based on landfill engineering criteria (decomposition temperatures, leachate contaminants). It was only the non-typical poor degradability of a large amount of organic contaminants in the leachate of the highly contaminated deposit from intensive decompositions that indicated industrial contamination (see details in Spillmann, 1986b). It was only possible to clarify the effects of chemicals on the stabilisation process and on the suitability of landfill systems to achieve degradation and immobilisation by a long-term direct comparison of the same wastes which had undergone the same treatment and by a detailed physical, chemical and biological investigation.

10.2.3.2 Effect of industrial residues on aerobic stabilisation of natural materials

The unfavourable influence of industrial contaminants on the stabilisation process and degradation performance of the waste body due to biological waste stabilisation can be proved by the characterisation of leachate contaminants based on municipal water management criteria (Chapter 5) and compared to the direct introduction criteria and to the eluates according to TASi (Technical Instruction for Municipal Solid Waste) (1993) (Section 6.1.1.4). The following results enable conclusions on the inhibition of biological stabilisation to be drawn:

- In contrast to the leachate from industrially non-contaminated material the contamination with easily degradable organic substances did not drop below the detection limit of $BOD_5 \ll 10$ mg/l and it exceeded the introduction conditions by more than one order of magnitude in the highly contaminated deposit.

- The contaminant level of the base discharges with difficult-to-degrade organic substances with a COD of 500 mg/l considerably exceeded the currently valid limiting values for discharge of leachates from landfills without a decreasing trend (limiting value: COD < 200 mg/l according to Waste Water Regulation 2002). The same conditions were found for TOC.

- The contaminant level of the leachates from the highly contaminated deposit with difficult-to-degrade organic materials exceeded the limiting values for discharge of leachates by more than an order of magnitude (COD $\approx$ 3000 mg/l $\gg$ 200 mg/l, TOC $\approx$ 1300 mg/l $\gg$ 70 mg/l).

- The ratio of organic nitrogen (org. N) to mineral ammonium-nitrogen (NH_4-N) increased and the ammonium-nitrogen was not nitrified in the highly contaminated deposit.

- The sulphate phase has not been reached.

- Even after the degradation of phenol and an extensive decay of cyanide the heavy metal contaminants were able to keep the level of organic contaminants, both in the leachate and in the eluate, clearly above the limiting values of discharge and/or of landfill classes as per TASi (Section 6.1.1.4) without declining.

It was proved that the biologically difficult-to-degrade leachate contaminants were, in fact, efficiently degraded rapidly in the non-contaminated base region when passed through the aerobically stable non-contaminated wastes:

- The high contamination of the leachate with extremely difficult-to-degrade organic compounds ($COD \gg BOD_5$) from the highly contaminated deposit was reduced by about an order of magnitude in the weakly contaminated bottom section. This performance exceeded the degradation performance of a biological sewage treatment plant using common operating technology by far.
- The incrustation potential (HCO_3^-) was substantially reduced in the bottom section by the nitrogen ammonification (pH change from acidity to alkalinity).

The cause of the reduced stabilisation was identified using detailed biological tests (Section 7.4.2):

- The number of (mainly anaerobic) organisms decreased because of concentrated industrial contaminants.
- Cyanide had a selective inhibiting effect particularly on actinomycetes.
- The degradation chain was sustainably impaired by the loss of entire groups of organisms.

The biological investigations during the aerobic reactivation of anaerobic waste bodies (Section 8.5.4) confirmed the results which were obtained by comparing the permanently aerobic waste body without industrial contaminants (Lys. 5) to the permanently aerobic waste body with a high level of industrial contaminants (Lsy. 6):

- A retardation of the activation was detected even in the landfill sectional cores with a moderate to medium contaminant level.
- The deposit with a high contaminant level exhibited ecotoxic behaviour. The luminous bacteria toxicity which is reversible in non-contaminated wastes was maintained in the contaminated waste despite an O_2 supply. Contrary to wastes with no or moderate contaminant level, Daphnia toxicity was also detected in the highly contaminated waste, which was not eliminated by aerobic degradation.
- The high level of heavy metal contaminants most likely sufficed to cause the unstable degradation processes because phenol was degraded and cyanide decomposed during the aerobic degradation (cf. Section 6.1.1.4).

10.2.3.3 Aerobic degradation of industrial residues polluting the environment

The advantage and limitations of the aerobic degradation method as identified in practice were confirmed. The following results have been

407

obtained for non-chlorinated phenols as representative substances for common industrial residues which are potentially aerobically degradable:

- Non-chlorinated phenols (measured as a phenol index) are almost completely degraded even in conditions of extreme contamination.
- Long-term leachate contamination by phenols is negligible in permanently aerobic landfills (Section 6.1.1.3).
- The process could reliably be reproduced during the aerobic reactivation of the anaerobic deposition despite extremely high heavy metal content in the highly contaminated landfill sectional core (Section 8.5.3.4).

The degradation and its limits concerning other test materials must be assessed in a differentiated way:

- Microorganisms existing in the tested waste were not able to degrade cyanide, normally degradable in nature, even in targeted degradation tests (Section 7.4.3.6). However, organisms were detected that can convert the cyanide into innocuous compounds by complexation. The almost complete complexation of cyanide found in the leachate (Section 6.1.1.3) was due to this microbial activity. The long-term decay was due to the instability of this compound.
- It was possible to degrade simazin to insignificant residues in a moderately contaminated waste.
- In the highly contaminated deposit, only the stored amount of simazin was reduced due to the degradation of the organic substance. It was emitted, but not degraded (Section 8.5.3.4).
- Lindane was degraded by aerobic degradation in the moderately contaminated waste at the same half-life as in an arable soil (1 year).
- In the highly contaminated deposit lindane degradation was inhibited, even under aerobic conditions (Section 8.5.3.2).

The negative influence of high contamination has been confirmed by ecotoxicity. Though it was substantially reduced by intensive aerobic degradation, the extremely low ecotoxicity of the non-contaminated, aerobically stabilised waste was not achievable.

10.2.3.4 Relocation of toxic organic industrial products illustrated by the example of the pesticides simazin and lindane

The properties of water-bound mass transfer of the pesticides simazin and lindane in municipal solid wastes have already been described by

Herklotz and Pestemer (1986) in a concentrated form and in the thesis of Herklotz (1985) in detail. The retention capacity of the following substrates was compared in the investigations:

- anaerobically extensively stabilised municipal waste (Lys. 1)
- sewage–sludge waste mix conserved in the acidic phase (Lys. 5)
- aerobically extensively stabilised, sewage sludge waste mix reacting like a forest soil (Lys. 5)
- parabraunerde: standard moderate-sorption soil of the Federal Research Centre for Agriculture and Forestry (Biologische Bundesanstalt)
- low moor soil (standard high-sorption soil 'Hinter Bruch', Gevensleben, organic content similar to the earth-like sewage–sludge waste mix).

The results can be summarised in the following points:

- The initial sorption is proportional to the organic content of the substrate. The degradation of organic wastes towards biological stabilisation is compensated for by the increase of sorption capability of the stabilisation products.
- The retention capacity of the wastes preventing an effective long-term desorption of simazin or other substances of the same solubility approaches the retention capacity of the moderate-sorption parabraunerde regardless of the degree of biological stabilisation of the waste: 98% of simazin was desorbed in ten desorption steps.
- Materials with low solubility such as lindane are retained over the medium term to the same extent as in the low moor. The breakthrough curves of the percolates of all three waste substrates keep on increasing progressively even after 1000 mm rainfall, so that desorption can also be expected over the long term.
- The sorption capacity of extensively stabilised sewage sludge waste mix did not reach the sorption capacity of a peaty field location despite the same content of organic substances. This was confirmed by the soil-science assessment (Sections 8.4.5 and 9.3) indicating that soil-like stable humic substances had not yet emerged.

The results of the long-term tests on the landfill sectional cores (Section 6.2) do not seem at first sight to be conclusive in themselves and contradict the laboratory-scale results in part. However, the relationships could be clarified in connection with the reactivation and degradation tests (Chapter 8) as well as isotope investigations (Chapter 4):

- Lindane and simazin emissions (Section 6.2.1), which were only mobilised after 5 years and detected in all four height measurement levels, were analysed in the leachates squeezed out by the consolidation processes following the compaction of the highly contaminated decomposed deposit (Lys. 10) in the bottom three stages (no, moderate and medium contamination, Chapter 6) (proved by isotope comparison).

- Unlike the substantially lower solubility and desorption capability of lindane in comparison to simazin, both materials were transported equally quickly and lindane was found at a higher concentration than expected according to solubility, because the organic substance transported by the leachate acted as a carrier for lindane (Section 6.2.3.2).

- The fact that lindane and simazin were not found in the leachate of landfill sectional cores with moderate and medium industrial contamination (Lys. 6) before adding the highly contaminated deposit (Lys. 10) can most probably be attributed to a degradation performance equivalent to that of an arable soil in the non-contaminated and/or low-contamination waste mixture (Lys. 6) during intensive decomposition. Degradation was blocked in the highly contaminated deposit (Lys. 10).

The shift of the contaminant centre (Section 6.2.2.3) corresponds qualitatively to the different solubility:

- The contaminant centre of simazin was detected in the initially non-contaminated base about 5 years after adding the highly contaminated deposit and the centre of lindane moved 1.5 m downwards into the moderate-contamination deposit over the same time.

- The fact that contaminants moved into the initially non or moderately contaminated zones over the years suggests that the degradation performance after the end of intensive decomposition is negligibly low for these materials in the non-contaminated and moderately contaminated waste mix in these zones and emissions cannot be prevented over the long term.

- The shift of the contaminant centres accelerated by biological degradation and consolidation was about 0.40 m per year for simazin and about 0.30 m per year for lindane.

It had been expected from the influence of recycling that the extensively separated collection of natural organic substances would reduce sorption. The opposite occurred because the relative content of plastic increased in the residual waste. The adsorption parameters for simazin and lindane (K_d values, Section 6.2.2.1) were at least doubled. Due to large cavities in the residual waste the relocation behaviour of pesticides was clearly differentiated in comparison to the total waste (Sections 6.2.2.2. and 6.2.2.3):

- Simazin with its high water solubility is transported hydraulically faster in the residual waste than in the total waste despite a higher sorption capacity, if the water regime makes hydraulic mobilisation possible.
- A simultaneous hydraulic mobilisation of lindane with simazin was not detected – in contrast to the total waste.
- A shift of the contaminant centre was primarily proved for simazin in the decomposed residual waste (about 1 m per year).
- A shift of the contaminant centre for lindane was not detected within 3 years.

The results of the detailed investigations into the shift of toxic organic industrial products can be summarised in three statements:

- Organic compounds with the same or similar behaviour as the tested model substances (multiple chlorinated circular hydrocarbons such as lindane and simple chlorinated triazine ring compounds such as simazin), which cannot be definitely degraded in a targeted biological treatment, are not likely to be degraded after compaction in the waste body, even if the basic conditions should be suitable for this according to our current knowledge.
- Sorption is intensive but limited in time. This results in a slow shift of the contaminant centre but not in a stable immobilisation.
- Even if the vertical shift is accelerated by a targeted extensive biological degradation of the wastes from high landfills (h > 50 m) the bulk of contaminants of this type will only reach the base by the time that the leachate treatment facility has been out of operation and the currently common technical long-term guarantee of 80 years for the basal liner has expired.

10.2.3.5 Aerobic stabilisation for immobilisation of toxic elements, particularly nonferrous heavy metals

Nonferrous heavy metals can be emitted from galvanic sludges, if they are not deposited in a geologically sufficiently stable form. One expected

411

a stable immobilisation of these elements by the organic substance in mixed landfills. To check this assumption galvanic sludge, which contained a particularly high nickel content, in the highly contaminated deposit was used in this research programme and, compared to the average nonferrous heavy metal content of other galvanic sludges, exhibited not only high toxicity of the elements but also particularly high mobility of nickel. It was therefore particularly well suited for the testing of mobility and immobilisation of heavy metals. It must be taken into account in the assessment of the results that the total content of trace elements in the earth's crust remains constant. However, they are not homogeneously distributed in the earth's crust, instead they are present either in an enriched or depleted form. Some of them are vital (e.g. zinc and copper) and only have toxic effects when their biological availability is increased beyond the natural level.

The measurement of transportation (Section 6.1.1.2) indicates that mainly nickel was discharged during the intensive degradation processes with a high contamination peak (160 mg/l) and nickel contamination in the leachate also exceeded the limiting values for discharge into the receiving stream during the consolidation phase by as much as two orders of magnitude. After consolidation, nickel contamination decreased by about two orders of magnitude and met the limiting values of discharge into the receiving stream despite an extremely high contaminant level within the landfill.

Solid and eluate tests (Section 6.1.1.4) proved that the metals contaminant centre had not been shifted – in contrast to the toxic organic substance. At least the eluate values of landfill Class II were met – with the exception of the highly contaminated deposit. The base met the Class I limiting values. Since the reactivation tests (Section 8.4) showed that biochemical degradation processes cannot be re-activated in an aerobically extensively stabilised material and an accompanied mobilisation of metals is impossible the observed stability state of the moderately and diffusely contaminated base must be considered as a long-term state. In more contaminated wastes, particularly in the highly contaminated deposit, natural stabilisation was limited due to the high level of heavy metal contamination. Though the advantage of biological stabilisation is obvious, this has been achieved at the disadvantage of a higher impact on the environment by organic compounds. Therefore the stability of immobilisation in highly contaminated deposits should also be considered as a proof of efficiency of the method, but not as a practicable removal method for toxic trace elements.

412

10.3 Estimation of long-term effects of wastes in standard landfills

10.3.1 Long-term trend of water and solid balance

The relationships between climatic water balance, water input, mass degradation and leachate discharge measured over the medium term (Spillmann, 1986b) have been confirmed and specified for the long term:

Leachate input

- Due to the mulch effect (interruption of capillary water ascent in the waste) only about 20 mm of a precipitation event evaporates from compacted waste without vegetation cover and with and without sand cover and only about 25 mm from cohesive covers (Chapter 3).
- High potential evapotranspiration does not prevent leachate from being discharged from landfills due to less frequent, but intensive precipitation events over the long term.

Leachate flow in major channels

Seeping water does not develop seepage fronts but flows in preferential channels from which the capillary stores will be filled over the long term. The percentage of direct run-off having short contact with waste is proportional to precipitation intensity. Up to 40% of precipitation input from heavy rain flows off in coarse channels even in carefully layered municipal solid wastes without bulky items (Chapter 4). This enables the following conclusions to be drawn:

- Mathematical models, based on the development of seepage fronts, are not suitable for the calculation of water balance models in waste.
- The fact that leachate flows off from a landfill does not prove that storage capacity is saturated even if up to 40% of an intensive precipitation event is measured in the discharge.

Leachate flow in capillaries

If storage capacity is saturated, intensive precipitation input, mainly from snow melts, disturb the force equilibrium in water-saturated capillaries and trigger a chain reaction in which capillary water stored over several years runs off as leachate (Chapters 4 and 6). Without knowing the isotope composition the process appears to be a fast-running seepage front. Therefore:

- the key transported masses of water-soluble polluting wastes are only expected after saturation of the storage capacity;
- the transported masses of high-solubility materials emerge in batches of contaminants, as soon as the capillary equilibrium is disturbed.

Storage capacity and its change

The water content of long-term storage in municipal waste, regardless of sewage sludge addition, was found to be about 45% by weight, i.e. approx. 0.8 t of water per 1 t of waste DS. Even minor stagnation effects, e.g. minor sewage sludge lenses, increased the water content to more than 50% by weight, i.e. 1 t water per 1 t DS waste (Chapter 3). Storage of residual waste differs from that of total waste not in its final value, but by a lower initial moisture content and thus a larger available storage capacity (Chapter 3).

Degradation of solids reduces the storage capacity to the same extent (1 t DS organic waste can store 1 t water). Since only wet organic substance is degraded biologically, the degradation of 1 t DS organic waste releases about 1 t stored water. In the case of biochemical oxidation of methane in the top landfill zones (poor or non-existing technical degassing) about 0.6 t water is additionally generated per 1 t degraded DS organic waste (Chapters 2 and 3).

The following relationships may be derived from the relationship of storage and degradation:

- With an initial moisture content of, e.g., 25% by weight, a degradation of 20% by weight DS waste and biological oxidation of methane, only about 0.2 t of precipitation water per 1 t of delivered waste dry substance (approx. 0.15 t per 1 t initial wet substance) is stored over the long term.
- Up to 0.5 t of precipitation water per 1 t waste dry mass can be stored until the activation phase, of which only 0.3 t of water is re-released during the degradation and mobilisation phases which can contribute considerably to the release of an emission batch.

Storage and mobilisation as a function of time

Under favourable conditions (no perched water, length of acidic phase about 1 year, methanogenic phase completed in 5 years) approx. 40% by weight of organic dry substance was degraded within 5 years. In this case the degradation corresponded to 20% by weight of the total waste DS, i.e. 0.2 t stored water per 1 t waste DS was released and possibly 0.1 t water was additionally generated anew. No extensive degradation was detected under anaerobic conditions over the long term (Chapter 3). The following forecasts can be established for the large-scale facility:

- If just a 5-m-high landfill activates stored capillary water under nearly ideal conditions after only 5 years, the capillary water from a similar, ideally operated 50-m-high landfill will only be released after 50 years (cf. model laws, Chapter 2).
- Practical landfills are operated less ideally than model landfills (delayed degradation by a longer acidic phase, coarse channels, local stagnation). In Germany, landfills often contain intermediate sealing soil covers (old guideline). Stagnation occurs before mobilisation batches of stored capillary water can flow off.
- Since the safe operation of engineering structures, sealing covers and basal liners are limited in time (large buildings, e.g. bridges, are currently designed for about 80 years), leachate from capillary storage will only reach the base in large quantities by then, when the safe operation of the sealing is no longer guaranteed and the treatment facility has not been operational for a long time.

10.3.2 Influence of water balance on the emissions from industrial deposits

Medium-term investigations (Spillmann, 1986b) on municipal solid waste have already indicated that the emission of water-soluble industrial wastes depends, to a large extent, on the water balance of the waste body, while the transport of materials of low water-solubility is primarily determined by adsorption and desorption. The question had remained open whether the reduction of storage capacity by degradation of organic wastes is decisive for mobilisation, or whether this process only strengthens the effect of storage saturation. This question has been answered by long-term observations, supplemented by measurements on residual waste and by the testing of the degree of degradation after 15 years of storage time.

Mobilisation of permanent contamination of soluble organic substances after storage saturation

Phenol was constantly mobilised from compact industrial deposits (phenol sludge) under anaerobic conditions, as soon as the storage of municipal waste (in this case: 1.5 m waste) was saturated after about 3 years (Chapter 6). The biochemical degradation was zero within the observed period of time of 15 years (Section 8.5, Lys. 9). Converted to a 50-m-high landfill of the same content the emission occurs at the same location after about 90 years at the earliest. In practice, the cover layer and vegetation cover extend the storage phase and thus saturation time is greatly postponed.

Transported masses released in batches

It has been proved for cyanide and other materials of this solubility class that disturbance of capillary equilibrium after saturation of the storage capacity is the key factor for mobilisation (Section 6.1.1.3). Hydraulic discharge peaks also emit the concentration maxima under these conditions (Section 6.1.2), so that the transported masses may increase by more than two orders of magnitude in comparison to emissions in dry weather. The environmental pollution generated can only be reliably measured if representative samples are taken from these batches. The same scale applies to the time-conversion to 50 m landfill height as to the example of phenol.

When there is no vegetation cover, retardation exerts its effect for about a century and it is extended proportionally to the reduction in precipitation input. Unstable compounds such as cyanide can break down under appropriate conditions in the retardation phase. The findings about retarded emission applies to materials of the same solubility and sufficient stability and/or conserving storage conditions for unstable materials.

The release of nonferrous heavy metals, to begin with mobile nickel (Ni), could also be attributed to the hydraulic discharge batch, which occurred in landfills at the end of the storage phase in connection with the biological degradation of organic substances (Section 6.1.1.2). Although the peak discharge batch from the compact lenses of galvanic sludge was substantially smaller than that from the biologically intensively degraded mix, it maintained a markedly higher level than the stabilised material mainly concerning nickel over the long term and substantially exceeded the discharge conditions for nickel into the receiving stream. A direct, reproducible relationship between discharge peaks and concentration increase analogous to cyanide has not been detected, but it was possible to establish one between mobility and biological activity (Chapter 8). This issue will be dealt with in connection with the assessment of biological stability (Sections 10.3.4 and 10.3.5).

Relocation of industrial products with low water solubility using the example of the pesticides simazin and lindane

The relocation of materials with low water solubility but easily adsorbed on organic substances can be characterised by the fact that materials

416

which are moderately dissolved in contaminated zones, are adsorbed when in contact with non-contaminated materials. If the process is repeated over a long time, the contaminant centre shifts despite a low solubility. In particular, the waste in standard landfills is characterised by degradation processes with different degradation rates occurring continuously in the unstable organic substance, coarse channels enable a part of the leachate to drain with little material contact and floating particles become available as contaminant carriers in the leachate. The effect of these waste characteristics on the displacement of difficult-to-degrade and difficult-to-dissolve materials has been cleared up in this test.

Herklotz (1985) published results on sorption behaviour of wastes in connection with medium-term processes (cf. Section 6.2.1). He suggests that the waste tested exhibits a very high sorption capacity for a time-limited storage regardless of sewage sludge content in the deposit. The reduction of organic substances due to anaerobic and aerobic degradation processes did not reduce sorption capacity because the sorption capability of stabilised organic substances increased to the same extent. Unlike a humose arable soil, substances of this kind were also gradually desorbed from the biologically stabilised waste, until the quantity of the percolating water was sufficient for the uptake of all materials with low solubility in the course of time. Under landfill conditions this process is affected by the interaction between the discharge in coarse channels and the release of capillary water as well as changes in adsorption capacities. The measurements on leachate contaminants and substance shift (Section 6.2) provided the following relationships:

- Lindane and simazin, and different materials with the same sorption and solubility behaviour are sorbed in a standard landfill, until the mobilisation phase starts after storage saturation and anaerobic degradation of key components of the saturated organic storage substance (in this case: approx. 20% by weight DS waste of total waste mass).
- Contaminant discharge batches – contrary to highly soluble materials such as cyanide – are not shown as pronounced peaks due to a low solubility, but can be detected by long-term sampling.
- Since leachate coming from coarse channels comes into contact with non-contaminated waste, the contaminant centre is gradually shifted into initially non-contaminated zones. The direct discharge to the base drainage through coarse channels is low in carefully compacted total waste.

417

- The shift of contaminant centres due to the interaction between the discharge in coarse channels and the contact with non-contaminated materials as well as the effect of carriers from the fine material of the contaminated and then degraded organic substances is not directly proportional to different solubility in the total waste of standard landfills. About 10 years after the beginning of deposition, both simazin and the considerably less soluble lindane were detected simultaneously in the initially non-contaminated base at a higher concentration than in the zone of placement (4 m relocation within 10 years).

- In the case of an anaerobic degradation process disturbed by chemical contaminants (acidic phase even 15 years after deposition), no shift was detected for either simazin or lindane in the total waste despite the fact that the storage capacity was saturated.

- The reduction of the natural organic substance fraction by intensive recycling increased the sorption capacity due to a higher percentage of plastics in the residual waste and increased the cavities available for water flow within the material. Thus simazin and materials of similar solubility are relocated more rapidly, while lindane and other substances of equally low solubility are adsorbed proportionally to the higher sorption capacity (no relocation was detected within 4 years).

The following forecasts can be made for the long-term emissions of industrial products with low solubility in water having the same or similar characteristics as simazin (simple chlorinated triazine ring) and lindane (multiple chlorinated hydrocarbon ring) based on the long-term measurements:

- Materials of this type are not immobilised permanently, only retarded for an incalculably long period.

- If these materials are not degraded, their removal from large landfills may take more than a half century.

- In the case of disturbed degradation processes the residence time up to a measurable removal is also incalculably long for flat landfills.

10.3.3 Operation-related biological instability as a cause of long-term emissions

10.3.3.1 Stability assessment of materials at rest

Stability tests on landfill sectional cores at rest without any direct effect by industrial contaminants represented extensively stable wastes

418

according to currently valid stability criteria based on medium-term investigations:

- Leachate contamination at the landfill base, judged by waste-water criteria (Section 5.1), suggests that emissions are low over the long term, regardless of operational technology and sewage sludge placement and only exceed the currently valid limiting values for natural organic substance to be discharged into the receiving stream.
- An extensive recycling, e.g. about 66% waste DS mass reduction, does not change the contamination in leachates per 1 t waste DS (cf. Section 5.4).
- Time-delays in stabilisation, which lead to contamination of leachate within the landfill (Section 5.1), are diminished by reduced oxygen influence – in nature in the porous aquifer near the surface – through a short cut (in this case: in the base crushed stone).
- The results of detailed microbiological investigations on the material at rest (Section 7.4.1) proved that, under favourable conditions (slow landfill construction with no stagnation and a very short acidic phase), biological stabilisation occurs up to the beginning of humic substance generation as early as within 5 years.
- A placement including sewage sludge lenses did not unfavourably affect the stabilisation according to biological criteria.
- The modern thin-layer placement (0.50-m lifts) without any earth cover extended the initial aerobic phase, however did not measurably accelerate the biological stabilisation in comparison to the 2-m-placement technique and applying mineral cover (sand).
- The influence of extensive recycling (Section 7.6) did not influence the process or final result of biological stabilisation.
- The 8-year extensive testing of biological stability of the material at rest (Section 7.7.3) confirmed biological stabilisation within the first 5 years.
- Mass reduction by biochemical degradation amounted to more than 20% by weight waste DS in basic anaerobic deposits and confirmed the extensive stabilisation for this landfill group.

Investigations of a potential methane generation (Section 7.4.1.4), after 5 years of storage, provided a first indication for a possible stabilisation by a re-activatable conservation after the methanogenic phase. The potential methane gas generation of the anaerobically degraded material in the methanogenic phase exceeded that of the aerobically stabilised waste by an order of magnitude, although all other stability indicators, including a mass reduction greater than 20% by weight waste DS approximated the values of aerobic stabilisation. This measurement did not qualify as verification of conservation, since an increasing stabilisation by degradation was still expected with increasing storage time.

A conservation which appeared stable to the outside, was found to be within the acidic range for highly compacted waste deposits, in this case the extremely compacted, anaerobically deposited sewage sludge waste mix (Lys. 3 and 4):

- The contaminants in the base discharges matched the low values of all other landfill types over the long term (Section 5.1).
- The high leachate contamination within the waste was degraded at a very low oxygen influence in the base, thus it was insignificant as an external effect.
- The detailed biological investigation (Section 7.4.1.1) indicated a stabilisation in the range pH < 7.
- The mass reduction was zero within 5 years (Section 3.1).
- The investigations of potential methane generation (Section 7.4.1.4) indicated that the anaerobic degradation up to methane generation was activated to a substantial extent by the change in deposition conditions.

It is known from tests of methane gas generation that the waste stabilised by organic acids is degraded anaerobically, as soon as stagnating acids are diluted and set in motion. This is a new finding that the waste may appear stable in the acidic phase in terms of common stability conditions such as organic leachate contaminants, gas generation of the waste within the waste body and microbial activity.

10.3.3.2 Conclusions from the reactivation tests

The activation tests on the extracted material following the observations on those landfill sectional cores being tested enabled the clarification of important reactivation processes on landfills appearing to be stable in practice and the estimation of the time to biologically irreversible stabilisation.

The reactivation tests (Chapter 8) proved that stability of a material degraded under optimum landfill conditions in the methanogenic phase can be reversed even if it has to be classified as stable according to currently recognised criteria:

420

- The investigations according to waste-management criteria indicated that mass reduction measured after 5 years remained constant within the measuring accuracy in the subsequent 10 years (Sections 8.3.2 and 8.5.2). Thus it follows that the organic mass, which still exhibited a major methane generation potential after 3 years of storage, remained unreduced in the subsequent 10 years.
- The deposit of municipal wastes with sewage sludge lenses classified as 'biologically stable' after 5 years of storage time (Lys. 7) had to be classified as 'stable' during the aerobic activation according to landfill engineering criteria (Section 8.3.2). However, a detailed chemical analysis concerning organic acids indicated that it had to be evaluated in part as a fresh material during hydrolysis (initial phase of landfilling) and was an extensively re-activatable material which was conserved by the lack of structure material as shown by the detailed biological investigation (cf. Sections 8.3.2–8.3.5).
- The sludge-free waste without and with a moderate levels of industrial contaminants, classified as extensively stable at rest (Lys. 2, bottom: no contamination and Lys. 2, top: low and medium contamination), reacted under aerobic reactivation conditions like a fresh material in every respect (cf. Sections 8.5.2–8.5.4).
- Biological stability, achieved by targeted aerobic stabilisation, was only detected in the domestic waste and only at a depth of 1–1.5 m maximum (related to the surface of landfill sectional cores) after approx. 15 years of storage time.

It can be concluded from the results of the reactivation tests that waste-management and in part biological criteria, which are currently considered suitably informative for stability assessment of wastes, provide necessary but not sufficient parameters. The new finding is that not only under acidic storage conditions can waste be conserved re-activatably. Degradation can be re-activated both aerobically and anaerobically even in such wastes where approx. 20% by weight DS has been degraded to methane. This explains why old, gas-pressure-free landfills appearing stable and exhibiting long-lasting intensive gas production endanger neighbouring facilities when building activity, e.g. pipeline construction, interferes with the waste body, and why a 40-year-old municipal wastes deposit reached a temperature of 60°C within 500 hours (temperature increasing), as soon as landfill gas was removed continuously and replaced by air (Chapter 11). The activation potential conserved by anaerobic conditions is maintained until diffusing oxygen creates the same state of aerobic stability that was detected in wastes that had gone through a targeted aerobic stabilisation (Section 8.4). Under the particularly favourable model conditions of Lys. 2 (no stagnation, no sealing layers, no compact sludge bodies, cavities free from sludge and earth, no internal gas pressure after about 5 years of storage time), only 1.5 m maximum of the municipal waste (measured from the surface) reached the stability of the targeted stabilised waste (Lys. 5) in about 10 years after the end of gas production. This finding yields the following minimum time for irreversible biological waste stabilisation based on the model law of biological stabilisation by diffuse gas exchange (cf. Section 2.8.3.2):

$$\frac{T_{stable}}{t_{stable}} = \left[\frac{H_{landfill}}{h_{model}}\right]^2$$

T_{stable} = time of stabilisation in nature [a]

t_{stable} = time of stabilisation in model [a]

$H_{landfill}$ = height of landfill [m]

h_{model} = height in model [m]

$$\min T_{stable} = \frac{10}{1.5^2}H^2 = 4.4\,H^2$$

Time of the acidic initial phase and active methanogenic phase $T_{methane}$ [a] have to be added to this period of time (internal gas pressure prevents O_2 access). Based on this the model body (Lys. 2) of about 5 m height would have reached complete biological stability of the targeted aerobically stabilised sewage sludge waste mix of Lys. 5 in

$$T_{methane} + \min T_{stable} = 5 + 4.4 \times 5^2 = 115 \text{ years}$$

A similar landfill structure of 50 m height with no disturbances, such as Lys. 2 needs at least

$$T_{methane} + \min T_{stable} = 50 + 4.4 \times 50^2 = 11\,000 \text{ years}$$

if the process is not extended by sealing layers, sealing covers and lack of water. The biological stabilisation of a landfill up to a state that excludes potential reactivation therefore requires geological periods of time for anaerobic landfills due to the consecutive conservation, even if methane gas generation is not impaired by disturbances, and it cannot be calculated based on technical scales.

10.3.4 *Influence of industrial deposits on biological stability*

It was proved in the long-term trend of natural leachate constituents of waste-water parameters (Section 5.3) that high industrial contaminations, previously fairly common, produced a stable acidic phase even in biologically very favourable flat and stagnation-free deposits (Lys. 9) whose end was not yet foreseeable after an observation period of 15 years.

The substantially lower contamination of the moderate and medium contamination stage (Lys. 2) had a clear inhibiting effect on biological mineralisation. The high biological degradation performance in the non-contaminated base of Lys. 2 compensated for a part of leachate contamination due to degradation inhibition. The disturbance was therefore particularly provable on the leachate constituents within the landfill: even a minor contamination clearly extended the dissolution of carbonate, which was then precipitated in undisturbed stable layers.

The detailed microbiological investigations (Chapter 7) indicated a decrease in the biological degradation activity parameters by several orders of magnitude for the highly contaminated deposit over the medium term (5 years). Concerning the long-term effects of chemicals they proved that the storage conditions have a larger influence on the necessary time for stabilisation than minor contamination of the first two contamination stages (Section 8.5.5). The diffusing atmospheric oxygen affected the stabilisation of moderately contaminated wastes so considerably that the inhibiting influence of chemicals could not be detected. The extensive inhibition of the highly contaminated deposit was proved in great detail chemically that the number of acid-producers was too small for the acidic phase. The high gas generation rate by aerobic degradation under ideal laboratory conditions was not reduced under practical decomposition conditions. In addition, the highly contaminated deposit – the only deposit in the test – exhibited considerable daphnia toxicity which a change to aerobic degradation conditions was only able to reduce but not entirely eliminate.

The findings on the influence of industrial deposits on the biological stabilisation can be summarised as follows:

- Minor industrial contaminants are also capable of retarding mineralisation.
- High, but common industrial contaminants can almost bring the degradation to standstill for an incalculably long time.

10.3.5 Assessments of different degradation inhibitions concerning future emissions

Natural conservation proved by the reactivation tests is based on the fact that microorganisms assume resting forms due to external conditions, e.g. stagnating water. As soon as the boundary conditions change, e.g. due to building activity intruding into the waste body or by diffusing oxygen after the end of gas production, the natural degradation starts with the associated natural emissions, which consist of minerals and humic substance-like materials on the water path. Both groups of materials are undesirable in the drinking water supply. The duration of emissions is incalculable, because even acidic conservation, which contains almost the entire degradation potential, may appear temporarily stable to the outside and because a considerable degradation potential remains even after the very extensive anaerobic degradation up to methane which cannot be recognised from the outside. Therefore deposits of this type do not fulfil the requirement of not putting a burden on future generations. However they do not pose a direct danger either.

Toxic industrial contaminants, even at minor concentrations, disturb the degradation chain and inhibit stabilisation. Deposits appearing temporarily stable cannot be recognised from the outside in this phase as potential sources of toxic materials. This remains firm to begin with, when zones of industrial residues are placed in a mixed landfill from municipal waste. The substantial difference to natural conservation consists of the fact that the degradation remains further disturbed by the reactivation which can be expected in each case by diffusing oxygen over the long term, even if to a small extent. The inhibition by toxic industrial wastes affects the activation negatively in two ways:

- First of all the disturbance of the natural degradation chain increases the emission of semi-degradable organic materials (residual COD) to a large extent (factor 2 to 10), and both deposited toxic materials and their degradation products are emitted by way of the gas and flow path (cf. e.g. Section 8.5.4).
- Particularly disadvantageous is the effect of the combination of conserving deposition conditions, e.g. compact sludge deposits, with degradation inhibition by toxic industrial products, which leads to a timely incalculable, but activatable halt in degradation. They are common as zones within mixed landfills and form an extremely long-term contamination potential for ground water which is not recognisable from the outside and by far exceeds the calculable life span of engineering safety facilities.

424

10.4   Conclusions from long-term instability of primarily stable plastics

Instability of PVC (Section 9.4) as a representative of plastics with high energy content and modified with additives was expected in large landfills (above 60°C) according to the manufacturer's data about material behaviour (materials sheets). However it is entirely new that under landfill conditions even extremely acid-resistant plastics may react as vehemently as fire (Section 9.5). A microbial initiation of chemical reactions has so far been believed to be impossible, since this group of plastics had proved biologically non-degradable under normal storage and test conditions. Thus it has been proved that an organic substance which is chemically stable in use is unsuited for final landfilling when there is a high energy gradient toward the surroundings. The law of entropy suggests that energy equalisation can always be expected.

These new findings enable the conclusion that fire safety must be tested by a complete simulation of storage conditions for intermediate storage facilities for plastics including potential disturbances. In the case investigated the fire was caused precisely by the microorganisms in the intermediate earth layers which are supposed to mechanically stabilise the plastic landfills and protect them against fire. The failure times of plastic structural components of landfill, mainly sealing membranes, will presumably be shorter than the approx. 80 years as specified by the manufacturers if they are not protected by constructional measures against the combined biological and chemical impact within landfills (cf. principle of 'composite liners' according to the inventor Prof. August, BAM Berlin, in Section 9.5.4 and Fig. 11-31).

11

Application of the results to waste management practice and drinking water protection

Timo Dörrie, Helmut Eschkötter, Michael Struve and Peter Spillmann

Preliminary remark: both landfills and contaminated aquifers were investigated in the research project supported by the Deutsche Forschungsgemeinschaft DFG (German Research Foundation). The final report on groundwater investigation has already been published by Verlag Chemie (Publishing House Chemistry, VCH) as a book entitled *Schadstoffe im Grundwasser, Bd. 2* (*Contaminants in Groundwater, Vol. 2*), edited by Spillmann *et al.* in 1995. The results of both parts of the research project have to be considered together for the practical application of the research.

11.1 Specification of targets and provisions for the methods

11.1.1 Assessment of 'fresh water' to be protected
The cost of fresh-water protection must be covered by the respective country's industries. Specifically countries in arid areas with limited precipitation possess very poor resources; these requirements must be particularly appropriate to the value of the protected water; in this case water with reduced salt content ('fresh water'). Its value can be assessed from its importance for human existence:

- Fresh water is the only 'foodstuff', which is absolutely necessary for life, must be available all times and cannot be replaced by anything else.
- Clean water is necessarily for the production of food products with no health implication and cannot be substituted.
- Clean, if possible low-mineral ('soft') water is irreplaceable for industrial production.

Fresh water is therefore the prime condition for human existence.

In the densely populated humid industrial areas of Central Europe, e.g. Germany, water balance calculations of water management authorities have indicated that groundwater supply is insufficient to supply the population reliably with clean water and to ensure the production of goods due to industry's high demands for quantity and quality of water, unless special measures are taken for the sustainable protection of fresh water. To maintain Germany's role as a reliable business location for industry, the protection of fresh-water resources was included in the Wasserhaushaltsgesetz (WHG, Water Protection Act) based on the 'cause for concern principle' (the danger need only be conceivable, it does not need to exist):

A permission to discharge substances into the groundwater may be only granted if no harmful pollution of the groundwater or another unfavourable change in its properties can be expected. (WHG, 2002, § 34(1)).

Substances may only be stored or deposited in such a way that no harmful pollution of the groundwater or another unfavourable change in its properties may arise. (WHG, 2002, § 34(2)).

All actions, in whose consequences an impairment of the groundwater can be expected, are thus forbidden. Similar laws apply to all states of Europe even if they have humid conditions on the basis of article 4, Paragraph 1(b) of the European Union's Water Framework Directive (WRRL, 2000), and it is also the basis of the European Union's Landfill Directive (EU-DepoRL, 1999). The precautionary principle applies to the current development in time: no burden on future generations.

A quantitative weighting of the influence of wastes as a cause of water pollution was carried out by the Swiss Federal Institute of Aquatic Science and Technology (Eidgenössische Anstalt für Wasserversorgung, Abwasserreinigung und Gewässerschutz – EAWAG)/Swiss Federal Institute of Technology in Zurich (ETHZ) using the example of Switzerland: Lichtensteiger and Zeltner (1994) showed that the amount of gravel deposited annually in Switzerland during the ice age was only about 10% of the current slag deposits from waste incinerators and only about 1% of waste deposits from the building industry. It also follows for other industrialised countries that human activity has represented the greatest geological event on the earth's surface since the ice age and must be evaluated even within the core of the precautionary principle according to geological criteria.

In the average sparsely populated but industrially developed humid areas of the USA or the CIS (Community of Independent States, former Soviet Union), the largest part of the population lives in industrially developed cities and metropolitan areas comparable to Europe, which are similar to Central Europe as far as water management is concerned. The value of water is then to be assessed in the same way as in Europe.

If fresh-water resources have to be protected in humid areas according to the 'cause for concern principle', this approach applies all the more to arid areas.

11.1.2 Assessment of leachate emissions based on the 'cause for concern principle'

Measurements on water balance in waste bodies (Chapters 3 and 4) come to the conclusion that, due to the 'mulch effect' of waste (no capillary rise of water), leachate generated by rare but intensive rainfall will spread, even in semi-arid areas. Leachate generation is much more intensive in less industrialised regions because wastes contain about 70% by weight water in these regions due to a high organic content. This water is removed to a large extent by the overburden pressure and degradation processes and directly contaminates the groundwater in permeable subsoil. Ponds of highly contaminated leachate may develop on low-permeability subsoil even under arid conditions (40°C, dry wind) (Fig. 11-1).

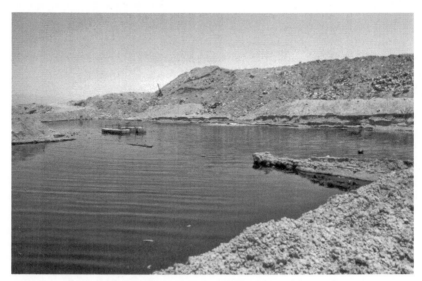

Fig. 11-1 *Highly contaminated leachate pond (1.5 hectare, 180 000 m³) in front of the Tehran Central Landfill, Iran (Körtel et al., 2003)*

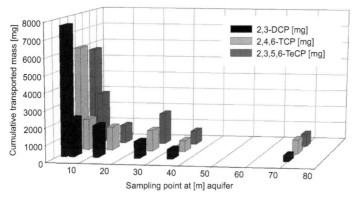

Fig. 11-2 Degradation behaviour of different chlorophenols. (Illustrated as cumulative contamination for dichlorophenol (DCP), trichlorophenol (TCP) and tetrachlorophenol (TeCP)) at six selected sampling points of a model aquifer (Herklotz and Rump, 1995)

The investigations of leachate components carried out in the preceding chapters showed that even such landfills that only contain biologically stabilised wastes of natural origin (elevated substance concentration) contaminate the drinking water resources. Therefore an impairment of these resources by leachate from any landfill may be expected. It has been proved that the extent and duration of contamination from different landfill types differ by several orders of magnitude. The extent is chiefly determined by the content inventory and the deposition conditions (Chapter 10). Groundwater contamination caused by leachate is eliminated only partly by natural cleaning processes (Spillmann et al., 1995; Schneider, 2005). This applies particularly to organic chemicals, whose degradation stops when the special degradation conditions are no longer available which also applies to salts. Herklotz and Rump (1995) proved, for instance, that a considerable degradation of chlorophenol and lindane took place over a 20-m flow distance in the porous aquifer, but the remaining contamination was transported unchanged for another 80 m (Fig. 11-2).

There is, therefore, a justified concern that groundwater contaminated by leachate cannot regain its initial quality simply through natural cleaning processes. Reliable results furnished by this research project on leachate emissions from landfills and their behaviour in groundwater allow the following conclusions for the protection of drinking water resources:

• Due to the extreme bandwidth of potential impairments of water resources a suitable investigation of the hazard potential of existing

429

contaminated sites is the pre-requisite for an effective water protection process using economically available means (Section 11.2.1).

- Technical protection measures only provide protection with a time-limited effect for unstable deposits. They have to be adjusted to the duration and type of emissions (Section 11.2.2).
- Existing waste deposits are unstable and as a rule must be converted after evacuation into materials which can be integrated into the environment and disposed of in a way suitable to the specific location (Section 11.3.2).
- Current and future wastes must be converted into materials which can be integrated into the environment and disposed of in a way suitable to the specific location to avoid an unlimited impairment of drinking water resources (Section 11.4.2).

11.1.3 Selection principle of application examples

The objective of protecting drinking-water resources complying with the cause for concern principle over the long term can be achieved in entirely different ways. Examples will be quoted in the following sections which prove that the targets can be achieved at economically reasonable costs. The authors do not suggest that the most favourable methods have been found in each case. The examples were selected to illustrate the latest technological state in industrialised countries on the one hand and provide potential solutions for the actual targets in a modified form for less developed countries, initially in arid regions on the other (Section 11.1.2).

11.2 Protection from emissions from highly contaminated landfills

11.2.1 Investigation of the risk based on information obtained from the research project

11.2.1.1 Industrial heritage and hydrogeological investigation
It can be concluded from the time-related model calculations of the test results that emissions from unstable landfills persist for centuries. Since extensive mining and metallurgical activity took place in prehistoric time which required considerable amounts of water, contaminated sites can often be found in areas of rising springs in mountains. They contain mobile toxic trace elements, e.g. non-ferric heavy metals which were ignored as associated elements at that time and are now

being washed out from anthropogenic material deposits. If this region is used by industry today, heavy metals can be mobilised by acid discharge from modern systems. Since these wastes were deposited in the proximity of manufacturing plants, they can be found with sufficient local knowledge and tested using simple tools. Thus, for instance, old and present deposits of the historical mining industry and metallurgical centre of the German Reich in the Harz mountains and their foothills were investigated by a team headed by the water management authority of the German State of Lower Saxony. This region is today the most important fresh water supplier of northern Germany. Considerable and, in some places, extremely high emissions of toxic trace elements were found in the foothills. The emissions in the meantime have been successfully stopped.

Since the invention of coal gasification, large quantities of tar and tar processing wastes seeped into the soil, often into gravel and sand, i.e. aquifers. These large accumulations of carcinogenic materials can also be found using industrial archives and historical maps, and the materials can be extracted and later converted. The same applies to wastes of the chemical industry on company-owned landfills and specified hazardous waste landfills.

The situation is different with industrially contaminated municipal domestic waste landfill sites, which were used to fill gravel pits, clay pits or quarries in order to restore the landscape. They are not near the production site and are 'remediated', thus no longer visible or even have buildings built on them. Due to the assumption that organic components of municipal wastes adsorb and 'decontaminate' industrial wastes, 'co-disposal' is still being practiced in this type of landfill worldwide today. (The recent Landfill Regulations 2002 in the UK arising from the EU Landfill Directive have outlawed the process of co-disposal in the UK.) In areas of light industry they may contain the full range of industrial waste, including extremely poisonous compounds such as dioxins (for example the Hamburg-Georgswerder landfill). The cumulative contents in mixed landfills can be calculated from the sum of the produced wastes minus locally known amounts of waste in company-owned and hazardous wastes landfills. Because of its importance, this type of landfill (Lys. 2 and 9) was investigated in the integrated research programme sponsored by DFG and the assumption of 'decontamination' was clearly disproved. In order to find these waste deposits with the aim of examining their effect on the drinking water and perform protection actions if necessary, precise local knowledge is required. The inhabitants and officials of a German industrial city known to the authors have

431

found more than 300 contaminated sites within the city, whose contents varied from industrially non-contaminated domestic waste to hazardous waste.

To be able to make appropriate decisions about the necessary extent of water protection measures, studies of the local inventory (following section) and hydrogeological information on the fresh water to be protected are necessary because of the extremely large bandwidth of the emissions.

The existing hydrogeological conditions have to be researched using industrial archives and historical maps. The necessary information is available in developed countries and can be retrieved from the national geological offices. Even in less developed countries, knowledgeable geologists can provide data about the most important drinking water resources and their supplies. The connection of water resources with the emissions of the deposits provides the picture of the potential risk. Austria, for instance, uses this principle in the remediation of the Mitterndorfer Senke, a 150-m-deep sand/gravel trough in the Vienna basin, one of the largest central European groundwater reservoirs.

11.2.1.2 Local determination of the current groundwater contamination
From the 'Schadstoffe im Grundwasser' (Pollutants in Groundwater) integrated research programme sponsored by DFG, the following information can be derived for the determination of mass transport which is important for measurement issues (Münnich, 1995):

- Salt-containing water sinks to the aquaclude even in a homogeneous silt-free fine to medium sand (typical northern German aquifers) with only approx. 1-m depth of water. The transversal dispersion is so low that even after a flow distance of 100 m a clear stratification can be measured.
- The organic contamination is the key factor in determining longitudinal dispersion of the salts: extensive dispersion of the Li^+ tracer in clean groundwater, a sharp Li^+ peak in connection with leachate from waste.
- Linear sampling is not satisfactory for the determination of local contamination even in a flow cross-section of $w \times h = 1.10\,m \times 1.00\,m$ – precise, simultaneous and flow-parallel raster sampling must be used (Fig. 11-3).
- The substance concentrations in the samples only agree with those in the groundwater when water is pumped slower from the sampling wells with cross flow than the flow velocity of water in the well.

432

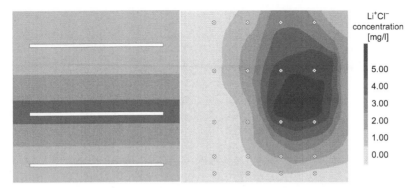

Fig. 11-3 Behaviour of a lithium chloride tracer in a model aquifer. Computer illustration of simultaneous and flow-parallel measurement of lithium concentration distribution in mg/l using linear (left) and punctiform raster (right) sampling (Lhotzky and Spillmann, 2002)

Complying with the currently recognised rules of groundwater control, water is pumped from control wells until the guide parameters – temperature, conductivity, pH value – become constant and then samples are taken. However, since the contamination also moves within an aquifer along path lines, this method only provides correct results in cases where the entire aquifer is homogeneously contaminated. This, however, is not the common case. Wells with sampling points at different heights (multi-level wells) are only used in special cases because of high building and operating costs.

Alternatively, sampling methods have been developed and used which rely on the results of this research project. They provide far better results with the same costs than more common well sampling techniques (Lhotzky, 1997). The different path lines, which traverse a control well, usually remain separated to such an extent within the casing that they can be selectively determined using profile measurement without a packer (Fig. 11-4). Temperature proved the most sensitive parameter (resolution $\pm 0.01°C$), followed by conductivity (resolution $\pm 0.1\,\mu S/cm$) and pH value (resolution ± 0.01). In order to ensure a turbulence-free measurement and sampling in 2-inch observation wells, hydrodynamically streamlined slim multi-parameter probes and sampling containers ($D \leq 30\,mm$) have been developed, which enable smooth measurement and sampling at 25-cm intervals without any judder down to a depth of 200 m (Fig. 11-4, right). First, the sampling depth is determined by a profile measurement, then a sample container rinsed with an inert gas and pressure is lowered under inert gas to the sampling depth causing as little turbulence as

433

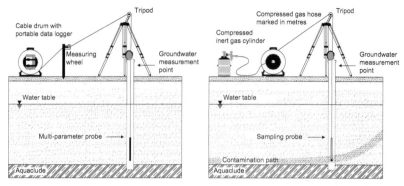

Fig. 11-4 Groundwater measurement system according to Lhotzky (1997): (left) packer-less profile measurement; (right) targeted measurement of individual contamination horizons including undisturbed sampling

possible. This method enables the taking of a representative sample, even with very volatile substances. To actually take the sample, gas pressure is reduced below the local pressure of the sampling point, which pushes the surrounding water into the sample container through a dirt filter. Then the inert gas pressure is increased again closing the intake valve and the sample can be removed. The sample container remains closed until the laboratory measurement starts.

In the following, results of a differentiated sampling will be presented as an example of a successful investigation into groundwater contamination from a large, old leaking landfill in the northern German lowland. In the area of the landfill – constructed without a liner on a loose rock – Quaternary overlies Cretaceous. The layers of the aquifer (moorland peat, fluviatile sand and gravel, glacial sand and blown sand) are altogether approx. 15 m thick and rest upon an aquaclude with very low permeability.

The results of a profile measurement of the guide parameters (temperature, conductivity, pH value) for a perfect well near the landfill are illustrated in Fig. 11-5 as an example. All three parameters exhibit a sudden change over the aquaclude, clearly indicating salt-containing leachate in an approx. 2.5-m-thick layer with sharp contours. A constant temperature increase from a depth of 10 m, however, allows the conclusion that the landfill also exerts its influence outside of the salt-containing zone, which must be tested separately.

It was the profile measurement that identified the layer above the aquaclude horizon as a layer of special interest. Figure 11-6 shows the conductivity and the chloride, sulphate and ammonium concentration distributions after layer sampling of the layer directly above the

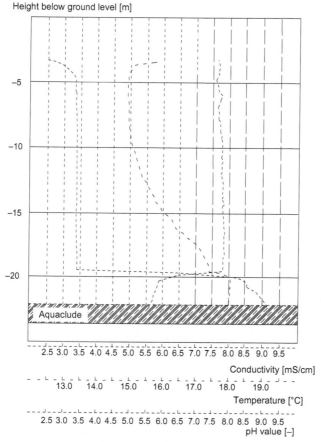

Height below ground level [m]

Conductivity [mS/cm]

Temperature [°C]

pH value [–]

Fig. 11-5 Profile measurement for risk assessment to groundwater by a leaky landfill in northern Germany. Illustration of the elevation profile of a downstream probe; continuously recorded conductivity, temperature and pH values (Lhotzky and Spillmann, 2002)

aquaclude surface. The conductivity isolines (Fig. 11-6(1)) – unlike the main flow of the groundwater – indicate a local transportation perpendicular to the main flow direction – induced by a groundwater pumping well. The area with the maximum chloride concentration (Fig. 11-6(2)) is equal to the old landfill area. The sulphate concentration (Fig. 11-6(3)) clearly differs from the distribution properties of chlorides and lies directly beneath the new disposal area of the landfill. Large amounts of demolition waste were deposited in this area. The ammonium distribution (Fig. 11-6(4)) near the landfill clearly deviates from the characteristics of the main groundwater flow and from the distribution of salt contamination. The spread coincides with the direction of the

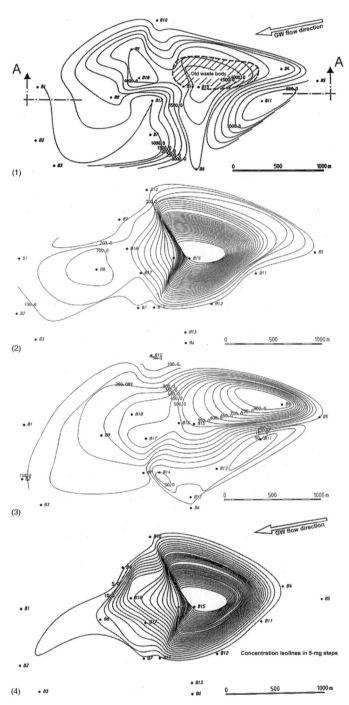

Fig. 11-6 *See caption opposite*

general groundwater flow only with increasing transportation and dilution; this means that considerable flow disturbances occur in comparison to the natural groundwater flow of the tested layer. The monitoring of the emissions can only provide proper results if these flows are measured accurately.

The following conclusions can be drawn for the practice from exact layered sampling of individual path lines using profile measurement:

- The transportation of toxic or carcinogenic trace substances does not evade the analysis by dilution in the sample.
- It is sufficient to monitor the location and direction of path lines already characterised chemically by regular profile measurements and to analyse the key contaminant flows once or twice annually.
- The high spatial and analytical accuracy of the investigations provides a proper contribution to 'natural attenuation' for long-term monitoring.

Thus a higher safety is obtained and guaranteed with a simultaneous substantial reduction in costs.

The profile measurement was further developed for fine to medium granular loose soils (flow valleys) by profile analysis to determine in situ volatile organic contaminants (Bracke, 2002). The analysis is based on a probe which is connected through a control and capillary hose system to a mobile laboratory. The measurement takes place online by means of a membrane interface probe (MIP) which can be coupled to a mobile gas chromatograph with several detectors (PID/FID/DELCD) (Fig. 11-7). Organic compounds in the saturated and unsaturated soil zone are thermodesorbed ($130°C$) by an integrated heating block, while the probe is vibrated into the soil. The mobilised evaporable compounds are adsorbed through a heated polymer membrane and transported into the interior of the probe. After diffusion through the membrane, the contaminants are transported by means of a high-purity carrier gas (N_2) through coupled capillary hoses to the detectors and measured online as total cumulative pollutant parameters (PID/FID/DELCD) semi-quantitatively and/or as single substances of

---

Fig. 11-6 Illustration of groundwater contamination around a leaking landfill in northern Germany. The concentration isolines (conductivity, Cl^-, SO_4^{2-} and NH_4^+) are shown for the plane over the aquaclude (aquaclude at $-20\,m$, cf. Fig. 11-5) (Lhotzky and Spillmann, 2002): (1) horizontal distribution of conductivity; (2) horizontal distribution of chloride concentration; (3) horizontal distribution of sulphate concentration; (4) horizontal distribution of ammonium concentration

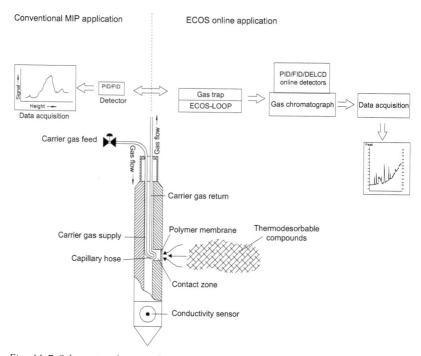

Fig. 11-7 Schematic of a membrane interface probe (MIP probe) for depth-oriented measurement of gaseous components with a conventional FID/PID detector (sum parameter) (Bracke, 2002). Left: conventional MIP measurement in comparison with the online application of the ECOS Company to determine single substances; Right: MIP GC ecosLOOP detector

very volatile substances (MIP GC ecosLOOP detector). Lithological conditions and soil classification can be simultaneously determined by the probe (conductivity by geoelectrical methods, compactness by penetration resistance measurement, thermal conductivity and thermal borehole resistance by heat energy flow in continuous thermal response measurement). A direct picture of the contaminant distribution is obtained by combining the detector measurements.

The advantage of the method is first that an almost complete picture of the organic contamination is obtained in the individual layers immediately during probing. The probing can therefore be carried out in the form of a result-dependent compression of a net instead of a fixed raster. Driven filter wells can be tested for characteristic layers. Fully equipped filtered drilled wells should only be sunk – if at all – in the key profiles. The method therefore reduces the investigation costs considerably in the subsoil in which probes can be driven, simultaneously increasing the reliability of the investigation.

The chemical analyses and above all their interpretation are always afflicted with the uncertainty that not all biologically active substances could be determined and that their interactions were not properly assessed (see Section 11.3.2.2). Therefore measurement methods are currently being developed that can directly show even low toxic or carcinogenic effects of current contaminations on living cells of warm-blooded creatures.

Tests are currently being developed within the 'Complex and Cellular Sensor Systems' Innovation College of DFG headed by the Institute for Cellbiology and Biosystems Engineering (Professor Weiss, Rostock University), to enable a direct characterisation of endogenous and exogenous neurotoxic effects. Changes in the electrical activity of biological neural networks will be determined quantitatively and qualitatively. Since communication in multi-cellular neural networks is the most important function of the nervous system, disturbances and/or changes in the development and function of the nervous system can be detected better and more susceptibly using an electrically active network which possesses all substantial transmitter receptors, than in cultures of single cells. This is substantiated by neurotoxicological investigations on heavy metals, organometallic compounds and pesticides, which, depending upon their mechanism of action, induce characteristic changes in the network activity. Clear impairments of the electrical activity occur at concentrations as low as 1/100 to 1/10 of cytotoxic concentrations. This system will be applied in the future for the characterisation of neurotoxic substances.

Carcinogenic materials induce changes of the genetic material, which can manifest themselves in the most diverse ways. Here, in principle, the mechanisms of gene mutations, chromosome mutations and genome mutations can be distinguished. These three mechanisms can, either in combination with each other or individually, cause malignant cell degeneration up to cell transformation and tumour cells. The determination of one of the three biological final stages does not yet give information about the carcinogenic potential of a foreign substance or mixture with sufficient safety. Therefore a micro core test for the assessment of carcinogenity has been modified within a landfill investigation headed by Professor Schiffmann, Institute for Cellbiology and Biosystems Engineering, Rostock University, to enable the recognition of changes of the genetic material on as broad a basis as possible (see Fig. 11-22). In addition to direct DNA damage, in essence, this test is also capable of indicating the effect of so-called 'non-gene-toxic' carcinogens. These are materials that exert their effect without any

439

direct influence on the DNA. The potential to determine such damage represents a substantial advantage with respect to conventional mutation tests (e.g. Ames test, DNA repair). Since this test has been well studied (Fritzenschaf *et al.*, 1993), it is possible to make first comparisons of the results which were obtained under identical physiological conditions (metabolism, bio-availability). This, in particular, applies to the investigation and comparison of very complex mixtures (e.g. Diesel particulate extract, landfill gas). Currently, the test is being adapted to the specific conditions of the preliminary exploration of contaminated sites.

The direct measurements of toxicity and carcinogenity expected in the future, coupled with profile measurements and profile analyses, require only low capital costs in relation to the information supplied by the investigation. Drilling of filter wells is, however, expensive. This method is not needed for investigations in a soil in which probes can be driven. The quality of the results remains the same when for instance perforated probes ($\frac{3}{4}$-inch dia. perforated PE pipes) are driven into the ground as water-level gauges instead of well casings (Fig. 11-8).

11.2.1.3 Determination of the contamination potential of the deposits (inventory)

It can be concluded from the results of the DFG long-term investigation that the emissions found in the groundwater do not need to correspond to the inventory whose emissions are expected in the future. Therefore the inventory, its state of activity and the re-activation capability of the degradation processes must be tested to be able to assess the contamination potential of a deposit. Drillings and/or probings necessary for this purpose correspond in principle to groundwater testing and may be supplemented by test pits. The authors believe, however, that in cases of high risk – such as the Hamburg Georgswerder landfill – a driven pipe system without material removal should be preferred to common drilling (Fig. 11-9).

Essentially, the gas has to be investigated analytically in the waste body. The main gases CH_4, CO_2, N_2 and CO_2 – measured in situ using portable multi-gas analysers – provide information about the state of activity in the deposits. Toxic trace gases enable conclusions to be drawn on the inventory of toxic or carcinogenic organic manufactured products. The currently available analysis technology provides the most detailed test results at minimum cost, when one proceeds in three steps (Spillmann, 1995b). In the first step, i.e. the orienting

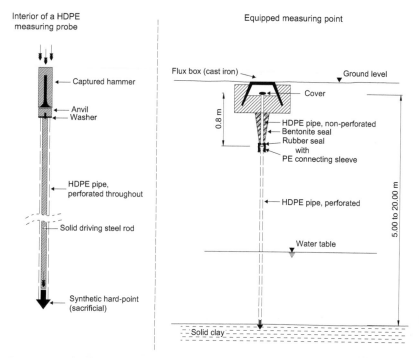

Fig. 11-8 Cost-efficient driven core test with casing performed using a pneumatic driven probe. Driving principle with a solid steel rod inside; a captured hammer hits a massive synthetic hard-point, not a PE pipe (left), fully equipped measurement point (right) (Spillmann and Meseck, 1986)

measurement, a rough raster is tested using measuring tubes, which detects the sum of common industrial chemicals and serves for the determination and first characterisation of contamination centres. In the second step, i.e. the identification measurement, all gases of the probes are determined, for which similar gas spectra were obtained in the first step, using the enrichment method (upgrading in one sample in each case). Adsorbent resins are used as enrichment systems for very volatile substances, activated carbon tubes for medium-volatile materials and high-purity, deep-freeze acetone for semivolatile gases. A susceptible trace analysis (e.g. GC-MS) then enables a full analysis of the large number of potential toxic trace gases due to the much reduced number of samples. Only in the third step, i.e. the localising measurement, the few key trace materials are identified by individual tests in a raster refined toward the contamination source in order to precisely locate the position of the contamination centres. Thus it is

441

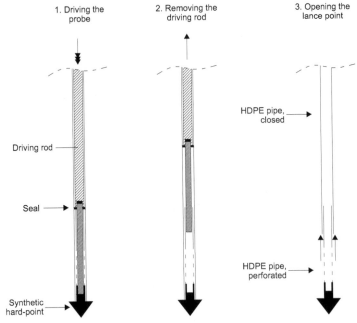

1. Driving the probe

2. Removing the driving rod

3. Opening the lance point

HDPE pipe, closed

Driving rod

Seal

HDPE pipe, perforated

Synthetic hard-point

Fig. 11-9 Closed driven probe for testing in highly contaminated areas (Spillmann and Meseck, 1986)

ensured that the extent of the expensive full analyses is reduced, without running the risk that certain contaminants remain undetected.

The tests on the groundwater under the landfill and on perched water in the landfill correspond to those on the surrounding groundwater (see above).

11.2.2 Technical immobilisation with time-limited effect

A short-term conversion of a large number of materials in landfills into materials which can be integrated into the environment, and the numerous contaminated sites (about 100 000 contaminated sites alone in Germany (UBA, 2000) raise high demands for a developed industrial nation (even more for a developing country). Therefore it is necessary to protect the drinking water resources (supply) against contamination from wastes until the conversion of these materials is complete. The manufacturers currently give a maximum period of guarantee of about 90 years for the technical measures needed for this purpose. Since immobilisation with a time-limited effect does not solve the problem, its costs for a practically ineffective measure must

442

be weighted against the costs of a material conversion (Section 11.4.2).

For the selection of material and design of the containment structures it has to be born in mind that although water contaminants are transported by water from the landfill, they can also spread from the leachate by diffusion through water-sealing structures. Furthermore, it has to be considered that structures that are waterproof in terms of hydraulic engineering always have a residual permeability and that a premature failure of the structures due to the not fully understood damaging capability cannot be excluded. A structure for a subsoil suitable for driven piles, which meets all requirements if the containment can be seated into the aquaclude, can be constructed with sheet-pile walls based on state-of-the-art techniques of civil engineering.

An annular space between two rings of sheet-pile walls seated into the aquaclude is divided into chambers by cross walls and the water levels drop from the outside inward (see hydraulic principle in Fig. 11-12). The precipitation is collected and drained over the enclosed area. The method has the following advantages: steel is impermeable to water flow and diffusion of all known substances. The gradients directed inward produce a flow in permeable spots, e.g. locks, from the outside inward. Therefore the technique is 'fail to safe'. Diffusion from the inside outward at faulty spots against the flow is impossible. Extensive failure, e.g. due to corrosion, can be located immediately by an internal rise of the water level in the respective chamber and repaired accordingly. Any flow in fissures and fault zones of the aquaclude is reversed, so that spreading contamination is also returned to the contamination centre. A contamination of the surrounding water is impossible with this method, as long as the system is properly maintained.

Weak points of the containment by steel sheet piles are the costs and corrosion sensitivity of steel. A corrosion-resistant and impermeable structure is obtained when the chamber system is built with cut-off walls and an HDPE geomembrane is included in the internal ring (Fig. 11-10). A diffusion-resistant version requires the inclusion of metal plates. Depending upon actual contaminant spectrum, the metal plate ranges from aluminium to high-alloy special steel.

If the contamination or the protected objects do not raise maximum demands from the diffusion barrier, the investigations of Münnich (1995) suggest that cut-off walls can also provide a barrier that is impermeable not only to convection but also to diffusion, if the grain size distribution of the sealing compound enables a minimal flow from

Fig. 11-10 Construction of a composite cut-off wall. Cut-off wall with an HDPE geomembrane for the containment of the Hünxe/Schermbeck household and hazardous waste landfill (Source: Archive of Bilfinger Berger AG, NL Spezialtiefbau (Division of Special Civil Engineering), Mannheim)

the outside inward. For sealing compounds with high porosity and stagnating pore water, he proved with metal ions that the diffusion flux outward is in the same order of magnitude as a convective transport. The containment with conventional cut-off walls without a diffusion barrier is therefore only successful if the environmental impact due to diffusion is insignificant.

A chamber system preventing convection was implemented for the first time in 1984/1985 by Kiefl[1,2,3] and Radl (1991) using cut-off walls. This design is known as the 'Wiener Dichtwand-Kammersystem' (Vienna chamber system) and is today a state-of-the-art construction practice (Figs 11-11 and 11-12).

If the aquaclude is at such a depth that a containment cannot be reached at reasonable costs, it is technically possible to reduce the vertical flow cross-sections by grout injection into the subsoil so that the level difference from the outside inward can be maintained without high pumping costs. The salt-containing vertical leachate flow is then reversed in the grouting zone. In deep groundwater, however, the leachate may have reached great depths before the containment was constructed, so that remediation by means of wells may be necessary even after an expensive and nevertheless imperfect containment. In such cases it has to be checked to see if it is better from the economic point of view to stabilise the situation by a system of wells and use landfill mining (see details in Section 11.3.2).

It is not reasonable in all cases to construct the containment alone using impervious structures. Applying a groundwater catchment and

444

Fig. 11-11 *Construction of the chambers of the Wiener Dichtwand Kammersystem (Vienna chamber system) using cut-off walls for the containment of the Rautenweg landfill, Vienna (Source: Archive of Bilfinger Berger Bauges. mbH (Building Company), Vienna)*

bypass system in conjunction with a barrier can often ensure a better result and cost reduction in comparison to solely impervious barriers (Spillmann, 1986a). For instance the most effective way of draining a backfilled pit can be the construction of an impervious capping and shedding rain water. The groundwater bypass system is easier to construct and maintain than a large-area liner which has to withstand water pressure (Fig. 11-13).

Also, in a fissured subsoil under a horizontal or slightly sloping terrain, e.g. in clay pits in mudstone filled with waste, artificially created drainage channels (bypasses) can provide a more effective protection from emissions than capping systems. A circular ditch equipped with a pipe, which cuts through the water-bearing cracks, can reroute groundwater almost without resistance along the deposit ('hydraulic short-circuit') thus preventing a flow through the waste (Fig. 11-14). If the surface of the waste is sealed against precipitation and the water level in the waste lowered, the fissure water flows both from the flanks and the base to the centre of the waste deposit, and the water pumped from there can be clarified. This arrangement prevents the spread not only by convection, but also by diffusion, as far as the

445

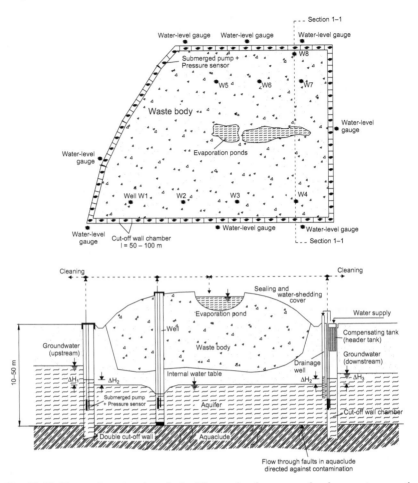

Fig. 11-12 *Plan and section through the Vienna chamber system for the containment of the Rautenweg landfill (adapted from Kiefl[1,2,3] and Radl, 1991)*

rock is a sufficient seal against diffusion. In addition to the circular ditch (French drain), the targeted sealing of some of the cracks may be useful, if the costs of dewatering and clarifying are substantially reduced in this way. However, this must be implemented in such a way as to prevent diffusion. Otherwise the full effect of the hydraulic barrier is lost. The hydraulic short circuit relieves the flow pressure even in the case of a large level difference.

The examples shown indicate that it is possible to construct a landfill cap even under extremely difficult conditions. An effective emission protection can even be achieved by simple means under the

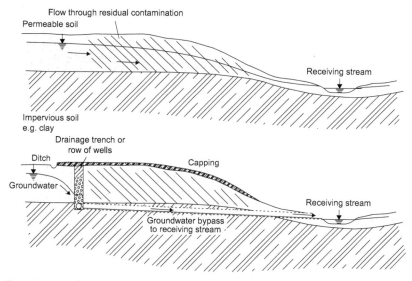

Fig. 11-13 Hydraulic containment of a landfill on a slope with water flow (Spillmann, 1986)

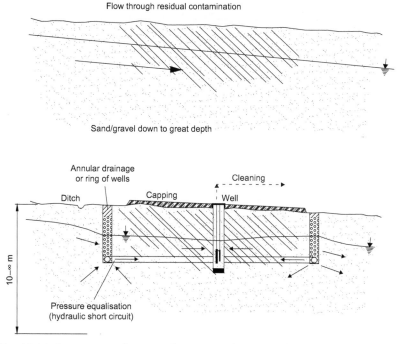

Fig. 11-14 Containment of a waste deposit in a deep permeable soil by constructing a 'hydraulic short circuit' (Spillmann, 1986)

requirements of municipal solid waste with no industrial contamination. Since this protection is time limited, the costs of capping – i.e. those of the postponement of the solution – should always be compared with the costs of the solution, i.e. material conversion of the wastes.

11.3 Long-term removal of contamination risks

11.3.1 In-situ stabilisation

11.3.1.1 Conditions of application
The reactivation tests (Chapter 8) have proved that degradation processes and accompanied emissions in a biologically stabilised waste cannot be activated any more, if a permanent aerobic state was reached in the deposit. Since the current state of sewage treatment technology enables environmentally polluting leachate emissions to be reliably eliminated, an extensive biological stabilisation of the waste under the protection of a technical immobilisation makes it possible for the emissions to be concentrated during the operational time of the sewage treatment plant (Section 11.2.2.). This protection measure – performed as an in-situ stabilisation – may be sufficient for deposits with a low toxic potential and low risk for usable waters, if the following conditions are taken into account:

- The sufficient supply of oxygen to the active zones by suitable technical systems is the key factor for a rapid and controlled conversion.
- The continuous, totally controlled removal of gaseous and liquid reaction products (e.g. discharged landfill gases, organic acids) and the released energy must be permanently guaranteed in order to prevent an unfavourable stagnation of the biological processes (acid preservation).
- It is necessary to evacuate accumulated water by pumping to remove and/or avoid a wet-preserving environment (moorland preservation, see Section 4.4.1, Lys. 4).
- Water evaporated by the biological autogenously heating process and removed by gas exchange must be replaced in order to avoid a dry preservation of the waste (mummifying).
- Nutrients that are in short supply and therefore lead to a stagnation of degradation (usually phosphorus) must be replaced.

In order to meet these basic requirements for controlled mass transport and evacuation, the following considerations must be applied for the successful execution of the in-situ reaction-to-entirety.

448

11.3.1.2 Fundamentals of mass transfer in the waste – a heterogeneous porous medium

Large-scale mass transfer for conversion and promotion of aerobic degradation occurs by convection in the large cavities of the wastes. Mass transfer to the biologically active cells takes place in capillaries and cells of the materials themselves, i.e. by slow capillary flow or diffusion. Here the materials that have been delivered and removed by the large-scale flow are distributed. Considering that up to 20% by weight of the dry mass of waste is converted into gas biochemically (see Section 3.1), a controlled, large-scale gas transport is a prerequisite for the conversion and acceleration of the degradation processes.

A convective gas exchange with a constant flow through the inlet and outlet lances produces a laminar flow with a parabolic velocity profile in water-free channels outside of the direct range of the lances. Since the flow velocity at the wall is always zero and it increases toward the axis of the channel proportionally to the square of the radius (parabolic distribution), gas exchange is concentrated in the centre of large channels. If a preferred discharge develops through macropores due to heterogeneity of the waste (bulky waste, demolition waste inclusions, commercial wastes), the effectiveness of the aeration measure is almost zero. In addition, oxygen-rich air is concentrated in the centre of the large channel due to the parabolic velocity distribution. Convection is equal to zero at the channel walls, in dead-end channels or in channels blocked by water. For the biochemical degradation, however, the diffusive gas exchange at the walls, in particular in small and dead-end channels, is the key factor. Its effectiveness depends exclusively on the concentration gradient of the gases according to Fick's law (Formula 2.12).

A gas exchange with a constant pressure difference and injected air (only 20.9% by volume of O_2) produces a small concentration gradient of oxygen at the surfaces that are decisive for the mass transfer and approaches zero even after a small O_2 consumption. An increased gradient can only be achieved by enriching the supply air with oxygen. The oxygen enrichment is however only efficient if the high concentration is not limited to the centre of large channels with its majority being expelled. If the supply air is compressed and introduced intermittently at the velocity of sound, the flow spreads out in all channels at the same velocity (principle of pressure surge). The flow is turbulent, so that the maximum oxygen concentration reaches the walls directly. Also, after reduction of the high initial velocity, turbulence maintains the advantage of wall contact over a long distance.

The surge-aeration therefore provides ideal conditions for the aeration of highly compacted wastes and an efficient enrichment with oxygen without large losses.

This principle was first implemented in 1990 by the first technical gas-conversion of a demolition waste and MSW (municipal solid waste) landfill for the purpose of landfill mining (see Section 11.3.2.4). Spillmann (1991) suggested various solutions for aeration and ventilation of old waste bodies (Fig. 11-15). The aeration measure was started with a modified continuously operational in-situ aeration and ventilation system used in soil remediation (Fig. 11-15(1)). Increasing inconsistency of the waste to be aerated – short-circuit flows through areas with low flow resistance – forced the system to rapidly reach its limits. Ranner (Spillmann *et al.*, 1992/1993) then developed a method which is capable of intermittent impulse injection of air or air enriched with oxygen (Fig. 11-15(4)). This pressure surge technique shoots air into the waste body like an explosion (2.5–6 bar). The pressure wave spreading spherically in the waste proved particularly suitable to penetrate the waste body homogeneously. This physically justified advantage of the pressure surge technique inevitably

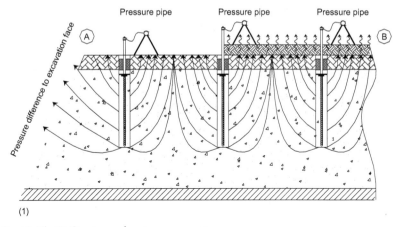

(1)

Fig. 11-15 (1) Continuous low-pressure aeration
Simple aeration for landfills without industrial contamination. Fresh air is pumped into the waste body through lances. Contaminated exhaust air is cleaned by a biofilter blanket (B), if necessarily, in combination with activated carbon filter blankets.
Advantages: Simple structure, relatively low power consumption.
Disadvantage: In addition to the disadvantage of laminar flow with regard to the gas exchange described, it is inevitable that air injected with a low pressure selects the path of least resistance. Compacted or heavily soaked areas therefore remain unaffected and do not take part in the rotting process. An improvement of the situation can only be achieved by increasing the amount of air injected. This however bears the danger of desiccation.

450

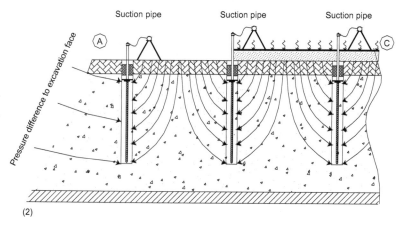

(2)

Fig. 11-15 (2) Continuous suction aeration
Simple implementation for the aeration of a less contaminated landfill. Gas exchange is achieved in such a way that the landfill is capped in a restraining manner but by no means imperviously (C = carbamide-base flexible solid foam layer) and the waste body is pumped out so intensively after the end of the gas utilisation phase that the oxygen supply exceeds consumption (Ryser's principle, applied on the Uttingen, Switzerland, landfill).
Advantages: Simple structure, inexpensive cap, relatively low power consumption.
Disadvantages: The disadvantages of the continuous low-pressure aeration are more expressed in the suction aeration technique, since the radius of action of the lances decreases greatly in the deeper layers of the waste body.

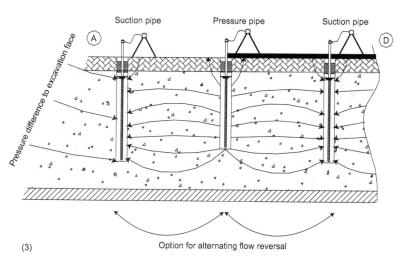

(3)   Option for alternating flow reversal

Fig. 11-15 (3) Continuous pressure-suction aeration (low-pressure aeration)
Combined pressure-suction aeration. Spillmann's first suggestion to the Vienna City Magistrate in 1990. For a better supply the flow can be alternatingly reversed. With (D) and without (A) a gas-tight protective canvas cover for the ventilation of contaminated sites with high contamination.

Advantages: Improved aeration results compared to the previous methods.
Disadvantages: Despite the considerably higher costs for installation and energy, the princi-pal problems of the continuous low-pressure ventilation cannot be eliminated.

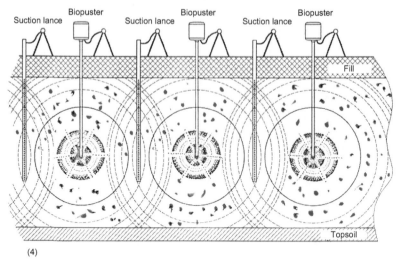

(4)

Fig. 11-15 (4) Pressure-surge aeration
System for a pressure-surge input of air/O_2-mixes, water and nutrients (BIOPUSTER®
system according to Ranner (Spillmann et al., 1992/1993)).
Advantages: The pressure surge is not sensitive to inconsistencies such as strong compaction
or high moisture content within the waste body. It can be used down to great depths without
any problems. The good penetration of the waste body ensures an efficient use of air and
an economical oxygen addition. The system is not sensitive to desiccation. Loss of humidity
can be compensated for by the injection of water aerosol.
Disadvantages: The method requires a high technological level. It needs compressed air asso-ciated with a high power requirement compared to the low-pressure technique.

Fig. 11-15 Principal schemes of various in-situ aeration methods

required a higher energy input than the continuous aeration. A consistent technological improvement of the method today enables pressures as low as 0.8–1 bar instead of the initial 2.5–6 bar while maintaining the same penetration effect. Thus the initially high power requirement was considerably reduced. Today, no significant disadvantage can be found in the energy balance in comparison to continuous aeration. The efficient aeration also enables a substantially reduced rotting time.

11.3.1.3 Determination of treatment time based on DFG research results
The extent of decay of readily decomposable organic materials is the key factor for organic leachate contamination (see Fig. 5-1). It depends

452

directly on the energy output (Fig. 2-13), thus it is proportional to a surface between the curves of waste and air temperature (in a diagram not shown here). If this energy release is compared in an expanded illustration (Fig. 11-16) with the total amount of the organic contaminants (BSB_5 and CSB) and that of mineral nitrogen compounds (NH_4-N and NO_3-N), the minimum aerobic treatment time of old wastes preserved in a fresh state can be determined (with optimum air supply without nitrogen and phosphate addition by, e.g., sewage sludge). Symbols characterise the energy release that must be produced as a minimum to achieve a significant reduction in leachate contamination. It can be concluded that exothermic processes for preserved old wastes are completed under favourable conditions in 15 months so much so that not only the biologically readily decomposable organic contaminants but the ammonium-nitrogen contaminants have been reduced as far as possible (far below $10\,mg\,NH_4$-N; i.e. the end of primary stabilisation has been reached).

To determine the aeration period for a waste aged under optimum anaerobic conditions (no preservatives) the estimated time period can be reduced by the amount of energy released as methane from the preceding anaerobic digestion process. This energy content thus serves as an input value in Fig. 11-16, top. The energy difference up to the 80% mark of total energy release (total energy release of primary stabilisation) then enables the remaining reaction time to be read off in units of weeks.

As a practical pre-design to determine the amount of oxygen input during this phase, the maximum respiration activity (RA) can be used. This can be estimated by a calculation using an RA_i analysis series (e.g. RA_4, RA_{21}, RA_{42}, with i = time [d], see Fig. 7-28) of the same waste sample. The general equation of state for ideal gases enables the determination of the oxygen volume necessary and the total air requirement by the O_2-content of the input air. The following points have to be taken into account:

- Only about a third of atmospheric oxygen (20.95%) is used up.
- Technical oxygen doped to air (20.95% by volume < O_2 content < 36% by volume) is fully used up.

Also, it has to be considered that the concentration difference is crucial where oxygen is needed (see Section 11.3.1.2). Furthermore, the sample mass 40 g of waste of RA_i is only partially sustaining. To compensate for this lack, the O_2-consumption must be regularly controlled and the O_2 supply optimally adjusted.

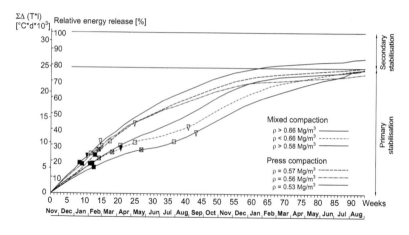

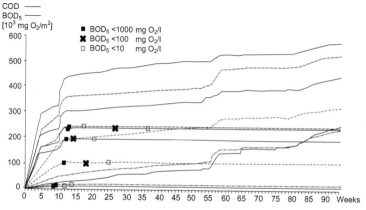

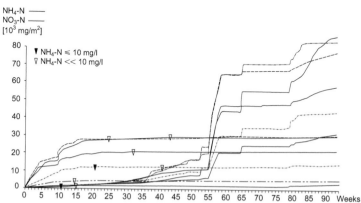

Fig. 11-16 *Comparison of the total energy release (top) for aerobic stabilisation of non-contaminated municipal solid waste (see Fig. 2-13) with the simultaneously released cumulative contaminant content of the two organic parameters CSB and BSB₅ (middle) and cumulative contaminant content of the mineral nitrogen parameters NH₄-N and NO₃-N (bottom)*

Secondary stabilisation to reach the fully completed reaction state (release of the 20% residual energy) takes decades (see Section 8.4), it is, however, so extensive that for maintenance purposes, only a small but necessary oxygen supply is sufficient. The objective of this phase must therefore be to establish a permanently aerobic landfill functioning over decades without using complex technical systems.

11.3.1.4 Technical feasibility of primary stabilisation

Research projects and practical implementation for the intensive release of energy of non-contaminated old landfills have been carried out for several years. One of the projects was terminated after primary stabilisation. After developing an aeration pattern adapted to the properties of the landfill material (see Section 11.2.1), pressure surge and low pressure aeration are currently being used to supply the old landfill with air or oxygen-air mixtures (Fig. 11-15). Technical modifications provide the best means of injecting water and nutrients as a spray directly through the system to meet the above requirement for optimal feeding of the microorganisms.

In the first world-wide in-situ stabilisation using the pressure surge method, an industrially non-contaminated household and commercial waste landfill underwent a reaction-to-entirety in a controlled way in Styria, Austria (Lorber and Erhart-Schippek, 2000). The Leoben University investigated five modifications of this aeration scientifically (Kaltenbrunner, 1999):

- standard operation (CO_2-enriched pressure surge)
- standard operation using water injection through the surge
- standard operation using water injection through injection lances
- standard operation using warm-water injection through injection lances
- standard operation using warm-water injection and doping microorganisms through pressure and injection lances.

In Germany the reaction-to-entirety of a landfill in Lower Saxony and two other old landfills is currently ongoing by applying low pressure aeration supervised scientifically by Professor Stegmann, Technical University Hamburg-Harburg (Heyer et al., 2000). The scientific analysis of all projects (gas behaviour, leachate characteristic at groundwater probes and wells, solids analysis...) will enable a comprehensive documentation of the reaction-to-entirety process. Results published confirm the above statements about primary stabilisation.

455

Since the aerobic degradation process induces a forced release of degradation products and pollutants within the aeration period (Figs 6-1–6-5), they must be collected and treated by suitable technical systems. Combinations of bio and activated carbon filters and systems for non-catalytic oxidation (Regenerative Thermal Oxidation, RTO) of gases with low calorific value have proved excellent as filters. In order to avoid an uncontrolled outburst of landfill gas due to the active injection of air or selected gases into the waste body, a suction capacity exceeding the injection capacity must be guaranteed, so that no migrant gases can escape the waste body. This requirement is fulfilled by pressure aeration and suction using a water ring vacuum pump. Gas containment is absolutely necessary for contaminated landfills, while containment can be omitted in the remediation of a non-contaminated landfill when surface emissions are continuously tested and the maximum emission concentration values are adhered to.

An intensive monitoring of the groundwater around the re-activated landfill is performed in all projects, but it is not possible to intervene in a controlled fashion when pollutant outbreaks are expected (Section 8.3.3.7). Considerations of this kind have been thoroughly integrated in the following projects. For example the contaminated sites have been contained downstream by a pump and treat system during the aerobic activation of a landfill in Lower Austria to establish the conditions for landfill mining (Budde *et al.*, 2002). Solutions of this kind for in-situ reactions-to-entirety combined with natural attenuation methods are being planned by the authors for the consecutive phase of secondary stabilisation.

11.3.1.5 Technical feasibility of secondary stabilisation

Secondary stabilisation is necessary if no landfill excavation is planned. The in-situ phase of secondary stabilisation is not current practice yet, however it is unavoidable if long-term control of the old landfill is to be guaranteed. Since the residual mineralisation and humus production require a low residual O_2 consumption over long periods of time, gas exchange for the secondary stabilisation must be guaranteed over long periods of time, so that no renewed stagnation occurs after the primary stabilisation. The objective is to establish a permanently aerobic landfill. The intensive aeration is uneconomical to maintain for this purpose, but it may be feasible at reasonable costs using the former field structure of the aeration system under the following considerations.

If the former aeration and ventilation lances are replaced with two non-perforated HDPE lances of different lengths and the previous

456

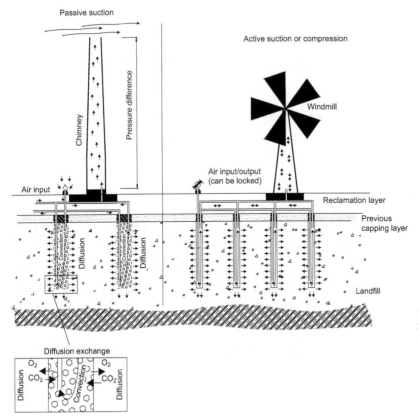

Fig. 11-17 Extensive aeration alternatives for long-term secondary stabilisation of old landfills

boreholes filled with high-permeability material (16/32 gravel) to prevent the hole from collapsing, aeration columns can be constructed, without high cost, able to supply oxygen to the landfill through diffusion (Fig. 11-17, left). The diffusion gradient between the aeration column and the waste body can be maintained actively by applying a wind mill to one of the double lances (forced draught) or passively by chimneys (chimney draught) (Fig. 11-17). This can ensure an almost maintenance-free, permanently aerobic landfill over a long period of time. If the area is intended to be used for landscaping purposes, the ventilation pipes can be covered with reclamation material. Destruction of the system due to subsidence of the contaminated site is not expected after the primary stabilisation (Lys. 6 has proved that no mass reduction took place during the secondary stabilisation, only conversion processes were observed). Thus there is no real danger of loosing the ventilation columns. Clogging of the gravel columns over the course of time is

457

compensated for by decelerating reaction processes, thus the two reductions cancel each other out.

If a country cannot afford a combination of intensive and extensive technical systems due to lack of financial resources, a modified form of secondary stabilisation can be applied: the injection pattern has to be refined, but it would be sufficient to use driven perforated probes (e.g. Fig. 11-8) with non-perforated dual interior pipes of various lengths instead of filtered wells (Fig. 11-17, right). Since the passive air exchange in the column would not be sufficient, one of the two HDPE interior lances can be closed and ambient air can be either actively compressed by the wind mill system or extracted through the chimney or removed by the wind mill system or a combination of simultaneous compression and extraction can be established (see principles on Fig. 11-17, top).

This alternative could be considered as a solution in countries with poor capital resources. However, the serious lack of stability in many deposits must certainly be taken into account. Not all emissions can be completely controlled, but they can be substantially reduced over the long term. The gains from emission trading rights by avoiding methane emission will provide financial support for the stabilisation measures, even if the positive water-related effects are not taken into account. The percentage of global methane emission from landfills is around 8% with a rising trend, exceeding 49% in certain places (Möller, 2001) and its greenhouse effect is twenty times higher than that of CO_2.

11.3.2 Landfill mining with targeted material conversion

11.3.2.1 Risk scenario for the excavation of landfills with low industrial contamination

The limits of an in-situ reaction-to-entirety are eventually reached when industrial wastes are present, which disturb the biological processes by substances toxic to microorganisms or cause autogenous heating or ignition. The long-term removal of the contaminants, which can be doubtlessly checked, can only be achieved in this case by a material conversion (Section 11.4.2) of the critical wastes following removal of the landfill (landfill mining). However before starting to excavate the landfill, the following health and safety considerations must be taken into account for protection of workers and residents.

Even if there is no direct contact with the materials for workers and residents, mobile materials, i.e. gases and dust, which enter into solution

458

through inhalation, may pose a risk. Therefore it is necessary to limit the landfill-specific risk before starting any digging.

Fire and explosion protection

It is still not fully understood by many engineers that inflammable land-fill gases and explosive gas mixtures released by anaerobic degradation of organic substances may pose a serious risk to inhabitants, residents and workers. To date, the necessary preventive measures originate from coal mining. Serious accidents occur due to landfill gas world-wide not only from landfill operation (Table 11-1).

Table 11-1 Overview of selected landfill gas explosion accidents with serious or fatal outcome

Landfill gas accidents
... 8-year-old girl was burned on her arms and legs when playing in an Atlanta playground. The area was reportedly used as an illegal dumping ground many years ago... (*Atlanta Journal-Constitution*, 1999)[1]
... on 28/04/1993 a gas explosion occurred on the Umraniye landfill in Istanbul burying nearby houses by mountains of waste due to landslides and killing 40 people... (Savar, 2003)
... while playing soccer in a park built over an old landfill in Charlotte, North Carolina, a woman was seriously burned by a methane explosion... (*Charlotte Observer*, 1994)[1]
... off-site gas migration is suspected to have caused a house to explode in Pittsburgh, Pennsylvania, 1987... (EPA, 1991)[1]
... landfill gas migrated into and destroyed one house near a landfill in Akron, Ohio, 1984. Ten houses were temporarily evacuated... (EPA, 1991)[1]
... an explosion destroyed a residence across the street from a landfill in Cincinnati, Ohio, 1983. Minor injuries were reported... (EPA, 1991)[1]
1975... in Sheridan, Colorado, landfill gas accumulated in a storm drain pipe that ran through a landfill. An explosion occurred when several children playing in the pipe lit a candle, resulting in serious injury to all the children (USACE, 1984)[1]
... the City of Ionia owns the landfill property, and operated the landfill from the mid- to late-1950s until 1969. During the early years of the landfill's operation, the trash was burned regularly,.... Nearby residents also reported occasional explosions at the site while it was in operation. In October 1965, one of these fires touched off an explosion that resulted in the death of a waste hauler. (MDEQ, 2002)
1969... Methane gas migrated from an adjacent landfill into the basement of an armoury in Winston-Salem, North Carolina. A lit cigarette caused the gas to explode, killing three men and seriously injuring five others (USACE, 1984)[1,2]

(1) Source: ATSDR, 2002

(2) http://www.hse.gov.uk/foi/internalops/hid/din/
530.pdf#search=%22methane%20from%20landfill%20gas%20explosions%22: An English source with a number of references. The 1984 Loscoe event was the turning point in English law

459

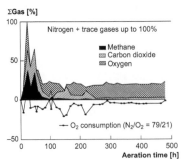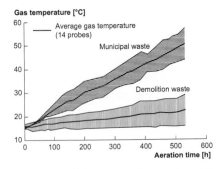

Fig. 11-18 *Typical gas release (left) and temperature (right) curves in the aeration of the over 30-year-old Donaupark-Bruckhaufen, Vienna, landfill containing domestic and demolition wastes (Dörrie et al., 2001)*

It was reported by citizens near the largest municipal waste landfill in Barsbüttel near Hamburg, Germany, that flames could be ignited from pipe culverts in the cellar walls. In particular if landfills are to be excavated, an explosion hazard is expected when there is an uncontrolled gas regime in the landfill, gas bubbles are present and spark-generating tools are used (an excavator shovel may do so). Gases from old landfills contain up to 80% by volume of methane, the flammable range $CH_4/CO_2/O_2$ is between 4–15% by volume CH_4, the stoichiometric point being 11%. This dangerous situation can be avoided, if before the excavation:

- biologically generated gases are pumped off in a targeted way and treated under controlled conditions (see filter technology in Section 11.3.1.2) (e.g. introduce the released gases into suitably designed active biochemical filters ('compost filter'));
- the landfill environment is aerated for the period of the excavation, degradation and transport in such a way that the generation of new methane is completely prevented (as a completely aerobic environment).

This procedure has been successfully applied in practice in more than ten projects (Budde *et al.*, 2002). Figure 11-18 shows the example of the Donaupark-Bruckhaufen, Vienna, landfill where the requirements for explosion prevention were fully met within as short a period as 100 hours, i.e. 4 days, of intensive aeration and the results were confirmed by subsequent projects. An air exchange which theoretically equals one pore volume per day ensured that there was no increased risk due to exceeding the LEL (lower explosion limit) over the entire excavation (see Fig. 11-18, left). It can clearly be seen that the 30-year-old

460

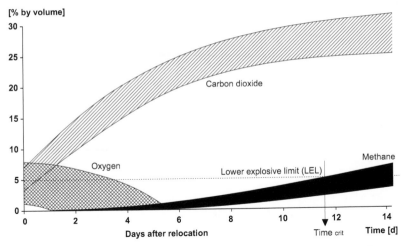

Fig. 11-19 Gas composition in a domestic waste landfill in Southern Germany after stopping the three-week pre-aeration (Source: ARGE Biopuster, 1998)

domestic waste area reacted to the injection of the air/oxygen mixture almost as intensively as fresh domestic waste (Fig. 11-18, right). This proves our statement about a re-activable preservation (Chapter 8). Another municipal waste landfill has shown that three weeks of intensive aeration made the old material so stable that in a week after stopping the ventilation there was sufficient time for excavation, transportation and processing (Fig. 11-19) before the landfill shifted back into an active anaerobic phase. This finding was also confirmed by subsequent projects.

Natural toxicity and odours
It has been proved in Section 8.3.4.2 that the anaerobic landfill environment (non-contaminated anaerobic Lys. 7) became extremely toxic as indicated by luminous bacteria. If a natural organic substance is degraded biochemically in an anaerobic way, intermediate products are produced from proteins, which are highly toxic for some single-cell aerobic organisms and most higher organisms: for instance the clostridium botulinum toxin is more than an order of magnitude more poisonous than the 'super poison' 3.-4.TCDD; hydrogen sulfide is toxic comparable to hydrocyanic acid. Humans therefore have a very sensitive warning system in form of their sense of smell, which reacts to the slightest traces of these degradation products or their associated substances, causing violent aversion and leading to high stress levels for

461

all exposed people. Contaminants of this kind are intolerable to both residents and workers.

The toxicity of the material can be avoided and the odorous substances from anaerobic degradation can be degraded in an aerobic way. The odour thresholds of the intermediate products from aerobe degradation are substantially higher and clearly cause far less repugnance (40–50°C: cellar smell, brewery smell; over 70°C also roasting substances). They are non-toxic (proof: the boundary area of Lys. 7 was affected by oxygen and the luminous bacteria toxicity disappeared) and people can be expected to tolerate them for a short time. However, the range over 70°C must be checked by a detailed gas analysis in order to be able to safely exclude any reaction products by thermophilic/hyperthermophilic bacteria. The following final conclusions can be drawn from this for smell emission control:

- a non-contaminated anaerobic deposit can be relocated without causing any stress and nuisance to residents and personnel, if it is first converted aerobically in-situ and mineralised into an 'earthen' odour;
- if a short-term odour nuisance (musty smell, silage) is acceptable, it suffices to change the anaerobic degradation to an aerobic degradation without waiting for the complete mineralisation and/or humification.

The easily decomposable organic substance that remain determine the aeration time since its anaerobic degradation products are the main source of unpleasant odours: in the form of organic acids (fatty acids characteristically smell like 'pigs' liquid manure' and determine the H_2S generation potential ('rotten eggs')) and in the form of ammonium (ammonia = 'horses' stable'). The measurement of BSB_5 (organic acids) and NH_4 in leachate/groundwater can provide information about the potential odour nuisance and the production of natural toxic gases. In normal anaerobic landfill technology the contamination is extended over 5–10 years for BSB_5 (see Lys. 3 and 7, Fig. 5-1) and over more than two decades for ammonium (see Lys. 3 and 7, Fig. 5-1). As has already been shown for the controlled aerobic reaction-to-entirety in Section 11.3.1.3, the length of an aerobic aeration for smell minimisation can also be estimated from the total energy release (Fig. 11-16).

As Fig.11-16 indicates, the development of organic fatty acids and aerobic degradation are in equilibrium if at least 20% of the total energy is released (Fig. 11-16, middle) for which 2 months are needed for freshly conserved waste in the most favourable case and 4 months under less favourable conditions. Ammonium is reliably nitrified if 30% to 40% of the energy is released, i.e. oxygen consumption is considerably reduced by an advanced degradation. Rapid degradation can achieve this point for freshly preserved waste within as little as 4 months or, under unfavourable gas exchange conditions, in about 1 year. For wastes aged anaerobically the processing times have to be reduced using comparisons of residual energy contents (see Section 11.3.1.3).

The above requirements have been implemented in practice by in-situ aerobisation using the pressure surge aeration procedure (Fig. 11-15) in over ten projects (Budde *et al.*, 2002). Olfactometric assessments took place in the most diverse projects. Figure 11-20 clarifies the effectiveness of an environmental conversion. After excavating an active anaerobic landfill (the objective being to repair an engineering component of the landfill), the odour load was determined in odour units (OU) following 3 weeks of intensive aeration (Fig. 11-20 top, Pr2i) and compared with a simultaneously performed measurement in a non pre-treated area (Pr1i). Even after this short period of aeration, a difference of two–three orders of magnitude is clearly recognisable between aerated and non-aerated landfill areas. It can also clearly be seen that the aeration fully eliminated the explosive conditions (Fig. 11-20, bottom).

Practice has proved that systems developed on the basis of this research project guarantee complete health and safety, resident and environmental protection for landfill mining of a non or slightly contaminated landfill and for the separation of the materials.

11.3.2.2 Risk scenario for the excavation of a landfill contaminated with industrial waste

The above statements for work on a slightly contaminated landfill apply. It was found in Section 8.5.4.2 for co-disposal that there is an extreme toxicity even for multicellular creatures (Daphnia toxicity), which causes substantial inhibition of micro organisms also under short-term aerobic reactivation conditions.

Basic understanding of the limiting value definition

The content of toxic and carcinogenic gases in modern deposits is known and recorded in detailed investigations (Poller, 1990). However, the main difficulties in protecting people are that not all out-gassing substances are measurable and more especially, their effects are not known. This applies even if all deposited materials are known, since their degradation products

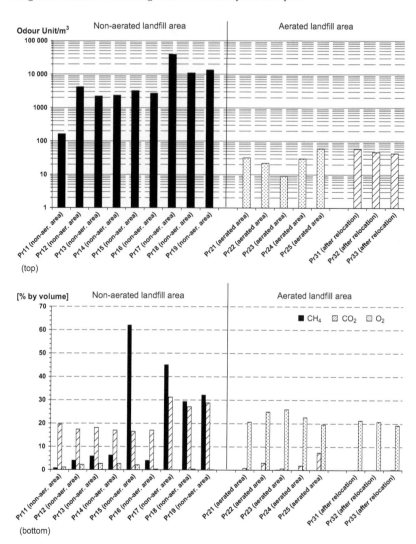

Fig. 11-20 *Olfactometric determination of odour units OU in connection with the relevant gas composition of an active anaerobic landfill in southern Germany (Source: ARGE Biopuster, 1998). Top: OU in a non-aerated reference area compared with an area aerated for 3 weeks and an area 1 week after completion of aeration and after relocation; Bottom: landfill gas composition in the non-aerated reference area compared with an area aerated for 3 weeks and an area 1 week after completion of aeration and after relocation (Pr = probe)*

are often unknown and may be far more poisonous than the initial materials: vinyl chloride (VC) from the degradation of perchloroethylene (PER) is carcinogenic; the OES (Occupational Exposure Standards) value (TRK, Technische Richtkonzentration) for VC = 2 ppm, the

464

MEL (maximum exposure level) value (MAK, Maximale Arbeitsplatz-Konzentration) for PER = 50 ppm. Even if it were analytically possible to determine all materials, no reliable statement could be given about their interactions. In an unfavourable case material combinations multiply the effect of single materials.

Eikmann *et al.* (1989) published investigations on the interactions of unknown toxic gases from landfills. Thorough chemical gas investigation of a cultivated mixed landfill indicated that the contamination was unproblematic. However, epidemiological studies reproducibly showed that a significant number of residents near the landfill showed symptoms which needed treatment after a very short time. The source of the unambiguously proved illnesses could be localised (the landfill), but even an extended analytical programme failed to find the cause. Findings from medical practice confirm the deleterious effect of gaseous emissions from landfills. König (1995) diagnosed toxic effects even in such cases when none were to be expected after careful chemical gas analyses. Health damage, including death, as a consequence of emissions from landfills occurred considerably more often than shown in the statistics (König, 1995). This finding was confirmed by König in 2005.

It is difficult to establish justified limiting values. Limiting values in air for workers on contaminated sites have been determined by the medical experts of the Health and Safety Executive (AMD, Arbeitsmedizinischer Dienst of TBG, Tiefbau Berufsgenossenschaft) as $\frac{1}{10}$ of the maximum exposure level (MAK values published by DFG) (see BGR 128 – TBG, 1997). These contamination limits apply solely to healthy, able-bodied people spending 8 hours a day in work and having regular health monitoring. Small children, old and ill people are also residents and are exposed to the contamination all day but are not protected by health and safety precautions or control tests. Therefore, for this group of people, logically substantially more stringent contamination limits should apply. The basic condition for a correct estimate is the execution of an emission prognosis for various scenarios, which would model the local topographic conditions. The use of such relevant models is state of the art.

Very volatile gases

It was shown in Section 8.5.3.1 that the group of the very volatile, synthetic substances – first of all wide-spread chain-like chlorinated hydrocarbons – produce extensive out-gassing when a landfill is excavated in an uncontrolled way. Since they contain the largest part of mobile xenobiotics and they exhibit the highest toxicological and carcinogenic risk, they must be accredited the greatest importance. The following requirements can be ascertained from the low vapour pressures and the resulting high mobility of the substances in order to remediate a landfill with industrial contamination – taking into account all requirements for the technical measures in excavating non-contaminated landfills:

- Very volatile substances (e.g. hydrocarbons, fluorinated or chlorinated hydrocarbons, BTX...) must be removed by the ventilation system; here a textbook technical gas containment is an absolute necessity (no gas entering the landfill may leak out) (see Fig. 11-15).
- The aerobic biological heating process must be intentionally induced for the thermal desorption and mobilisation of pollutants (see Figs 8-8–8-10) – in so far as the toxic potential permits.
- Contaminated gases must be collected in a controlled way and treated according to the pollutant.
- Extremely serious local contaminant sources – such as drums, etc. – must be separately removed and treated while simultaneously maintaining a protective ventilation for personnel protection measures.
- It must be tested to see if inflammable gas mixtures are present.

These basic requirements have already been implemented for the excavation of mixed landfills as Fig. 11-21 indicates. Within a short aeration period of a few weeks the content of very volatile halogenated hydrocarbons and BTX in the raw gas was reduced by one to two orders of magnitude. In order to cope with the above combined effects, a measuring system, modified for landfill gas, was installed for the first time to determine the carcinogenic potential (see Section 11.2.1). First results show a significant positive effect of aeration (Fig. 11-22). The extracted gas was cleaned by the combination of a bio filter and activated carbon filter.

Semivolatile gases

Semivolatile compounds cannot be fully eliminated with the maximum achievable temperatures of 70°C. As trace gases, however, they pose a potential risk both for residents and personnel during excavation.

They may emerge as natural intermediate products initially through an incomplete anaerobic degradation (e.g. phenol), but then are present only to a small extent. They can be further degraded aerobically, and are toxicologically insignificant as gases after a chiefly aerobic biochemical

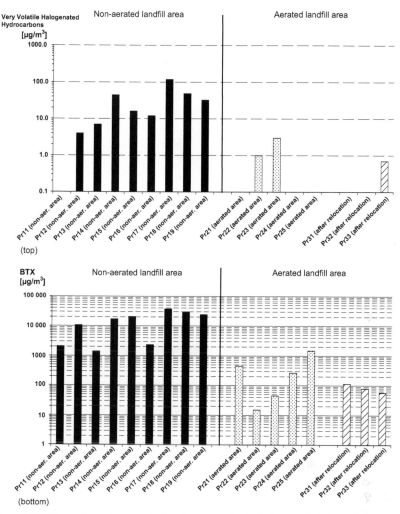

Fig. 11-21 Illustration of contamination by very volatile halogenated hydrocarbons and BTX of the landfill gas in an active anaerobic landfill. The landfill was first pre-aerated and then dug up to repair a landfill component (Source: ARGE Biopuster, 1998). Top: concentration of very volatile halogenated hydrocarbons in landfill gas of the non-aerated reference area compared with an area aerated for 3 weeks and an area 1 week after completion of aeration and after relocation; Bottom: BTX concentration in landfill gas of the non-aerated reference area compared with an area aerated for 3 weeks and an area 1 week after completion of aeration and after relocation (Pr = probe)

mineralisation. They occur as anthropogenic deposits in larger quantities. Non chlorinated manufactured products of this substance group are often non-degradable due to an asymmetric material composition and high concentration. Chlorinated compounds are not available for

467

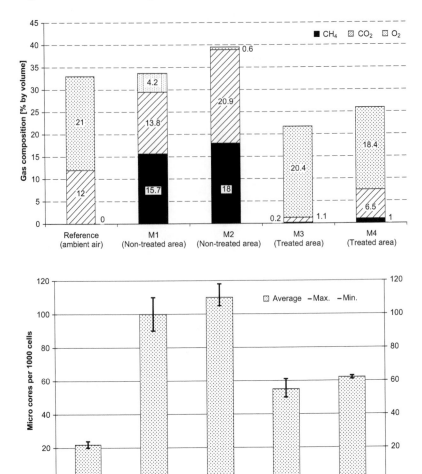

Fig. 11-22 *Result of the investigation of landfill gas from a south German landfill for its carcinogenic potential with the help of the micro core test shown in Section 11.2.1.2. An aeration of 3 to 4 weeks reduced the cancer risk by half (Source: Schiffmann, 1998)*

aerobic biochemical attack even under favourable conditions. Apart from the above aspects of protection, it follows for the remediation of such sites:

- The spread of these pollutants must be calculated by an emission forecast.
- A technical containment by fluid mechanical means must prevent the exposure of residents to semivolatile substances through trace gas release.

468

- Workers must be suitably protected by the correct personal protective equipment (PPE), especially during excavation.

or

- Industrial contaminants of these zones must be uncovered section-wise under the protection of vacuum, isolated as islands and, if necessary, sealed with metal-plastic composite sheets against a diffuse escape of the remaining gases.
- These substances must be degraded and/or expelled to a considerable extent by further targeted actions.
- Inertisation is needed by freezing or targeted injection.

It has to be noted, however, that in Germany according to § 19 GefStoffV (Dangerous Goods Ordinance) (1999), the technical protective systems apply in principle according to the legally specified order of rank which must always be used before personal health protection. Thus, generally the following logical priority applies for work on landfills:

- Procedure for the prevention of development or for the degradation of toxic and carcinogenic gases.
- Procedure for the controlled collection and cleaning of released gases.
- Technical devices to guarantee a high degree of freedom in work.
- Personal protection equipment (respirators, protective clothing).

The priority of necessary actions indicates that sufficient protection for work on landfills can be influenced by the choice of the remediation method. Avoidance of an uncontrolled escape of pollutants must be the highest principle of the procedure!

For the control of semivolatile pollutants it must be first tested to see whether extra precautions are needed for in-situ degradation of contaminants. Pecher's (1990) research results suggest that this is quite possible if sufficient processing time is available for the biochemical degradation of the potentially toxic substances before excavation of the site. Even highly chlorinated phenols can be biochemically degraded within a short time, if a suitable environment is adjusted (addition of a deficient substance may be necessary). Substances not susceptible to an aerobic attack can be rapidly dechlorinated in the methanogenic phase anaerobically. The (usually cyclic) dechlorinated hydrocarbons can be degraded more easily aerobically. Particularly effective is the rapid alternating change of the redox potential from aerobic to anaerobic degradation and back, analogous to the processes in the upper soil

469

layers after rain (Hanert, 1991; Hanert *et al.*, 1992). The difference of the in-situ pre-treatment for excavation as opposed to in-situ remediation is the reasonable length of time: the degradation performance decreases proportionally to the diminishing nutrient for microorganisms with progressive degradation. The degradation needed for the excavation is therefore reached considerably faster than for example, a remediation target. The periodically changing oxygen supply provides good conditions to achieve sufficient pollutant degradation needed for the excavation at reasonable costs.

The change of environment was successfully implemented in practice for groundwater remediation, the procedure is being tested for an old landfill.

Deposits that cannot be remediated microbiologically (either due to quantity, e.g. drums, or due to their extreme high toxicity) must be loaded into special containers and transported to a facility to be processed under special safety precautions, e.g. in an aerated enclosure or by nitrogen freezing or sealing with injection lances. If the effect of the pollutant scenario on the human organism is not fully known, isolation protective aeration alone suffices both for the items and the workers with regard to personal safety according to TBG. It has to be noted that it may be dangerous to use filter protection alone, even if filter-penetrating substances are not present because, if an additional ventilation system is completely omitted, the main landfill gases may mix with the ambient air and produce non-breathable, suffocating air-gas-mixtures. Based on this important finding, the excavation of the Bachmannig, Austria, landfill was carried out under an enclosure with additional aeration of the wastes (Budde *et al.*, 2002).

11.3.2.3 Risk scenario for the control of plastics in landfill

Based on the findings of Chapter 9 and operational experiences in two landfills in northern Germany, it is highly recommended to consider carefully whether plastics should be disposed of in landfills. The Cause for Concern Principle is already met by the findings; because it cannot be excluded that existing landfills containing large amounts of plastics may ignite spontaneously despite complex state-of-the-art lining and capping systems. The aftercare of landfills now receives a new aspect, which must be considered in the risk assessment during final completion of the landfill and before starting reclamation. As long as the temperatures in the landfill do not reach 50°C, all activity is accompanied with a substantially lower risk which is easier to control.

The risk rises with the temperature, and then bacteriologic processes are overtaken by chemical processes. If further influences are added by pressure – both weight and wind pressure – the risk increases considerably. This danger can only be avoided or reduced by early and targeted action.

In the following, the case history of two landfill fires is described which finally triggered research into spontaneous heating of plastic landfills, which is described in Chapter 9. Initially extensive measures had to be devised to control the phenomenon of spontaneous ignition for acute hazard control. Even if the first landfill fire was caused by an ignition in a slag tip beneath the landfill, the second case shows a conventional spontaneous ignition within a plastic mono landfill – as simulated and proved in Chapter 9.

Up to 1995 only a few insiders knew that there are mono landfills for plastics in Germany and Europe. In September 1995 a hidden fire developed in a landfill containing 130 000 tons of plastic. A targeted investigation using driven core sampling discovered temperatures of 800°C. These tests were performed with water pipes in such a way that the temperature was controllable in the inserted measuring points. It was found that metallurgical residues with high carbon content under the plastics were spontaneously ignited by exothermic processes of a neighbouring slag tip. The risk investigation between operators, authorities and experts led very quickly to the result that the use of water as a fire extinguishing agent was to be rejected; since water on hot coal supports the generation process of coal gas (the raw gas developing in oxygen deficient atmospheres is highly poisonous and explosive).

To prevent endangering the area, liquid nitrogen freezing was used for the first time in the world (editors' information) to cool and neutralise the fire source. For this purpose the nitrogen was introduced through copper lances into the seat of fire and the temperatures were measured at the measuring points by data logger. Within 3 days, the seat of fire was cooled from 700°C to −100°C in the lower landfill range. Since more than 800 litres of gas can develop from a litre of liquid nitrogen, a worst case scenario for the risk of gas release to the neighbourhood was investigated by means of an emission prediction. The authorities only gave permission to use nitrogen after experts excluded any risk to the neighbourhood. More than 220 tons of liquid nitrogen was delivered in tankers for the test. Based on the enormous success, it is suggested that liquid nitrogen should be considered as a potential variant for fire fighting in similar acute situations (Struve, 1996).

471

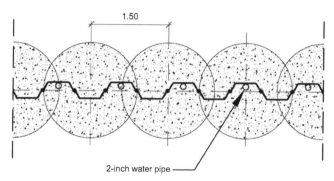

1.50

2-inch water pipe

Fig. 11-23 Coolable fire wall. Large bored-pile cut-off wall filled with sand, diameter 1800 mm, with integrated steel sheet piles and 2-inch water pipes to provide cooling (Lhotzky and Struve, 1996)

The slag tip, next to the plastic landfill, will continue produce heat due to chemical reactions over decades and thus represents a risk for the plastic landfill. To prevent further heat transfer, the first coolable fire wall was constructed in the world using large bored piles with an integrated sheet-pile wall (down to a depth of 22 m) (Fig. 11-23). Since the sheet-pile wall has been constructed in a gas-tight fashion, heat transfer has been sustainably prevented so far (10 years).

The risk of heat accumulation in the landfill still existed however, since several impervious horizons were present. As several thousand tons of material with temperatures between 400°C and 800°C contain enormous amounts of heat which migrates toward the external embankment, three ballast columns of 1.80 m diameter were bored to a depth of 25 m in the landfill and filled with crushed stone to ensure a controlled heat dissipation (Fig. 11-24).

After small localised fires developed suddenly on the external embankment, excavation of the fire nests was started and the nests were filled with cohesive soil. A few weeks later this phenomenon emerged again 10 m further along on the embankment. Again, it was excavated and filled. After a few repetitions of the procedure the problem was finally solved by targeted injections of a total of 10 000 tons of grouting material (containing quick-setting cement) through vertical and horizontal drillings (Fig. 11-25). The drillings were directed by temperature measurements to channel the grout under and behind the smouldering fire.

All special work ran without accidents and without problems for the residents, who were kept up to date about the problems and their

Fig. 11-24 Large bored piles filled with crushed stone to prevent heat accumulation by con-trolled heat dissipation. The flickering air and the plumes of steam indicate heat removal. Bright spots in the area of the steam plume are sulphur precipitations

solutions by regular meetings. The measurement by the temperature probes was carried on as progress control.

About 2 years after completing the grouting work, a slow heating appeared in an area of the landfill and smouldering processes emerged again on the external embankment. The peculiarity of these processes was that they always occurred at the embankment toe.

Material was taken from these areas to be investigated in the simulation reactors of Rostock University (see Chapter 9). It was proved that the reason for the smouldering processes was that thermophilic/hyperthermophilic bacteria were active in the plastics. Strains were separated up to 95°C where these specialists are active only in narrow temperature spectra – sometimes as narrow as 3°C – and they were only identified as spores outside of these areas. It was also shown that, in addition, chemical processes such as chemical cracking of the plastics occur at temperatures around 70°C, generating very inflammable gases in the landfill.

473

Fig. 11-25 Horizontal drillings injecting insulating material into the smouldering toe of the landfill embankment

The phenomenon that the smouldering processes always emerged at the embankment toe of the landfill was explained as follows. It was proved that the processes in the landfill produced the very volatile CS_2 (carbon disulfide) gas. Combined with air, this gas produced a very explosive and inflammable mixture, which is 2.6 times heavier than air. This explains the path to the embankment toe. It was more difficult to prove the existence of CS_2, since it is used as an extraction agent in the laboratory and is thus suppressed in the analysis. It is assumed that this material was first a reaction product of landfill bacteria. The fact that there is a gas fire however, can be clearly seen in Table 11-2: the sharp temperature maximum remains within the range of a few decimetres between 230 and 232 m without significant material consumption – which could have been seen by an indentation of the embankment.

Meanwhile there is a second landfill in Germany, in which exactly the same phenomena have occurred. This landfill was closed 20 years earlier and lies near an industrial site without raising much attention. In the summer of 2004 a fire was observed. After a short time more fires emerged, always near the embankment toe. Here, too, bacteria were proved to be the cause of the fire. After 1 year two localised smouldering processes were observed in the landfill of 60 000 m^3 (Struve, 2005b). These smouldering processes produce thiophens, which are a nuisance for the neighbourhood due to their unpleasant smell (very low smell threshold value). Damage control is pending.

Based on the results of Chapter 9 and the problems discussed here about landfills containing plastic which are badly affected with

Table 11-2 Temperature [°C], driven core probe 434 (Struve, 2005a)

Depth [m] AOD*	Feb 03	June 03	Oct 03	Feb 04	June 04	Oct 04	Feb 05	June 05
235	8	19	6	3	19	16	3	20
234	18	39	27	22	27	21	9	22
233	46	57	48	49	36	29	20	26
232	83	85	86	96	44	34	27	29
231	157	114	66	133	53	43	36	37
230	37	37	38	51	36	38	35	38
229	33	30	31	34	29	32	33	33
228	31	28	29	31	29	29	28	31
230.50	–	173	–	–	–	–	–	
230.40	219	–	–	140	53	–	–	
230.80	–	–	98	–	–	–	–	

* (Above Ordnance Datum), benchmarked for the North Sea Level (mNN)

smouldering fires caused by bacteria, risk assessment of existing landfills must be extended by including spontaneous combustion. Mixed landfills, even pure plastic landfills, contain sufficient material for bacterial metabolism, so that unfortunately the two described landfills will not remain individual cases. This problem is also known in temporary storage facilities for commercial wastes and must be looked into. Excavation of such landfills is one of the greatest challenges that health and safety protection and excavation technology face, requiring the elimination of inflammable gas mixtures in each case. An intensive programme is currently ongoing to solve this problem.

11.3.2.4 Landfill mining by material-specific extraction of wastes
After ensuring good working conditions for the excavation of a landfill, the objective is to remove the waste as cleanly as possible so that, after a simple separation, it can be channelled directly to utilisation or an extensive material differentiation, as described in Section 11.4.2. That this is no utopia and has been performed several times in practice is shown by the following two examples which have been carried out according to the excavation scheme illustrated in Fig. 11-26 (Dörrie et al., 2000).

Separation of a demolition waste/municipal waste landfill with no industrial contamination
The possibilities of material utilisation were implemented as early as in 1990 when an aeration system was first used for the controlled landfill

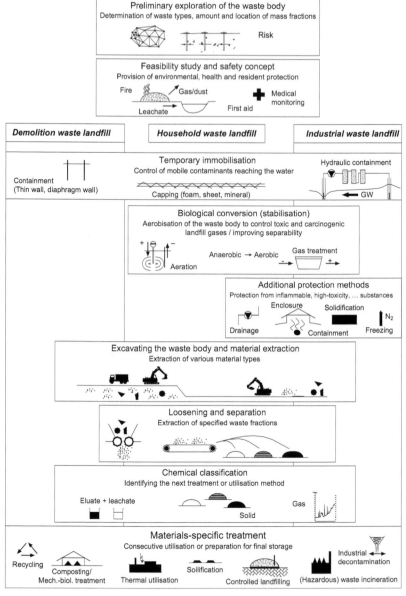

Fig. 11-26 Basic flow chart for landfill mining of a mixed landfill (Dörrie et al., 2000)

mining of the Donaupark landfill in Vienna. After a 4-week stabilisation 15 000 tonnes of fairly clean demolition waste was recovered from 1 100 000 tonnes of landfill material and in such a price-efficient way that profitable use was possible (Fig. 11-27). It was also possible to

Fig. 11-27 Landfill mining produced 15 000 tonnes of demolition waste from the aerated Donaupark landfill (Source: Archive of Bilfinger Berger Bauges. mbH (Building Company), Vienna)

stabilise the organic substance to an extent that the sieve overflow >40 mm could be classified one landfill class higher than the initial material. The relocation enabled such a huge increase in compactness of the waste that only 30% of the original landfill volume was needed on a municipal waste landfill at the new standard. This confirmed our findings in Section 8.5.2.3 that relocation and renewed compaction alone can provide a large gain in landfill volume.

Separation of a municipal waste/industrial waste landfill with industrial contamination

A new quality level of the differentiated, controlled landfill mining system was reached while performing landfill mining in the Berger landfill in Wiener Neustadt, Austria. After a deep biological stabilisation (12 m) the waste was excavated in two terraced stages (Fig. 11-28). Large quantities of industrial plastic waste were separated during the excavation procedure using a backhoe designed as a fork. Waste batches that could clearly be distinguished visually (various industrial influences) were stored separately, sampled, analysed for the expected pollutants and kept in temporary storage facilities until the material was released for transport. The transport destination depended on the chemical contamination and ranged from Professor Husz's soilification technique (Langes Feld, Vienna, 2002), to demolition waste processing, inert material landfills, domestic waste landfills, hazardous waste landfills and, possibly, hazardous waste incineration plants.

A ten-fold hazardous waste contamination was revealed by the landfill mining as compared to what had been expected after the preliminary

Fig. 11-28 Terraced excavation, sieving the plastics by a backhoe with a special fork-shovel (left). Computer and video controlled temporary storage of single batches in boxes until classification of the contamination by chemical analyses, followed by controlled transport to the specified facility (right) (Source: Archive of Bilfinger Berger Bauges. mbH (Building Company), Vienna)

exploration. About 5500 drums were recovered under increased health and safety precautions and annealed in a rotary kiln before being transferred to scrap yards. Nevertheless, the material utilisation could be carried out: 68% by weight was used for soilification (Husz, 2002) and 4% by weight (aluminium slag) was utilised thermally (Figs 11-29 and 11-30). Only 22% by weight was disposed of in expensive municipal solid waste landfills and 6% by weight on mineral material and/or demolition waste landfills. A thermal utilisation of the 12% by weight of light fraction (plastics with high energy content, with about 50% of the

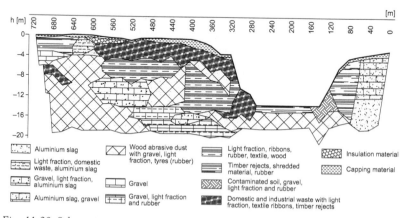

Fig. 11-29 Schematic cross-section of the Berger landfill, Weikersdorf, Lower Austria based on the findings of preliminary exploration pits: mixed municipal waste/industrial waste landfill (Dörrie et al., 2000)

478

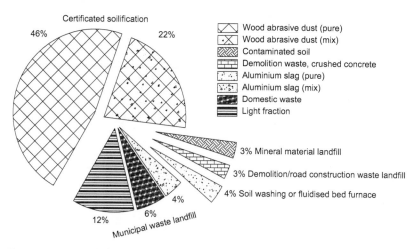

Fig. 11-30 Material flow diagram after separating 900 000 tonnes of landfill material in the Berger landfill. The material fluxes of metal and road planing's and the recovery of 5500 drums are not indicated (Dörrie et al., 2000)

energy content of the landfill) could not be carried out due to existing contracts with landfill operators. Also, the transport of at least half of further 10% by weight of waste to municipal waste landfills can be avoided by extensive separation (see Section 11.4.2), so that after a consistent material conversion, a maximum of 10% by weight of the initial mass was left to be landfilled.

That landfill mining is reasonable not only in highly industrialised regions for land remediation (property development) or for purely ecological reasons, has been proved in practice: a current inquiry of the power generating industry (Greece, Turkey) to the authors suggests the removal of high-quality substitute fuels from the landfill to pay for the remediation. This method has already been developed as Spencer (1990) and Brammer (1997) report on completed projects in the USA. Another inquiry aims at landfill mining to recover the reclamation substrate – again performed more than 50 years ago in Tel Aviv (Savage et al., 1993; Savage, 2003). At that time the contractor company paid for the excavation and was allowed to market the material profitably (cultivation material for citrus groves). A consistent blending of further potential options (Table 11-3) will enable a permanent solution for groundwater protection by eliminating the issue of old landfills with an full range of gas- and potential fire-related problems (Section 11.3.2.3).

Table 11-3 Reasons for major excavations in landfills. Evaluation of more than 77 landfill excavations performed world-wide for >10 000 m³ waste volumes (Budde et al., 2002)

Reasons for intrusions in landfills

Higher-ranking reasons
- Building project
- Planned land use
- Disturbing landscape

Risk prevention
- Ground- and drinking-water protection
- Resident protection
- Mechanical stability
- Fire fighting

Use
- Gain of landfill void space
- Extraction of valuable materials
- Fuel/energy production

Improvement of landfill safety
- Repair measures
- Landfill expansion
- Embankment shaping

Costs
- Reduction of aftercare costs

11.4 Avoiding polluting leachate emissions from present and future wastes

11.4.1 Immediate actions

The most obvious and immediately effective water protection measure consists of the deposition of currently produced wastes on a sealed base, collection of the water and only discharging it into the environment after cleaning. The clarification technology is at such an advanced level today that its consequent application prevents any water contamination. The seemingly easy-to-solve problem of applying a basal barrier and collecting leachate soon resulted in complex solutions which still exhibit fundamental shortcomings. The invention of the composite liner comprising a mineral layer and an HDPE geomembrane (Professor August, Federal Institute for Materials Research and Testing, BAM Berlin) solved the problem both of diffusion and convection through large areas because the properties of the different barriers complement each other:

- Nonpolar materials (e.g. very volatile halogenated hydrocarbons) diffuse through nonpolar barriers (e.g. HDPE geomembrane) but are blocked by polar (e.g. mineral) barriers.
- Polar substances, i.e. water, permeate through polar barriers (mineral layers), but are stopped by nonpolar materials (HDPE geomembrane).
- Thick mineral layers are unsusceptible to local damage; but they exhibit a residual permeability on a large liner surface.
- HDPE geomembranes are impervious over large areas, but very sensitive to local damage.

However, state-of-the-art solutions still fail to exhibit the following properties:

- fail to safe
- stable protection of HDPE geomembranes from penetration by 16/ 32 mm gravel of the drainage layer
- preventing the mineral layer from desiccation under the HDPE geo-membrane.

According to the state of the art, all provisions can be fulfilled (Fig. 11-31, left) if the HDPE geomembrane is embedded into the mineral layer, the gravel is placed to prevent penetration according to the rules of concrete technology (pumped concrete, vacuum compression) and a controlled

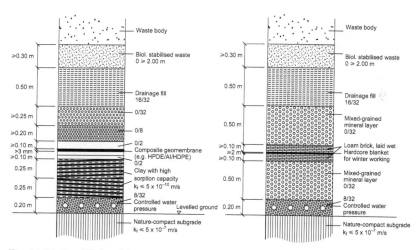

Fig. 11-31 Landfill basal liner systems. Left: basal liner system for municipal solid waste landfills with high industrial contamination, principle according to the inventor Prof. August (BAM Berlin) (Spillmann, 1993b); Right: modified landfill basis for municipal solid waste in underdeveloped countries

external water pressure is applied from beneath adjusted to the overburden. The basal structure according to Fig. 11-31 meets all provisions and can also be used as a controllable barrier against high chemical contamination (Spillmann, 1993b). Reliable basal liners established on the same principle of function can also be constructed for less developed countries at substantially reduced costs (Fig. 11-31, right). The authors' experience of several years shows that the principle is also suitable to collection and settling ponds.

The long-term and large-area drainage of untreated waste has proved extremely unreliable in the past, because the crushed stone can become completely clogged by lime deposits due to biological activity in the anaerobic phase at an increased pH due to micro organisms (Ramke and Brune, 1990). Even large-diameter drainage pipes (200 mm) may cease to function if they are not regularly cleaned with high-pressure water jet equipment. Otherwise they give rise to serious damage such as uncontrollable springs on the slope or severe landfill slope failure (Kölsch, 2003a; 2003b; 2003; 2007). Prerequisite for the reliable operation of a landfill with a basal liner is therefore biological stabilisation prior to disposal. This stabilisation has been carried out in its simplest form as flat windrows for over 25 years in Germany (Spillmann and Collins, 1981; Haschemi, 1998; Breuer, 2000). Air supply to the fresh waste is ensured by the chimney draught principle: the waste is placed on pieces of bulky material (e.g. old pallets from bulky waste) to a height of approx. 2 m (Fig. 11-32). Fresh air flows into this area through drainage pipes (100 mm). The gas mixture in the waste is heated up by aerobic degradation processes and then experiences buoyancy due to the temperature difference to the colder external air – as in a chimney – and leaves the waste heap through its surface and a

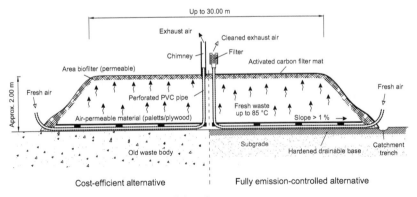

Fig. 11-32 Schematic of static natural draught aeration

chimney. Cold external air flows from below and re-supplies the decom-posing materials with oxygen. Before constructing the windrows, the material can be sieved, and only the reject is landfilled. On the compacted decomposed material, aeration is developed anew and fresh waste is arranged for new decomposition.

In the meantime, the basic principle has been modified in various ways, successfully tested for many years in climates of the most diverse countries – from the tropics (Brazil and Thailand, Maak, 2003) to arid regions (Iran, Körtel *et al.*, 2003; Körtel and Spillmann, 2005) – and also permanently used in practice (Fig. 11-33). It is available as a cost-efficient alternative (Fig. 11-32, left windrow half) and a fully emission-controlled alternative (Fig. 11-32, right windrow half).

11.4.2 Material conversion of wastes

11.4.2.1 Fundamentals of material differentiation of wastes

The oldest model which proved the most successful in practice was aimed at killing pathogenic organisms by controlled biological degrada-tion and producing a substrate for humus production which can be integrated into the environment (composting). This natural integration was stopped when the wastes contained increasingly toxic trace elements and persistent organic compounds from industrial wastes which were unable to be integrated into the environment. Switzerland was the first industrialised country to draw a conclusion and pursued the objective of fully oxidising all organic carbon compounds from non-recyclable wastes and landfilling the minerals as rock like materials. This model made it *de facto* compulsory in Switzerland that all residual wastes are incinerated. Other industrial countries – e.g. Germany – are currently implementing this principle.

The practice of thermal treatment of unspecified waste mixes has in the meantime shown that the minerals post-react intensively even after the best incineration, soluble salts are removed by way of the leachate and the inventory of trace elements in the slag exceeds that of the natural top soil by more than one to two orders of magnitude (see details in, e.g., Baccini and Gamper, 1994). This also applies to vitrified resi-dues. It has not been taken into account yet that in real engineering applications neither '0%' nor '100%' can be achieved. If for instance only 1% organic carbon is permitted in the residue (the current limiting value for landfill class I in German regulations), 1 tonne of incineration residue contains 10 kg organic coking residues which may include the entire spectrum of halogenated and non-halogenated organic

Fig. 11-33 Different examples of natural draught aeration for various climatic zones. German variants: 'chimney draught method' (top left, Spillmann, 1978) and modified 'Schwäbisch Hall model' (top right, Haschemi, 1998/Breuer, 2000). 'AMBRA' version of Faber Umwelttechnik GmbH, Alzey, for Brazil (2nd row left, Sao Sebastiao), Mexico (2nd row right, Atlacomulco) and Thailand (3rd row, Phitsanulok) (Münnich et al., 2001; GTZ, 2003; Maak, 2003), chimney draught system 'Tehran model' according to Körtel et al. (2003) and Körtel and Spillmann (2005); here on the landfill of Tehran (bottom)

pollutants. The objective, not to put any burden on future generations, completely oxidize the organic compounds and limit the mobility of trace elements to an environment-compatible extent on a long-term basis, cannot be achieved by the currently applied thermal treatment.

The perfect oxidation of organic hydrocarbon compounds and the production of minerals which can be integrated into the environment are results of chemical processes with high quality provisions. They can only be met if the reactants are known and the reaction conditions can be held within the tolerances permitted by technology. The conversion of the individual material components of municipal waste is not particularly difficult. Due to seasonal and regional changes, however, the reactants are not consistent in the continuously changing municipal waste mixture, because the mass ratios of the component fluxes vary considerably (e.g. different amounts of tree loppings in winter and summer). However, the goal of neutralisation can be achieved without high costs if the main waste groups are separated before thermal treatment. Although the mass of the different component fluxes may vary, their chemical-physical characteristics change little or may even remain constant, thus the treatment can be carried out even with a refined material differentiation.

Separation into the main components is possible and economically reasonable if the adhesive effect of the organic compounds (starch, protein, fats) is eliminated by their biological degradation into CO_2 and H_2O or by changing them into water-insoluble, not sticky humic substances. The biological conversion (at approx. 80°C) expels the volatile pollutants which can then be collected separately. After conversion, the material can be differentiated into three main fractions using wet separation without producing waste-water: minerals (up to coarse sand), an earth-like fraction and plastics. Metals, wood and rubber emerge as side groups (patent DE 19734565.4-09, Professor Spillmann).

Fine differentiation of the main groups in special treatment plants is current state-of-the-art practice. The washed mineral fraction can be used if needed; in any case it can be landfilled. The soil-like fraction can be thermally treated in two stages: the organic substance is burnt substoichiometrically (gassed) in the gasification stage without changing the minerals, and the carbon contained in the gas is only then completely oxidised under specified conditions without producing carbon black. Since mineral after-reactions are not expected from the minerally unchanged material of the first stage in this procedure, heavy metals can be adsorbed using clay additives – if necessary – and deposited as geologically stable sediments in a compacted form. Plastics can be used as a sulphur-free, high-quality fuel if the utilisation plant is designed to chemically bind chlorine from PVC. This is also state-of-the-art practice. The industry has suitable ways of utilisation for the other sub-groups (patent DE 19706328.4, Professor Spillmann).

485

11.4.2.2 Practical example for a material-differentiated treatment of fresh wastes

In principle it is insignificant which separation procedure is used for biological stabilisation. It is crucial, however, that the gaseous emissions are collected and the extent of the engineering action is appropriate to the success. A procedure was selected as an example that is particularly robust and cost-efficient and has low space requirement due to a great windrow height. It has been state-of-the-art practice for 15 years.

There were ideas of producing specific material fractions from a non-uniform residual waste in order to be able to utilise them. This initiated the concept of a material specific treatment system for the residual waste with biological stabilisation as its focus (Fig. 11-34) (Spillmann et al., 1998).

First separation stage: mechanical pre-treatment

The objective of the mechanical pre-treatment is to condition the residual waste in such a way that the conditions for a biochemical degradation of the organic constituents are optimised (Fig. 11-35). The major tasks are in particular granulisation and wearing down of fibres (enhancing the chances for biological reaction), homogenisation, adjustment of an optimum water content for the (aerobic or anaerobic) degradation processes, admixture of nutrients (e.g. adding sewage sludge), separation of specified high-calorie fractions and valuable materials (e.g. metals) and, finally, separation of visible unacceptable objects and pollutants (e.g. batteries, paint boxes). The mechanical processing is thus a first stage separation process and is a preventive measure both for personnel and the granulisation equipment on the one hand and helps avoid contamination in the composting process on the other. Granulisation using jet cutting proved suitable (Fig. 11-36), whose performance has already been proved under operational conditions (Nassour et al., 2002). The advantage of this processing method is that no wearing parts come into contact with the waste and dust generation is prevented. Granulisation, homogenisation and water enrichment can be performed simultaneously using this technology within the same equipment.

Second separation stage: biological stabilisation

The tasks of biochemical stabilisation are the conversion by oxidation of natural organic substances contained in the residual waste to carbon dioxide and water and/or into insoluble minerals and humic substances. It is the second separation stage in this concept, in which carbon is separated from the organic compounds on the one hand and volatile

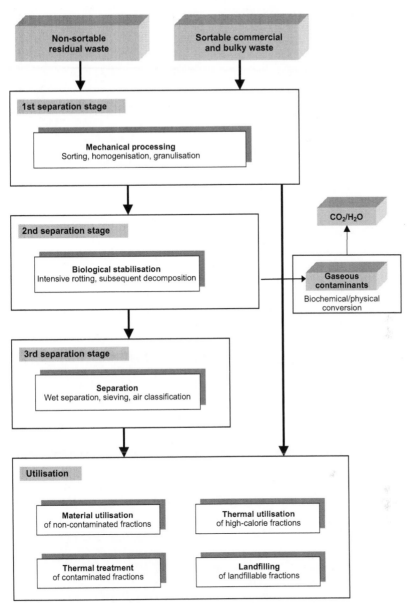

Fig. 11-34 Flow chart of a material-flux-specific waste treatment using material separation (Eschkötter, 2004; Eschkötter et al., 2004)

synthetic hydrocarbon compounds are removed, collected and treated on the other. Up to 70% of organic substances (oDS_{bio}) contained in the residual waste can be degraded under practical conditions in mechanical-biological treatment plants (MBTs) within about 10–20

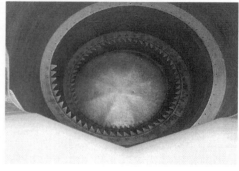

Fig. 11-35 Transfer station for the incoming dust carts to the windrows or boxes (windrows of the Bad Kreuznach, Germany, landfill). The plant was designed in such a manner that the waste can be transferred without a waste bunker directly into a tipping drum (diameter: 3.20 m; capacity: 30 tonne). Domestic waste can be mixed with sewage sludge and further additives can be added in the drum. An optimum mixture is obtained through a minimum of one-hour mixing in a slowly rotating drum – a principle used in dust carts for 50 years. Connected bundles can be disentangled by teeth arranged in a spiral line (principle Noell). External view (left) and interior (right), teeth after 5 years of use

weeks of intensive decomposing (Müller *et al.*, 1999). The fraction of the mass to be considered in a non-uniform material can be markedly reduced. The biological after-stabilisation using the natural draught of air ('chimney draught method', see principle in Section 11.4.1) further reduces the mass fraction to be separated and provides stability, which can meet the eluate limiting values for the disposal of Municipal Solid Waste Incineration (MSWI) material of landfill class II for $TOC_{eluate} < 100$ mg/l and for all others for landfill class I (Table 11-4).

Fig. 11-36 Mixing drum with integrated jet cutting device. Cooperative research project for the integration of jet cutting in the drum. Successful industrial application (Nassour et al., 2002)

In the example shown (Fig. 11-37), the waste was arranged in 8-metre-high windrows and aerated by pressure surges using high-pressure air enriched with oxygen through vertical compressed-air lances. Contaminated air was removed through suction lances and conveyed to suitable filter technology. The windrows were covered with gas-tight HDPE protective sheets. After 10 weeks of intensive active aeration, the subsequent decomposition took place in chimney-draught windrows.

Third separation stage: separation into material fractions

The Department of Waste Management of Rostock University (Spillmann *et al.*, 1998; Dohme and Eschkötter, 2001) carried out large-scale pilot tests aimed at the treatment of non-uniform material with regard to the German landfill limiting values. The best cost-benefit ratio was to be achieved by sieving the coarse fraction in a disk separator with >20 mm disk distance (not sieve aperture!) and cleaning adhesive materials before thermal utilisation (Fig. 11-38). The remaining mass can then be disposed of on a Class II landfill. In this simplest case about 0.3 tonne would be thermally utilised, 0.4 tonne landfilled and 0.3 tonne completely degraded biologically from 1 tonne dry waste matter of the raw material without any re-cleaning. Further differentiations have shown that dry sieving does not yield any further substantial improvement concerning the specific calorific values or the energy content of the <20 mm fractions.

If one considers the energy contents of the sieved fractions (Fig. 11-38, top), then one obtains about 30% of waste dry matter of the waste mass with a calorific value comparable to that of brown coal and the inventory of accompanying substances of the waste. This process cannot be used to produce a profit (status 2005). If the stabilised waste is differentiated into specified material types (in this case using wet separation technology, Fig. 11-39), then one obtains high-calorific fuels with specified characteristics (Fig. 11-40, bottom). A prerequisite for an extensive wet separation producing no waste water is stabilisation up to a TOC eluate $\ll$ 100 mg/l. This quality has been the state of the art for 30 years (Götze *et al.*, 1969; Kahmann, 1971; Spillmann, 1978, 1979, 1981, 1989, 1993a). The wash water can then be recycled. The wet material separation provided the material types shown in Fig. 11-41 with the relevant utilisation possibilities, as related to 1 tonne of initial dry matter.

Approximately two-thirds of the initial organic dry mass is converted by the biological conversion processes into CO_2 and water. Depending upon the fraction of the organic materials of the total waste, the part to

Table 11-4 Solid and eluate parameters compared with the TASi limiting values (1993) and recommended target values for biological stability, obtained after 25 weeks of intensive rotting and 3 months of subsequent decomposition (solid criteria) and 8 months of subsequent decomposition (eluate criteria)

Parameter	Unit	Residual waste after intensive rotting and subsequent decomposition				TASi limiting values (1993)	
		<8 mm	8–40 mm	>40 mm	Average	LC* I	LC* II
Solid[1]							
Ignition loss	% by weight DM**	20.3	20.3	33.5	23.5	3	5
Ignition loss corrected	% by weight DM**	12.8	18.9	16.7	16.0	–	–
TOC	% by weight DM**	13.3	12.2	14.7	13.2	5	3
Respiration act. in 96h	mg O_2/g DM**	<1	<1	4.5	<1.9	5	Target values for biological stabilisation
Gas production in 35d	l/kg	<1	2	3	<1.9	20 (in 21 d)	
Eluate[2]							
pH value	–	7.7	7.9	7.6	7.7	5.5–13.0	5.5–13.0
Conductivity	µS/cm	2420	1940	2160	2182	10 000	50 000
TOC	mg/l	44	44	61	48	20	100
Phenols	mg/l	0.01	0.02	0.024	0.017	0.2	50
As	mg/l	<0.04	<0.04	<0.04	<0.04	0.2	0.5
Pb	mg/l	<0.02	<0.02	<0.02	<0.02	0.2	1
Cd	mg/l	<0.002	<0.002	<0.002	<0.002	0.05	0.1
Cr-VI	mg/l	<0.05	<0.05	<0.05	<0.05	0.05	0.1
Cu	mg/l	0.053	0.044	0.053	0.05	1	5
Ni	mg/l	<0.02	<0.02	<0.02	<0.02	0.2	1
Hg	mg/l	<0.0002	<0.0002	<0.0002	<0.0002	0.005	0.02
Zn	mg/l	0.07	0.05	0.12	0.07	2	5
Fluoride	mg/l	0.12	0.15	0.12	0.13	5	25
Ammonium-N	mg/l	0.8	0.25	0.02	0.41	4	200

Cyanide (easy to release)	mg/l	<0.01	<0.01	<0.01	<0.01	0.1	0.5
AOX	mg/l	0.2	0.33	0.2	0.25	0.3	1.5
Water-soluble fraction	% by weight	0.19	0.16	0.2	0.18	3	6
COD	mg O_2/l	111	118	195	134	–	–
BOD$_5$	mg O_2/l	<3	<3	<3	<3	–	–

* LC: landfill class; ** DM: dry matter; [1] 3 months of subsequent decomposition; [2] 8 months of subsequent decomposition

Fig. 11-37 *Full-scale test rig for mechanical-biological waste treatment with separation of non-uniform material into specified fractions (Eschkötter, 2004)*

Fig. 11-38 *Separation of aerobically stabilised residual waste. Shredder (background) with attached disk separator (front) of the EuRec Technology GmbH (Eschkötter, 2004)*

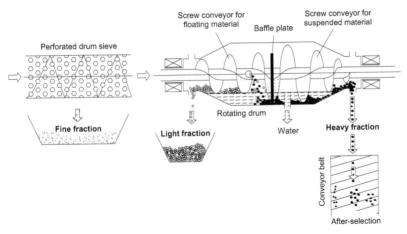

Fig. 11-39 *Wash drum; schematic set-up for wet separation of mineralised residual waste*

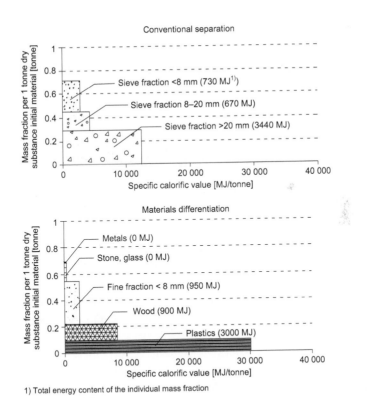

1) Total energy content of the individual mass fraction

Fig. 11-40 *Comparison of energy contents between dry sieve separation and material differentiation by wet separation (Spillmann et al., 2002)*

493

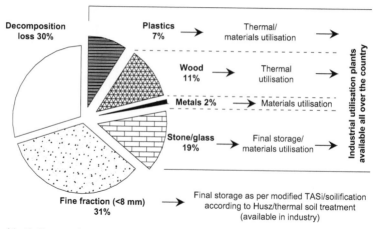

Fig. 11-41 Separated material groups and their utilisation possibilities (Eschkötter, 2004)

be further treated is reduced by around 20% to 40%, in this by 30%. The fraction of stones and glass (14%) is freed from adhering material as far as possible and can undergo a profitable use (e.g. in road construction) or can be landfilled as an inert material according to current TASi regulations. The metal fraction can also be used. The fine material is submitted to a chemical-physical analysis and can be further utilised depending on its properties and/or treated thermally like the soil. Numerous utilisation plants exist enabling thermal utilisation of old timber. The landfilling properties of the fine material <8 mm fulfil the parameters for landfilling of mechanical-biologically pre-treated residual waste and can also be landfilled on Class II landfills after 2005. The substantial difference to simple sieve separation is in the vigorous differentiation and the suitability for the utilisation of the material groups:

- Only about 10% of the mass is plastics in the current example which however contains at least half of the entire energy (Fig. 11-40, bottom) and can be processed to yield high-quality, specified fuels for high temperature processes or diesel (in the meantime the catalytic depolymerisation has become an economical technology for the conversion of plastics into diesel or heating oil).

- About the same amount of energy is contained in the materially undefined sieved calorie-rich fraction of the threefold mass, which has to be classified as a waste material due to the impurities from the process engineering. Due to its low specific calorific value it is not suitable for high temperature processes.

494

- The wood fraction with 10% of waste dry matter is unusually high for residual waste, but it is insignificant in comparison to the thermal utilisation of contaminated wood in industry and can be utilised without any problem.
- As opposed to the wood fraction the 8–20 mm sieve fraction cannot be thermally utilised as a waste according to current regulations, but must be treated thermally with provided energy. This, together with the soil-like fine fraction, yields an amount for thermal treatment of about 40% of waste dry matter in the case of sieve classification, while only 30% of waste dry matter has to be treated thermally under the same conditions if material differentiation is used. This mass can still be reduced if a sand fraction can be separated.
- About 14% of waste dry matter can be used as a coarse mineral fraction in commercial plants without any further thermal treatment using material differentiation.

The main advantage of the presented simple material differentiation of unspecified waste mixes is that reducing the need for thermal treatment and improving the utilisation possibilities can minimise the costs. In this consideration, German guidelines are the basis of an evaluation which does not apply everywhere.

Channelling the pollutants

In addition to the actual separation, wet separation also enables a channelling of pollutant fractions with regard to their calorific value. For the example under consideration the pollutant fractions shown in the following figures apply for very volatile gaseous pollutants (Fig. 11-42) and heavy metals (Fig. 11-43). The non-mobile heavy metal contents of the fractions stone/glass with 24.6% and metal with 1.7% of the stabilised residual waste were not taken into account.

Aerobic degradation drives out the volatile toxic substances such as benzene in the high-temperature phase (approx. 80°C) towards treatment (Fig. 11-42). The solid content remaining after decomposition is wet-separated and analysed for heavy metal contents (shown as content related to the total mass of the stabilised residual waste). The heavy metal content of the hard plastics fraction is remarkably high, although the mass proportion of this fraction is only 3.9% of the stabilised residual waste (Fig. 11-43). Initially the heavy metals Cu and Cd are strongly represented, since they are used as colorants (these heavy metals fall out with the catalytic converter in catalytic

495

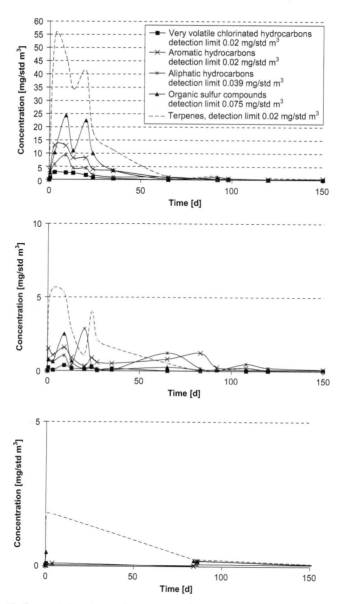

Fig. 11-42 Gas emissions in aerobic intensive decomposition of a mechanical-biological waste treatment (Eschkötter, 2004)

depolymerisation of plastics). The wood fraction also exhibits a high heavy metal content. The toxic elements arsenic and nickel dominate here. More than half of the heavy metal content apart form Cd is concentrated in the fine fraction. Lead and zinc make up about 70%

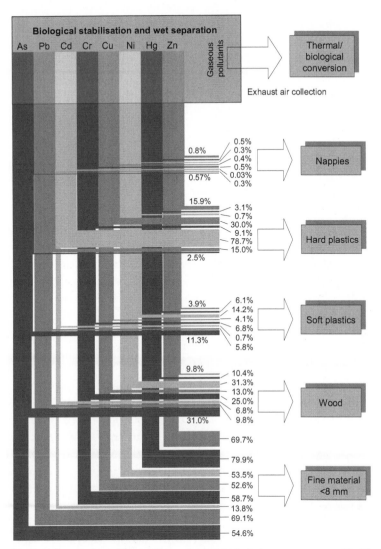

Fig. 11-43 Pollutant channelling within mechanical-biological waste treatment (Eschkötter, 2004)

and mercury about 80% of the respective total content. The majority of the heavy metal content could be enriched in the fine fraction by washing the adhering materials during the wet separation.

The differentiated constituents of the heavy metal fraction show the advantages of the material-fraction-oriented treatment. Utilisation methods of the individual fractions can be adjusted to the specific heavy metal content. Thus nearly 90% of the Cd content can be

497

removed through the hard plastics fraction and a majority the arsenic content through the wood fraction. The majority of the heavy metal contamination is concentrated in the fine fraction. If utilisation of this fraction is impossible due to an elevated contamination, then about one third of the entire input remains after biological stabilisation and wet separation, which must be further classified as a waste in the sense of intensive waste management.

Production of high-quality fuel

Separation of a washed plastic fraction provides no fuel so far. The plastic mix contains chemically and physically different materials.

Fuels for chemically high requirements can be obtained according to the state of the art only with a granulisation. This could be ultrasonic re-cleaning and separation into $\rho > 1.0\,g/cm^3$ and $\rho < 1.0\,g/cm^3$. The fraction $\rho < 1.0\,g/cm^3$ contains the bulk of the plastics (PE; PP) without any metal (approx. 80%). This material is free from sulphur and chlorine, thus reaches the minimum quality for desulphurised heating oil.

Constant physical conditions must be established in the second step. The fuel must either be atomised and burn similarly to a fuel vapour in suspension, or it must be supplied in well-defined pieces (agglomerates) with a homogeneous burning profile. The form of delivery is decided by the customer and the price.

The $\rho > 1.0\,g/cm^3$ fraction, in addition to halogenated compounds and metals, also contains heavy halogen- and metal-free plastics such as non-foamed polystyrene. This fraction is therefore also suitable for thermal utilisation if the composition is homogenised and constant physical characteristics are ensured. A fluidised bed reactor was developed by Rostock University for instance (Steinbrecht and Neidel, 2001), in whose fluidised bed the radicals are bound by additives in status nascendi and whose temperature can be controlled with a specified fuel so precisely that the mineral halogen compounds do not disintegrate. The prerequisite for this form of utilisation is a specified agglomeration. In addition to dust-removal from the exhaust gases, a 'police filter' (e.g. coke) is only necessary.

11.5   Summary of applications in practice

Justification of the actions

It can be concluded from the results of the investigations presented that groundwater contamination can only be avoided if the materials can be integrated into the environment and the disposal suits the local conditions. All substances which fail to fulfil these conditions are likely to impair the environment over an in-calculably long period. Research results suggest periods of more than a hundred years for landfills to keep 'bleeding'. It has been proved that the self-cleaning capability of the aquifer is not sufficient to degrade or immobilise these substances sufficiently in a stable form. Assumptions prevailing in current practice concerning the extent, duration and range of toxic and carcinogenic substances emitted by the leachate from landfills are therefore far on the unsafe side. The discussions about their action threshold values and lately insignificance threshold values are being carried on (GICON, 2004).

The pollutants often move on very narrowly limited path lines with low transverse dispersion. A density deviating from that of groundwater induces a vertical flow. Therefore, the currently applied well tests only have limited suitability to determine the groundwater contamination. Depth-oriented sampling and measurements using a MIP (membrane interface probe) are considered more effective remedies.

Since available fresh water in arid areas sets limits to the possibilities of human existence and it is not in abundance in humid areas either, it is necessary both in industrialised and less developed countries to take effective precautions against this serious and long-lasting risk on drinking water resources.

Protection of drinking water from emissions from existing deposits

Even rich industrialised countries cannot afford to carry out all immediate material conversions of the numerous existing old landfill sites and their location-relevant integration into the environment. Therefore drinking water resources and their catchment areas must first be protected. It is necessary for this purpose to assign the drinking water resources (usually well known geologically) to historical and current contamination sources by knowledgeable specialists (toxic trace elements from mining and metallurgical industry; polycyclic aromatics from coal coking and tar processing; extremely toxic organic substances from military contaminated sites of the 1st and 2nd World War; municipal mixed waste landfills with toxic industrial wastes of all kinds in today partially unknown locations).

For investigation of the current groundwater contamination it is necessary to find the actual contaminated path lines by profile measurement and layer sampling for which small driven filter wells are sufficient. The programme of analysis is focused on the toxicity and carcinogenicity of the material mixes measured directly on living warm-blooded creature cells.

For the assessment of the inventory of unknown mixed landfills the following procedure can provide the most reliable results at minimum costs. First the phase of biological activity and/or re-activation should be estimated by analysing the main gases in certain areas of the landfill. Then the type and position of industrial wastes have to be determined by toxic trace gas analysis using the enrichment method. In the case of perched water the samples have to be supplemented analogous to groundwater investigation. Toxicity and carcinogenity in contamination centres will be directly measured in the future similarly to groundwater. Only a few of the expensive material samples are then necessary to perform informative solid analyses to find the contamination centres.

The allocation of current and potential risk to drinking water resources is the basis for choosing the preventive measures. Three possibilities are available:

1. Material conversion of the deposit with consecutive utilisation or integration of the products into the environment; no aftercare is necessary.
2. Material conversion with in-situ integration into the environment if necessary using engineered groundwater protection during the conversion; long-term control of aftercare necessary.
3. Technical immobilisation of the deposit to gain time until the material conversion of the wastes; intensive aftercare up to the conversion stage is necessary.

Current state of the art technology does not enable an infinite technical immobilisation which is, however, still the object of valid legislation (TASi, 1993). Fortunately, landfilling without any pre-treatment has been forbidden in Germany since June 2005.

Based on the state of the art, the material conversion combined with utilisation or disposal of the products suitable to the specific location can be achieved in two steps. In the first step the landfill is sufficiently stabilised in-situ by engineered rearrangement of gas and water management that it can be safely excavated and the material separated with regard to its composition and chemical contamination. In the second step the main groups are processed and polluting industrial materials and fractions contaminated with toxic materials are converted based on available technology and depending on the material. For mixed landfills the conversion of anaerobic to aerobic degradation using pressure surge aeration (BIOPUSTER® method), including controlled exhaust and cleaning of the reaction gases, has been proved a the suitable first step over the last 15 years. The subsequent steps have been state-of-the-art separation technology for a long time. If the biological stabilisation is continued so that water-insoluble materials similar to humic substances are produced in situ or in separate equipment, a very selective wet separation can be performed without producing waste water. The products can then be specified exactly and used in price-efficient industrial utilisation and processing methods.

Biological in-situ stabilisation uses the findings proved by our research so that landfills with minor industrial contamination can be converted into permanently aerobic landfills where their emissions can no longer be activated. During the active phase the groundwater is protected, until contamination is no longer expected. If this goal cannot be achieved against expectation, the landfill can be excavated safely and its material utilised. An anaerobic stabilisation results in possible reactivation and is therefore not sufficient for the protection of drinking water, specifically since extremely dangerous leachate and gases develop during this long period.

Immobilisation by encapsulation requires an engineering structure, which – as any engineering structure – only has a limited service life. Its sole purpose can only be to gain time until a material conversion. Regarding immobilisation it has to be born in mind that not only convection transports contaminants, but diffusion can also contribute considerably to emissions. Also, it has to be noted that 'technically impervious' can leave residual permeability in the subsoil which is acceptable in hydraulic engineering but not for protection from toxic and carcinogenic trace substances. The 'Wiener Dichtwand Kammersystem' (Vienna chamber system) is an example for a doubtless perfect solution of the convection problem because flow is directed through the residual leaks from the outside inward. Gross permeability and damage can be recognised by the changing water balance in the chamber, and they can be localised and repaired ('fail to safe'). If the diffusion problem must also be solved in addition to the convection problem, a plastic-metal composite plate can be inserted into the inner ring of the cut-off wall. On slopes, in water-bearing cracks and in aquifers with a large water level gradient, it is often more effective to combine impervious structures with an artificial drainage ('hydraulic short-circuit') instead of insisting on a solution based exclusively on barriers. The cost of containment is additional expenditure to gain time for material conversion in each case but containment cannot replace conversion.

Groundwater protection against contaminations from present and future wastes

An immediately effective protection of groundwater against contamination from wastes can be achieved by a disposal of waste on top of an engineered barrier which is impervious to convection and diffusion, is kept reliably drained and is situated on a low-permeable subgrade at a great distance from the groundwater. Environmental contamination can be excluded based on the level of sewage treatment and gas treatment technology, as long as the engineered barrier is regularly maintained and the treatment equipment is properly operated. The same design principle applies to the engineered barriers as for the encapsulation of an old landfill. The drainage layer made of coarse, low-lime crushed stone (e.g. 16/32 mm granulation) only remains operational if the initial acid phase (hydrolysis) of the degradation processes is safely avoided by technical acceleration, e.g. by aerobic degradation of organic acids before disposal. Due to the proven re-activation, anaerobic deposits of this kind can be regarded as equal to immobilised old deposits. As proved, the emissions last substantially longer (>100 years) than can a guarantee be provided for the barriers. The necessary material conversion must therefore be performed at a later time at the expense of future generations.

If wastes requiring special monitoring are collected separately and converted according to their material properties into compounds which can be integrated into the environment, the materials of the residual waste mix can be converted using the same simple means as for an old landfill.

Protection of the atmosphere from emissions from existing landfills

The research was focused on groundwater protection and provided wide ranging findings. The results of the emergence of landfill fires by thermophilic/hyperthermophilic bacterial activity, obtained as a side effect, must trigger a re-consideration of the concerns about existing landfills and valuable material repositories. An alarming result is that acid-proof plastics believed to be inert provide sulphur as a foodstuff for the biogenic process. Furthermore it is startling to find that the warm environment created by the bacteria is sufficient for the degradation process or the chemical breakdown (cracking) of plastics, i.e. between 70 and 80°C – or even at lower temperatures if suitable pressure is available – the biogenic processes are already overtaken or even displaced by chemical processes. The landfill then becomes a reactor to be controlled at high cost which can produce gas fires over years (see Section 11.3.2.3). The bacteria alone produce highly reactive carbon disulfide CS_2, aliphatic hydrocarbons with chain lengths of C 3 to C 12 and BTEX aromatics.

The cause of landfill fires now moves to the forefront of public interest, particularly since the aftercare of 300 landfills in Germany now begins. Today it is already certain that the engineered landfill cap fails to prevent this biogenic process and, under certain conditions, even intensifies it by heat accumulation. Nevertheless large quantities of plastic wastes were still deposited in landfills in 2005, even though being a valuable resource, since plastics – instead of causing problems in landfills – can be used to produce oil using a simple technology. For instance Israel possesses no primary energy and stores 1.0 million tonnes of plastics on a landfill on the outskirts of Tel Aviv (Struve, 2003) – equivalent to 800 millions litres of diesel oil.

Concluding remark of the editors

In our daily activities we see that the results published here have international consequences. In many third world countries, the industrial revolution is only now beginning, i.e. many countries remain at the beginning of the problems which industrialised countries have already faced. For the majority of the world's population the quality of life of a clean environment is a dream objective. Therefore the results and techniques presented here should be implemented both in practice and in technical regulations not only in the industrialised countries, but also world-wide for the improvement of the environment.

The increasing frequency of natural catastrophes raises many questions which are concentrated on global warming and the effect of greenhouse gases. The globalisation of the environment moved into a crucial phase by passing the Kyoto Protocol on climate control in 2005. This protocol also embodies the promotion of co-operation between industrialised and developing countries in the field of climate protection. For this purpose the Clean Development Mechanism (CDM) ('mechanism for an environmentally compatible development') has been established. In view of the enormous, unsecured and partially burning mountains of waste in the world, the Kyoto Protocol can be found to a large degree in this book, so that the topicality of the 30 years of research and engineering tests presented could not be more explosive.

Our knowledge and its globalisation compel us to act.

References

Abbt-Braun, G. (1987): Untersuchungen zur Struktur isolierter Huminstoffe: Allgemeine Charakterisierung und massenspektrometrische Bestimmungen. München, Technische Universität, Inst. f. Wasserchemie u. Chemische Balneologie. (Investigations into the structure of isolated humic substances: General characterisation and mass spectrometric determination. Institute for Water Chemistry and Chemical Balneology, Munich Technical University). Thesis.

AbfAblV (2001): Verordnung über die umweltverträgliche Ablagerung von Siedlungsabfällen (Abfallablagerungsverordnung – AbfAblV) – Artikel 1 der Verordnung vom 20. Februar 2001 (Regulations on Environmentally Compatible Deposition of Municipal Waste (Waste Deposition Regulations – AbfAblV) – Article 1 of the Regulations of 20 February 2001), BGBl (Federal Gazette). I p. 305, amended by Article 2 of the Regulations of 24 July 2002, BGBl (Federal Gazette).

AbfKlärV (1992): Klärschlammverordnung vom 15. April 1992 (Sewage Sludge Regulations of 15 April 1992), BGBl (Federal Gazette). I p. 912, last amended Art. 2 Verordnung zur Änderung abfallrechtlicher Nachweisbestimmungen vom 25. April 2002 (Regulations for the Amendment of Waste Regulatory Proof Provisions of 25 April 2002), BGBl (Federal Gazette).

AbwV (2002): Verordnung über Anforderungen an das Einleiten von Abwasser in Gewässer (Abwasserverordnung – AbwV) – vom 15. Oktober 2002 (Regulation on the provisions of releasing waste water into receiving waters (Abwasserverordnung – AbwV, Waste Water Regulations) – of 15 October 2002, BGBl (Federal Gazette).

Alajberg, A., P. Arpino, D. Deur-Siftar and G. Guiochon (1980): Investigation of some vinyl polymers by pyrolysis-gas chromatography-mass spectrometry. J. Anal. Appl. Pyrolysis 1, 203–212.

ARGE Biopuster (1998): Erfolgsnachweis der aeroben Stabilisierung eines verdichteten Deponiekörpers zur Gewährleistung des Arbeitsschutzes während der Reparatur des Entwässerungssystems (Proof of aerobic stabilisation of a compacted waste body to guarantee work safety while repairing the drainage system). Archive of Arbeitsgemeinschaft Biopuster (Biopuster Working Group), Wien (Bilfinger Berger Baugesellschaft mbH, Porr Umwelttechnik GmbH, G. Hinteregger Baugesellschft mbH) (unpublished).

ATSDR (2002): Reports of the Agency for toxic substances and disease registry. USA, http://www.atsdr.cdr.gov.

Atwater, J., S. Jasper, D. Muvinic and F. Koch (1983): Experiments using daphnia to measure landfill leachate toxicity. *Water Research* 17, 1855–1861.

Audisio, G. and F. Bertini (1992): Molecular weight and pyrolysis products distribution of polymers. I Polystyrene. *J. Anal. Appl. Pyrolysis* 24, 61–74.

Baccini, P. and B. Gamper (eds) (1994): Deponierung fester Rückstände aus der Abfallwirtschaft (Landfilling of solid residues from waste management). vdf Hochschulverlag AG a. d. ETH Zürich.

Ballin, G., P. Hartmann and P. Spillmann (2004): Ermittlung der Ursachen zur Selbstentzündung von Kunststoffmonodeponie (Determination of the causes of spontaneous ignition in plastic mono landfills). In: DepoTech 2004, A.A. Balkema/Rotterdam/Brookfield.

Ballin, G., P. Hartmann, D. Steinbrecht and P. Spillmann (2005): The spontaneous combustion of apparent inert plastics – problem, research method, solution. 9th CEST International Conference on Environmental Science and Technology, Rhodes Island; Global NEST and Dep. of Environment Studies, Univ. of the Aegean.

Bracke, R. (2002): Stand der Entwicklung bei der Durchführung von Geomonitorings mit kombinierter in-situ Direct-Push-Feldmesstechnik (State of the art in the implementation of geomonitoring using combined in-situ direct push field measurement technology). BEW/LUA NRW Seminar: Gefährdungsabschätzung und Sanierung von Altlasten (Risk assessment and remediation of contaminated sites), 02–03 July 2002.

Brammer, F. (1997): Rückbau von Siedlungsabfalldeponien – Schrittfolge und Entscheidungskriterien bei Planung und Ausführung (Mining of MSW landfills – process sequence and decision criteria in planning and execution). Papierflieger (Flyer), Thesis at the Carolo-Wilhelmina Technical University of Braunschweig, FB Bauing- und Vermessungswesen (Faculty of Civil Engineering and Surveying).

Brammer, F. and H.-J. Collins (1995): Aufnehmen – Verwerten – Deponieren von Müll aus alten Kippen – A Abfallwirtschaftliches Teilprojekt. Abschlußbericht zum interdisziplinären Verbundprojekt (Mining – re-using – depositing landfill wastes from old sites. Waste Management Subproject, Final Report on an Interdisciplinary Integrated Project) AZ. II/67336 commissioned by the Volkswagen Foundation, Hannover.

Brammer, F., M. Bahadir, H.-J. Collins, H. Hanert and E. Koch (eds) (1997): Rückbau von Siedlungsabfalldeponien (Mining of MSW landfills). B. G., Teubner Verlagsges., Stuttgart, Leipzig.

Breuer, W. (2000): Optimierung des Schwäbisch Haller Verfahrens zur Belüftung von Müllrottemieten (Optimization of the Schwäbisch Hall method for aerating waste decomposition windrows). Diplomarbeit an der Universität Rostock, Inst. f. Landschaftsbau und Abfallwirtschaft (Final thesis at the University of Rostock, Institute of Landscape Engineering and Waste Management).

Budde, F., P. Chlan and T. Dörrie (2002): Landfill restoration with the BIOPUSTER® – System – Aeration as prerequisite for occupational-, residential and environmental safety. EUROARAB 2002, 10–12 October 2002, Institute of Landscape Engineering and Waste Management, University of Rostock.

Bundesgesetzblatt BGBl. I (2001): Verordnung über die umweltverträgliche Ablagerung von Siedlungsabfällen und über biologische Abfallbehandlungsanlagen (Regulations on environmentally compatible deposition of municipal solid wastes and biological waste treatment facilities), BGBl (Federal Gazette).

Chammah, A., H.-J. Collins, H.-G Ramke and P. Spillmann (1987): Einfluss von Recycling-Maßnahmen auf den Wasser- und Stoffhaushalt von Hausmülldeponien (Influence of recycling measures on water and materials balance of municipal waste landfills) – *Müll und Abfall (Waste and Refuse)*, 19(9), 353–358.

Chefetz, B., P. G. Hatcher, Y. Hadar and Y. Chen (1998a): Characterization of dissolved organic matter extracted from composted municipal solid waste. *Soil Sci. Soc. Am. J.* 62, 326–332.

Chefetz, B., Y. Hadar and Y. Chen (1998b): Dissolved organic carbon fractions formed during composting of municipal solid waste: properties and significance. *Acta Hydrochim. Hydrobiol.* 3, 172–179.

Chiavari, G. and G. C. Galletti (1992): Pyrolysis-gas chromatography/mass spectrometry of amino acids. *J. Anal. Appl. Pyrolysis* 24, 123–137.

Christy, A. A., A. Bruchet and D. Rybacki (1999): Characterization of natural organic matter by pyrolysis/gc-ms. *Environ. Intern.* 2–3, 181–189.

Collins, H.-J. and P. Spillmann (1990): Lagerungsdichte und Sickerwasser einer Modelldeponie von selektiertem Hausmüll (Bulk density and leachate of a model landfill of selected municipal waste). *Müll und Abfall* 6, 365–373.

Collins, H.-J., D. Maak and C. Reiff (1998): Das Kaminzugverfahren als entscheidende Aktivität einer stoffstromspezifischen Restabfallbehandlung In: Neues aus Forschung und Praxis, Bio- und Restabfallbehandlung II, biologisch, mechanisch, thermisch (The chimney draught method as a key component of a materials flow-specific treatment of residual waste. In: New from research and practice, Bio and residual waste treatment II, Biological, mechanical, thermal methods). Witzenhausen: Baeza-Verlag, 557–579.

Cunliffe, M. and W. T. Williams (1998): Composition of oils derived from the batch pyrolysis of tyres. *J. Anal. Appl. Pyrolysis* 44, 131–152.

Degener, P. (2006): Sickerwasserkreislauf zur Behandlung von Sickerwässern der aerob-biologischen Restabfallbehandlung – Diss. a.d. Agrar- u. Umweltwiss. Fak. d. Universität Rostock. (Leachate re-circulation for leachate treatment in aerobic biological treatment of residual waste. Thesis at the Faculty of Agriculture and Environmental Protection, University of Rostock).

Degener, P., M. Franke, P. Spillmann and B. Sprenger (2004): Rottesimulationsreaktoren zur Optimierung der aerob biologischen Restabfallbehandlung (Decomposition simulation reactors for optimization of aerobic biological treatment of residual waste). *Müll und Abfall (Waste and Refuse)* 8, 373–377.

De Man, J. C. (1975): The probability of most probable numbers. *European J. Appl. Microbiol.* 1, 67–78.

Derenne, S. and C. Largeau (2001): A review of some important families of refractory macromolecules: composition, origin and fate in soils and sediments. *Soil Sci.* 11, 833–847.

De Smedt, F. Wauters and J. Sevilla (1986): Study of tracer movement through unsaturated sand. *Journal of Hydrology* 85, 169–181.

DFG (ed.) (1982): Rückstandsanalytik von Pflanzenschutzmitteln (Residual analysis of pesticides). VCH Verlag Chemie, Weinheim-Deerfield Beach-Basel.

Dinel, H., M. Schnitzer and S. Dumontet (1996): Compost maturity: extractable lipids as indicators of organic matter stability. *Compost Sci. Util.* 2, 6–12.

Dohme, M. and H. Eschkötter (2001): Stoffgerechte Restabfallbehandlung am Beispiel der MBA-Demonstrationsanlage 'Am Langen Feld' in Wien. In: Tagungsband 4. Dialog 'Abfallwirtschaft M-V'., Universität Rostock, Institut für Landschaftsbau und Abfallwirtschaft. (Material-related treatment of residual waste using the example of the MBT demonstration facility in 'Am Langen Feld' Vienna. Proceedings 4th Dialogue 'Waste Management Mecklenburg-Western Pomerania'. Institute of Landscape Engineering and Waste Management Rostock University). 121–140. Rostock, June 2001.

Dörrie, T. (2000): Kombinierte Ablagerung von MVA-Schlacken und MBA-Abfällen, Diplomarbeit an der Universität Rostock, Institut für LBAW. (Co-disposal of waste incineration slags and MBT wastes. Final thesis at the University of Rostock, Institute of Landscape Engineering and Waste Management).

Dörrie, T., P. Spillmann and M. Struve (2001): In-situ-Stabilisierung von Altdeponien ohne und mit Rückbau. 3. Symposium Natural Attenuation – Umsetzung, Finanzierung, Perspektiven (In-situ stabilisation of old landfills with and without landfill mining. 3rd Natural Attenuation Symposium – Implementation, financing, perspectives). 4–5 December, DECHEMA Haus, Frankfurt a. M.

Dörrie, T., W. Breuer, A. Nassour and P. Spillmann (2000): Dokumentation zum Rückbau der Deponie 'Helene Berger' in Niederösterreich. In: Wasserwirtschaftliche Sanierung von Bergbaukippen, Halden und Deponien (Records on the mining of the 'Helene Berger' landfill in Lower Austria. In: Water management reclamation of mining dumps, waste tips and landfills). 51. Berg- und Hüttenmännischen Tag 2000, in Freiberg, TU Bergakademie (Mining and Metallurgical Conference 2000, University of Mining and Technology), Freiberg/Saxony, Germany, 278–293.

Ehrig, H.-J. (1980): Beitrag zum quantitativen u. qualitativen Wasserhaushalt von Mülldeponien (On the quantitative and qualitative water balance of landfills). – Veröff. d. Inst. f. Stadtbauwesen (Publications of the Institute of Town Planning), Braunschweig Technical University, 26, 1st edition. 1978, 2nd extended edition 1980.

Ehrig, H.-J. (1986): Chemische Untersuchungen, Abwassertechnische Summenparameter (Chemical investigations, waste water management sum parameters). In: Spillmann, P. (ed.) (1986), Chapter 5.1 and 9.

Eikmann, Th., S. Michels, Th. Krieger and H. J. Einbrodt (1989): Medizinische Aspekte bei der Untersuchung und Bewertung von Altlasten. Forum Städte-Hygiene (Medical aspects of the investigation and assessment of contaminated sites. Town Hygiene Forum) 40(4) 239–244.

Eschkötter, H. (2004): Die mechanisch-biologische Restabfallbehandlung als Bestandteil eines verwertungsorientierten Stoffstrommanagement – Abfallwirtschaft in Forschung und Praxis (Mechanical-biological treatment of residual wastes as a component of an utilization-oriented material flow management – Waste management in research and practice). Vol. 131, E. Schmidt-Verlag, Berlin.

Eschkötter, H., A. Körtel and P. Spillmann (2004): Senkung der statistischen Entropie der Siedlungsabfälle durch biologischen Degradation (Reduction of statistic entropy of municipal solid wastes using biological degradation). In: DepoTech 2004, A.A. Balkema/Rotterdam/Brookfield.

EU-DepRL (1999): Council Directive 1999/31/EC of 26 April 1999 on the landfill of waste 399L0031 OJ L 182, 16/07/1999 pp. 0001–0019.

European Commission DG XI.E3 (ed.) (2000): The Behaviour of PVC in Landfill – Final Report by ARGUS (Germany), Prof. Spillmann (University of Rostock), Carl Bro (Denmark) and Sigma Plan (Greece). Website of the European Commission, http://www.europa.eu.int./comm./environment/waste/report7.htm.

Faix, O., D. Meier and I. Grobe (1987): Studies on isolated lignins in woody materials by pyrolysis-gas chromatography-mass spectrometry and off-line pyrolysis-gas chromatography with flame ionization detection. *J. Anal. Appl. Pyrolysis* 11, 403–416.

Filip, Z. (1983): Beurteilung von Stabilisierungsvorgängen in deponiertem Hausmüll mit Hilfe eines Respirationstests (Assessment of stabilisation processes in landfilled municipal waste using a respiration test). *Forum Städtehygiene (Town Hygiene Forum)* 34, 139–143.

Filip, Z. and J. Berthelin (2001): Analytical determination of the microbial utilization and transformation of humic acids extracted from municipal refuse. *Fresenius J. Anal. Chem.* 371, 675–681.

Filip, Z. and E. Küster (1979): Microbial activity and the turnover of organic matter in municipal refuse disposed of in a landfill. *Europ. J. Appl. Microbiol. Biotechnol.* 7, 277–280.

Filip, Z., W. Pecher and J. Berthelin (2000): Microbial utilization and transformation of humic acid-like substances from a mixture of municipal refuse and sewage sludge disposed of in a landfill. *Environmental Pollution* 109, 83–89.

Filip, Z. and R. Smed-Hildmann (1995): Huminstoffe im Feststoffmaterial der Modellporengrundwasserleiter (Humic substances in the solids of model porous aquifers). In: Spillmann et al. (1995), Chapter 5.3.

Fischer, J. (1996): Bestimmung organischer Schadstoffe im Abfall und Sickerwasser (Determination of organic pollutants in waste and leachate), Thesis. Technical University of Braunschweig 1995, Papierflieger (Flyer), Clausthal Zellerfeld.

Franke, M. (2003): Wasserhaushaltsprognose in Deponie mit Sickerwasserk-reislaufführung. In: Inst. f. Landschaftsbau u. Abfallwirtschaft d. Universität Rostock (Hrsg.): 6. Dialog Abfallwirtschaft M-V: Betrieb, Stilllegung und Nachsorge- von Deponien, S. 65-85; Rostock, Universitätsdruckerei (Forecast of water balance in landfills with leachate re-circulation. In: Institute of Landscape Engineering and Waste Management, University of Rostock (ed.): 6th Dialogue 'Waste management Mecklenburg-Western Pomerania': Operation, closure and aftercare of landfills, pp. 65–85; Rostock, University Press).

Franke, M. (2004): Massenspektrometrische Untersuchungen an Prozesswässern und Feststoffen aus der aerob mechanisch-biologischen Restabfallbehandlung. Universität Rostock, Inst. f. Umweltingenieurwesen, Diss. 150 S. (Mass-spectrometric investigations on process water and solids from aerobic mechanical-biological treatment of residual waste. University of Rostock, Institute of Environmental Engineering). Thesis.

Franke, M. and P. Degener (2003): Wasserhaushalt in der offenen Nachrotte von Restabfällen (Water balance in open post-decomposition of residual wastes). *Müll und Abfall (Waste and Refuse)* 8, 406–409.

Franke, M., G. Jandl and P. Leinweber (2005): Organic compounds in re-circulated leachates of aerobic biological treated municipal solid waste. *Biodegradation* 17, 473–485.

Franke, M., G. Jandl and P. Leinweber (2007): Analytical pyrolysis of recircu-lated laechates: towards an improved municipal waste treatment. *Journal of Analytical and Applied Pyrolysis* 79, 16–23.

Frimmel, F.-H. and M. Weis (1991): Aging effects of high-molecular-weight organic acids which can be isolated from landfill leachates. *Water Sci. Tech.* 1–3, 419–426.

Frimmel, F.-H. and M. Weis (1995) Heavy metals complexed by humic substance-like substances, in P. Spillman, H.-J. Collins, G. Matthess, W. Schneider (eds) *Schadstoffe im Grundwasser, Bd. 2: Langzeitverhalten von Umweltchemikalien und Mikroorganismen aus Abfalldeponien im Grundwasser – Deutsche Forschungsgemeinschaft – DFG. (Contaminants in groundwater, Vol. 2: Long-term behaviour of environmental chemicals and microorganisms from landfills in groundwater – German Research Foundation)*. VCH-Verlags-gesellschaft mbH, Weinheim.

Fritzenschaf, H., M. Kohlpoth, B. Rusche and D. Schiffmann (1993): Testing of known carcinogens and noncarcinogens in the Syrian hamster embryo (SHE) micronucleus test in vitro; correlations with in vivo micronucleus formation and cell transformation. *Mutat. Res.* 319, 47–53.

Garcia, C., T. Hernandez and F. Costa (1992): A chemical-structural study of organic waste and their humic acids during composting by means of pyrolysis gas chromatography. *Sci. Total Environ.* 119, 157–168.

GefStoffV (1999): Verordnung zum Schutz vor gefährlichen Stoffen – Gefahr-stoffverordnung – vom 15. November 1999 (Regulations for protection from hazardous substances – Hazardous Substance Regulations – of 15 November 1999), BGBl (Federal Gazette).

Gertloff, K.-H. (1993): Setzungsanalyse und Setzungsprognose für eine Hausmülldeponie (Subsidence analysis and forecast for a municipal solid waste landfill). *Müll und Abfall (Waste and Refuse)* 10/93, 752–766.

GICON (2004): Inventarisierung von Grundwasserschäden und deren Beurteilung in Großprojekten 'Ökologische Altlasten' der neuen Bundesländer. – Abschlussbericht an das UBA (Inventory and assessment of damage to groundwater in large-scale projects 'Ecological contaminated sites' of the new German States. Final Report to the Federal Environment Agency), Berlin/Dresden.

Götze, K., M. Budig and E. Homrighausen (1969): 'Giessener Modell', gemeinsame Beseitigung fester und flüssiger Abfallstoffe ('Giessen model', co-disposal of solid and liquid wastes). Der Städtetag, (4) and (5)/69, pp. 202–205, pp. 251–255.

Grainger, J. M., K. L. Jones, P. H. Hotten and J. F. Rees (1984): Estimation and control of microbial activity in landfill. In: Microbiological Methods for Environmental Biotechnology. J. M. Grainer and J. M. Lynch (eds), Academic Press, London, 259–273.

Greim, H. and H. Sterzel (1990): Bewertung von Altablagerungen aus der Sicht der Humantoxikologie. In: Collins , H.-J. and Wolff J. (eds) (1990): Erfassung und Bewertung von Altablagerungen. Zentrum für Abfallforschung, TU Braunschweig (Assessment of old deposits from the viewpoint of human toxicology. In: Collins, H.-J. and Wolff (eds): Determination and assessment of old deposits. Waste Research Centre, Technical University of Braunschweig).

Grewe (1987): Wasseraufnahme und Entwässerung von Papier (Water absorption and drainage of paper). Oral communication of the Institute of Paper Production of Technical University of Darmstadt to the Author of Chapter 2.

Grün, I. (1984): Cadmium und Bodenmikroorganismen (Cadmium and soil micro organisms). Thesis. University of Gießen, Institute for Agricultural Microbiology.

GTZ (2003): Sektorvorhaben Förderung der mechanisch-biologischen Abfallbehandlung, Deutsche Gesellschaft für technische Zusammenarbeit GmbH (Sector project promotion of mechanical-biological waste treatment. German Society for Technical Co-operation), Eschborn.

GUfA (1970): Untersuchung der mechanisch-biologischen Aufbereitung eines Hausmüll/Sperrmüll/Klärschlamm-Gemisches. Arbeitsgemeinschaft Gießener Universitätsinstitute für Abfallwirtschaft (Investigation of mechanical-biological processing of a municipal waste/bulky waste/sewage sludge. Working Group of the Institutes for Waste Management, Giessen University). Report to the Federal Ministry of the Interior.

Gunschera, J., J. Fischer, T. Dartsch, W. Lorenz and M. Bahadir (1995): Aufnehmen – Verwerten – Deponieren von Müll aus alten Kippen. – Chemisch-Analytisches Teilprojekt. Abschlußbericht zum interdisziplinären Verbundprojekt (Mining – re-using – depositing landfill wastes from old sites. Chemical-analytical Subproject. Final Report to Interdisciplinary

Integrated Project.) Az. II/67336 i. A. d. Volkswagenstiftung (Volkswagen Foundation), Hannover.

Hanert, H.-H. (1991): Überlegungen zu einer zweistufigen (anaerob-aerob) mikrobiologischen Aufbereitung des Restmülls. In: Collins, H.-J. Aufbereitung fester Siedlungsabfälle vor der Deponierung. Zentrum für Abfallforschung (ZAF), d. TU Braunschweig. (Deliberation on two-stage (anaerobic-aerobic) microbiological processing of residual waste. In: Collins, H.-J. Processing of solid municipal wastes before landfilling. Waste Research Centre of the Technical University of Braunschweig).

Hanert, H., P. Harborth, M. Kucklick, E. Lang, R. Rohde, Ch. Waschke and M. Wittmaier (1992): Möglichkeiten mikrobieller Stoffumsetzungen bei der Abfallentsorgung. – Zentrum für Abfallforschung (ZAF), d. TU Braunschweig. H. 7, 'Ist die thermische Behandlung von Abfallstoffen vermeidbar?' (Possibilities of microbial material conversion in waste disposal. Waste Research Centre of the Technical University of Braunschweig, No. 7 'Is the thermal treatment of wastes avoidable?').

Haschemi, H. (1998): Verbessertes Verfahren zur Verrottung von Haus- und Gewerbemüll (Schwäbisch-Haller-Modell) (An improved method for decomposition of municipal and commercial waste (Schwäbisch Hall model)). In: *Müll und Abfall (Waste and Refuse)* 8, 502–511.

Haschemi, H. (2002): Die bakteriologische Kontamination von Insekten auf einer Abfalldeponie und ihre Abhängigkeit vom Entsorgungsverfahren. Habilitations-Schrift an der Universität Rostock, Agrar- und Umweltwissenschaftliche Fakultät (Bacteriological contamination of insects on a landfill and their dependence on the disposal method. Habilitation Thesis at the Faculty of Agriculture and Environmental Protection, University of Rostock).

Hatcher, P. G., K. J. Dria, S. Kim and S. W. Frazier (2001): Modern analytical studies of humic substances. *Soil Sci.* 11, 770–794.

Haude, W. (1955): Zur Bestimmung der Verdunstung auf möglichst einfache Weise. – Mitteilungen des Deutschen Wetterdienstes (A simple determination of evaporation. Publications of the German Weather Service) No. 11. Vol. 2, Bad Kissingen.

Henseler-Ludwig, R. (ed.) (1993): Technische Anleitung zur Verwertung, Behandlung und sonstigen Entsorgung von Siedlungsabfällen. Bundesanzeiger Verlags-Ges. (Technical Instructions on utilization, treatment and other disposal of municipal solid wastes. Federal Legal Gazette Publishing House Co.), Köln.

Herklotz, K. (1985): Sorptions- und Mobilitätsverhalten von ausgewählten Pestiziden in Hausmüll, Böden und Porengrundwasserleitern (Sorption and mobility behaviour of selected biocides in municipal wastes, soils and porous aquifers). Thesis. University of Hannover.

Herklotz, K. and W. Pestemer (1986): Physikalisch-chemische Untersuchungen zur Sorption und Mobilität von Pestiziden in Abfällen (Physical-chemical investigations into the sorption and mobility of biocides in wastes). In: Spillmann, P. (ed.) (1986), Chapter 6.

Herklotz, K. and H. H. Rump (1995): Ausbreitung und Elimination von ausgewählten Chlorphenolen in einem mit Müllsickerwasser belasteten Porengrundwasserleiter (Spread and elimination of selected chlorophenols in a porous aquifer contaminated with leachate). In: Spillmann, P. (ed.) (1995), Chapter III.

Herrmann, A., P. Maloszewski and W. Stichler (1987): Changes of oxygen-18 content of precipitation water during seepage in the unsaturated zone. Intern. Sympos. on Groundwater Monitoring and Management, Dresden.

Heyer, K.-U.; L. Andreas and U. Brinkmann (1997): Standardarbeitsvorschrift SAV 3: Beprobung von Abfallstoffen in Deponiesimulationsreaktoren (DSR). In: Umweltbundesamt (Hrsg.): Verbundvorhaben Deponiekörper, 2. Statusseminar . Im Auftrag des Umweltbundesamtes, Projektförderung: BMBF, (Standard working regulation SAV 3: Sampling of waste materials in landfill simulation reactors (LSR). In: Federal Environment Agency (ed.): Integrated project Waste Body, 2nd Seminar. Commissioned by the Federal Environment Agency, Project Management: BMBF Federal Ministry of Education and Research). Project No: 1460799 and 1460799A to G. 345-358.

Heyer, K.-U., K. Hupe and R. Stegmann (2000): Die Technik der Nieder-druckbelüftung zur in situ-Stabilisierung von Deponien und Altablagerun-gen (The technology of low-pressure ventilation for in situ stabilisation of landfills and old deposits). *Müll und Abfall (Waste and Refuse)* 7/2000, 438–443.

Hoffmann, G. (1968): Eine photometrische Methode zur Bestimmung der Phosphatase-Aktivität in Böden. Z. Pflanzenern. Bodenk. (A photometric method for the determination of phosphatase activity in soils. *Journal of Plant Nutrition and Soil Science*) 118, 161–172.

Hughes, E. D., C. Ingold and R. Pasternak (1953): Mechanism of Elimination Reactions. – Part XVIII. Kinetics and Steric Course of Elimination from Isomeric Benzene Hexachlorides, *Journal Chem. Soc.* 3832–3819.

Husz, G. (2002): Vererdung von Abfallstoffen Band I – Teil 1 Bodenwis-senschaftliche Grundlagen für die Vererdung und Anwendung von Erden (Soilification of waste materials. Volume I – Part 1 Soil science fundamen-tals for the soilification and use of soils), ÖKO-Datenservice, Wien.

IAEA (1983): Guidebook on Nuclear Techniques in Hydrology. IAEA Wien, Techn. Rep. Ser. No. 91.

Irwin, W. J. (1982): Analytical pyrolysis: a comprehensive guide. In: Chroma-tographic science series 22. New York: Marcel Dekker-Verlag, 578 S.

Isermeyer, H. (1952): Eine einfache Methode zur Bestimmung der Bodenat-mung und der Carbonate im Boden. Z. Pflanzenern. Düngung und Boden-kunde (A simple method to determine soil respiration and carbonates in soil. *Journal of Plant Nutrition, Fertilisation and Soil Science*) 56, 26–38.

Jourdan, B., P. Spillmann, H. Münz, E. Britzius, J. Stritzke, H. Koch, G. Holch and A. Rothmund (1982): Hausmülldeponie Schwäbisch Hall – Homoge-nisierung und Verrottung des Mülls vor der Ablagerung. – Bundesmin. f. Forsch. u. Technolog. (Schwäbisch Hall Municipal waste landfill –

Homogenization and decomposition of waste before deposition. Federal Ministry of Research and Technology). Research Report. T 82-180, Fachinformationszentrum Karlsruhe.

Jungbauer, A. (1994): Recycling von Kunststoffen (Recycling of plastics). Würzburg: Vogel-Verlag.

Jury, W. A., W. R. Gardener and W. H. Gardener (1991): Soil physics. John Wiley & Sons Inc., New York.

Kahmann, L. (1971): Physikalisch-chemische Untersuchungen zur Beurteilung von Rohgemenge der Giessener Abfallbeseitigungsanlage. Diplomarbeit am Inst. f. Bodenkunde u. Bodenerhaltung (Physical-chemical tests for the assessment of raw mixtures of the Giessen waste disposal facility. Final thesis at Institute of Soil Science and Soil Preservation) Justus-von-Liebig University, Gießen.

Kaltenbrunner, W. V. (1999): Biologische in-situ-Sanierung mit Hilfe des Bio-Puster-Verfahrens am Fallbeispiel Feldbach. Diplomarbeit am Institut für Entsorgungs- und Deponietechnik (Biological in-situ remediation using the Bio-Puster method on the Feldbach case history. Final thesis at Institute for Disposal and Landfill Engineering), Montanuniversität (University of Mining) Leoben.

Keeling, A. A., J. A. J. Mullett and I. K. Paton (1994): GC-mass spectrometry of refuse-derived composts. *Soil Biol. Biochem.* 6, 773–776.

Kern, M. (2000): Potenziale zur stofflichen und energetischen Verwertung im Hausmüll. In: Wiemer, K.; Kern, M. (eds): Bio- und Restabfallbehandlung IV. Biologisch-mechanisch-thermisch (Potential for materials and energetic utilization of municipal waste. Bio and residual waste treatment IV. Biological, mechanical and thermal methods). Witzenhausen: Baeza-Verlag, 986–1006.

Kettern, J. (1990): Untersuchungen zur biologisch-chemisch-physikalischen Behandlung von Deponiesickerwässern. In: Gesellschaft zur Förderung der Siedlungswasserwirtschaft an der RWTH Aachen (Investigations into biological-chemical-physical treatment of leachate. In: Society for Promotion of Municipal Water Management at RWTH Aachen University) (ed.). GWA-Bd (GWA Volume) 113. Also RWTH Aachen University, Thesis.

Kiefl, M. and F. Radl (1991): Altlast Rautenweg. In: Die Sanierung von Altlasten in Wien. Band II, Magistratsabteilung 45 – Wasserbau (The Rautenweg contaminated site. In: Remediation of contaminated sites in Vienna. Volume II, Magistrate's Department 45 – Hydraulic Engineering). Herold, Wien.

Kölsch (2003a): Standsicherheitsuntersuchung auf der Deponie Hildesheim-Heinde (Investigating the stability of the Hildesheim-Heinde landfill). In: Witt, Katzenbach (eds.): 1. Symposium Umweltgeotechnik, Bauhaus-Universität Weimar, Schriftenreihe Geotechnik, No. 10, Vol. 2, Weimar. Download at http://www.dr-koelsch.de

Kölsch (2003b): Standsicherheit und Entwässerung von Deponien (Stability and drainage of landfills). In: Morscheck, G. (ed.) (2003): 6. Dialog Abfallwirtschaft Mecklenburg-Vorpommern: Betrieb, Stilllegung und Nachsorge von Deponien (6th Mecklenburg-Pomerania Waste Management Dialogue: Operation, closure and aftercare of landfills). Institut für Landschaftsbau und Abfallwirtschaft (Institute for Landscape Engineering and Waste Management), Universität Rostock.

Kölsch (2003c): Monitoring of landfills. 18th International Conference on Solid Waste Technology and Management. Philadelphia (USA).

Kölsch (2007): Stability analysis according to different shear strength concepts. Proceedings HPM 2. Southampton 2007.

König, K. (1994): Ruhender Punkt – Expertenmeinung zum Deponie-Rückbau. Teil II. (Resting point – expert opinion on landfill mining. Part II). *ENTSORGA-Magazin* 9.

König, K. (1995): Arbeitsmedizinische Anforderungen bei den verschiedenen Arbeiten an Deponien. – Vortrag zur Fachtagung 'Sicherheitstechnische Aspekte und Arbeitsschutz während der Betriebs- und Nachsorgephase von Abfalldeponien' (Occupational medicine requirements for working on landfills. – Presentation at the conference 'Safety engineering aspects and occupational safety during operation and aftercare phase of landfills'). 06–07 Nov. 1995 at the University of Rostock.

Körtel, A., H. Haschemi and P. Spillmann (2003): Das Teheraner Modell – Die biologische Stabilisierung extrem wasserreicher Siedlungsabfälle unter ariden Klimabedingungen mit minimalem technischen Aufwand (The Tehran model – Biological stabilisation of extremely water-rich municipal solid wastes under arid climate conditions applying minimum expenditure). *Müll und Abfall (Waste and Refuse)* 2.

Körtel, A. and P. Spillmann (2005): The 'Tehran Model' – A large scale introduction of a static composting process for the high water content municipal waste in arid region. 9th CEST International Conference on Environmental Science and Technology, Rhodes Island; Global NEST and Dep. of Environment Studies, Univ. of the Aegean.

KrW-/AbfG (1994): Gesetz zur Förderung der Kreislaufwirtschaft und Sicherung der umweltverträglichen Beseitigung von Abfällen vom 27. September 1994 (Promotion of recycling management and assurance of environmentally compatible disposal of wastes Act of 27 September 1994), BGBl (Federal Gazette).

Kucklick, M., P. Harborth and H. H. Hanert (1995): Aufnehmen – Verwerten – Deponieren von Müll aus alten Kippen. – B Mikrobiologisches Teilprojekt. – Abschlußbericht zum interdisziplinären Verbundprojekt (Mining – re-using – depositing landfill wastes from old sites. B microbiological subproject. Final report on the interdisciplinary integrated project.) Az. II/67336 Commissioned by Volkswagenstiftung (Volkswagen Foundation), Hannover.

Kucklick, M., P. Harborth and H. H. Hanert (1996): Aussagekraft von Sickerwasseranalysen zur Beurteilung der biologischen Stabilität von Deponien. Zentrum für Abfallforschung TU Braunschweig, Heft 11: Nachsorge von Siedlungsabfalldeponie (Informative power of leachate analyses for the assessment of biological stability of landfills. Waste Research Centre,

Technical University of Braunschweig. No. 11: Aftercare of municipal solid waste landfills).

Küster, E., W. Neumeier and H. Rötlich (1989): Mikrobiologische und physikalisch-chemische Untersuchungen an Sickerwasser aus einer Mülldeponie. Forum Städtehygiene. (Microbiological and physical-chemical investigations on leachate from a landfill. Town Hygiene Forum) 40, 38–42.

Küster, E. and S. T. Williams (1964): Selection of media for isolation of streptomycetes. *Nature* 202, 928–929.

LAGA (1998): Anforderungen an die stoffliche Verwertung von mineralischen Reststoffen/Abfällen + Technische Regeln. Länderarbeitsgemeinschaft Abfall. (Provisions on materials utilization of mineral residual materials/ wastes + Technical regulations. State Waste Working Group), No. 20, 4th extended edition.

Lehmann, W. D. and H.-R. Schulten (1976a): Physikalische Methoden in der Chemie: Allgemeine und Elektronenstoß-Massenspektrometrie I (Physical methods in chemistry: General and electron impact mass spectrometry I). *Chemie in unserer Zeit 5 (Today's Chemistry 5)* 147–158.

Lehmann, W. D. and H.-R. Schulten (1976b): Physikalische Methoden in der Chemie: Massenspektrometrie II – Chemische Ionisations-, Feldionisations- und Felddesorptions-Massenspektrometrie (Physical methods in chemistry: Mass spectrometry II – Chemical ionization, field ionization and field desorption mass spectrometry). *Chemie in unserer Zeit 6 (Today's Chemistry 6)* 163–174.

Leinweber, P. and H.-R. Schulten (1998): Nonhydrolyzable organic nitrogen in soil size separates from long-term agricultural experiments. *Soil Sci. Soc. Am. J.* 2, 383–393.

Leinweber, P., A. Wehner and H.-R. Schulten (2002): Qualitätsbeurteilung von Komposten aus Bioabfällen mit klassischen biologischen und chemischen sowie mit modernen spektroskopischen Methoden. In: ATV-DVWK (Hrsg.): Mechanische und biologische Verfahren der Abfallbehandlung. (Quality assessment of compost from biowastes using conventional biological, chemical and modern spectroscopic methods. In: ATV-DVWK (ed.): Mechanical and biological methods of waste treatment). Berlin: Ernst & Sohn-Verlag, 499–521.

LfU B.-W. (ed.) (1992): Der Deponiegashaushalt in Altablagerungen – Leitfaden Deponiegas – Handbuch Altlasten, Bd. 10. Landesanstalt für Umweltschutz (Landfill gas balance in old deposits – Landfill gas manual – Handbook of Contaminated Land, Vol. 10, State Environmental Office). Karlsruhe.

Lhotzky, K. and M. Struve (1996): Hazard analysis on the HMG landfill. Unpublished report for HMG; Archive of Prof. Dr. W. Hartung + Partners Consultants, Braunschweig.

Lhotzky, K. (1997): Milieusondenmessung. In: Bundesanstalt für Geowissenschaften und Rohstoffe, Ref. B 3.15, (Hrsg.) Handbuch zur Erkundung des Untergrundes von Deponien und Altlasten, Bd. Geophysik, Kap. 13 (Environment probe measurement. In: Federal Institute for Geosciences and Natural Resources, Ref. B 3.15, (ed.) Handbook for the investigation of the underground of landfills and contaminated land), Vol. Geophysics, Chapter 13, Springer-Verlag.

Lhotzky, K. and P. Spillmann (2002): Risk assessment of groundwater contamination – 3-dimensional monitoring with profile measurement and layer sampling – 2nd Intercontinental Landfill Research Symposium at Asheville NC, USA, 13–16 October.

Liang, B. C., E. G. Gregorich, M. Schnitzer and H.-R. Schulten (1996): Characterization of water extracts of two manures and their adsorption on soils. *Soil Sci. Soc. Am. J.* 60, 1758–1763.

Lichtensteiger, Th. and Ch. Zeltner (1994): Wie lassen sich Feststoffqualitäten beurteilen? (How can solid qualities be assessed?) In: Baccini and Gamper (1994).

Lorber, K. E. and Erhart-Schippek, W. (2000): Erkenntnisse interdisziplinärer Begleituntersuchungen der Sanierung der Altablagerung Feldbach (Results of interdisciplinary accompanying tests of the remediation of the Feldbach old deposit). 5th DEPOTECH 2000, 21–23 November 2000 in Leoben. A.A. Balkema, Rotterdam.

Lott-Fischer, J., A. Albrecht and P. Kämpfer (2001): Mikrobiologie der Kompostierung von Abfällen. In: Kämpfer, P.; Weißenfels, W.D. (Hrsg.): Biologische Behandlung organischer Abfälle. (Microbiology of waste composting. In: Kämpfer, P.; Weißenfels, W. D. (eds): Biological treatment of organic wastes). Berlin: Springer-Verlag, 1–43.

Maak, D (2003): Vom open dumping zur emissionsarmen Deponierung – Umweltfreundliche Abfallentsorgung auf der Deponie Sao Sebastiao, Brasilien –TRIALOG, Heft 77, Vereinigung zur wissenschaftlichen Erforschung des Planens und Bauens in Entwicklungsländern e.V. (From open dumping to low-emission landfilling – Environmentally friendly waste disposal on the Sao Sebastiao, Brazil, landfill – TRIALOG, No. 77, Society for scientific study of planning and building in developing countries).

MAK- and BAT-Werte-Liste (1994): Deutsche Forschungsgemeinschaft (DFG), Senatskommission zur Prüfung gesundheitsschädlicher Arbeitsstoffe, Mitteilung 30 (German Research Foundation, Senate's commission for the testing of health-endangering materials), No. 30.

Maloszewski, P. and A. Zuber (1982): Determining the turnover time of groundwater systems with the aid of environmental tracers. 1. Models and their applicability. *Journal of Hydrology* 57, 207–231.

Maloszewski, P., W. Rauert, W. Stichler and A. Herrmann (1983): Application of flow models in an Alpine catchment area using tritium and deuterium data. *Journal of Hydrology* 66, 319–330.

Maloszewski, P. and A. Zuber (1985): On the theory of tracer experiments in fissured rocks with a porous matrix. *Journal of Hydrology* 79, 333–358.

Maloszewski, P., A. Zuber, W. Stichler and A. Herrmann (1990): Bestimmung hydrologischer Parameter in Einzugsgebieten mit Kluftaquiferen unter Verwendung von Umweltisotopen und mathematischen Fließmodellen. Freiberger Forschungshefte (Determination of hydraulic parameters in catchment areas with fissured aquifers using environmental isotopes and mathematical flow models. – Freiberg Research Papers), C 442, 11–21.

McCrady, M. H. (1918): Tables for rapid interpretation of fermentation tube results. *Can. J. Public Health* 9, 201–209.

MDEQ (2002): Report of the Michigan Department of Environmental Quality at http://www.michigan.gov.

Moldoveanu, S. C. (1998): Analytical pyrolysis of natural organic polymers. In: Moldoveanu, S. C. (ed.) Techniques and instrumentation in analytical pyrolysis, Vol. 20. Amsterdam: Elsevier.

Möller, D. (2001): Zur Klimarelevanz von Deponiegasen. Sitzungsberichte Thüringische Akademie der Wissenschaften (On climate relevance of landfill gases. Minutes of the meetings of the Thuringian Academy of Sciences).

Mortimer, C. E. (1987): Chemie: das war das Basiswissen der Chemie (Chemistry: that was the basic knowledge of chemistry). Georg-Thieme-Verlag, Stuttgart.

Moser, H. and W. Rauert (1980): Isotopenmethoden in der Hydrologie (Isotope methods in hydrology). Verlag Bornträger, Berlin, Stuttgart.

Müller, W., R. Wallmann and K. Fricke (1999): Technische Anforderungen an die Mechanisch-biologische Restabfallbehandlung. Tagungsbeitrag zum Verbundvorhaben Mechanisch-Biologische Behandlung von zu deponierenden Abfällen. Beiträge der Ergebnispräsentation. (Technical provisions on mechanical-biological treatment of residual wastes. Paper to the integrated project on mechanical-biological treatment of wastes to be deposited. Contributions to result presentation) pp. 85–116. Potsdam, 7–8 September.

Münnich, K. W. (1995): Reduktion des Stoffdurchganges durch mineralische Abdichtungen bei inverser Strömung. Dissertation Leitweiss-Institut-Mitteilungen, Nr. 130 der TU Braunschweig. (Reduction of mass transfer through mineral liners due to inverse flow. – Thesis. Reports of Leichtweiss Institute), No. 130. Technical University of Braunschweig.

Münnich, K., C. F. Mahler and D. Maak (2001): Mechanical-biological pre-treatment of residual waste in Brazil. In: 8th Intern. Landfill Symposium Cagliari CISA, Grafiche Galeati, Imola/Italy.

Nagasawa, S., R. Kikuchi, Y. Nagata, M. Takagi and M. Matsuo (1993): Stereochemical Analysis of γ-HCH Degredation by Pseudomonas Paucimobilis, UT 26. *Chemosphere* 26, 1187–1201.

Nassour, A., P. Degener and K. Kraase (2002): Hochdruck-Wasserstrahlen zur mechanischen Aufbereitung von Restabfällen – Auswirkungen auf die nachgeschaltete Rotte sowie die Ablagerungskriterien. Müll und Abfall, Heft 12, Jhrg. 34, (High-pressure water jets for mechanical processing of residual wastes – Effects on the downstream decomposition as well as storage criteria). *Waste and Refuse* 12(34), 659–662.

Neumann, U. (1978): Untersuchungen von Vegetationen an geschlossenen Müllablagerungen in der Bundesrepublik Deutschland. – Umweltforschungsplan der BDI (Investigations on vegetation on closed waste deposits in the Federal Republic of Germany. – Environmental Research Programme of BDI), UBA.

517

Neumann, U. (1981): Anleitung zur Rekultivierung von Deponien, Teil 2 (Guidance for landfill reclamation, Part 2), UBA-FB80-103.

Neumeier, W. and E. Küster (1981): A method for measuring the potential activity of methanogenic bacteria in waste disposals. *Europ. J. Appl. Microbiol. Biotechnol.* 12, 231–233.

Neumeier, W. and E. Küster (1986): Mikrobiologische Untersuchungen zur Kennzeichnung von Stabilitätsvorgängen in Abfalldeponien (Microbiological investigations for the marking of stability processes in landfills). In: Spillmann, P. (ed.) (1986).

Niedersächsisches Umweltministerium (MU) (1988): Durchführung des Abfallgesetzes; Abdichtung von Deponien für Siedlungsabfälle, RdErl. d. MU v. 24.6.1988 – 207–62812/21–8 S., 2 Abb.; Hannover (Niedersächsischer Dichtungserlaß) (Implementation of the waste act; Lining of landfills for municipal solid wastes, Decree of the Ministry for Environment of 24 June 1988 – 207–62812/21–8. Hannover – Lower Saxony Liner Decree).

Niese, G. (1963): Versuche zur Bestimmung des Rottegrades von Müllkomposten mit Hilfe der Selbsterhitzungsfähigkeit (Tests for the determination of the decomposition degree of waste composts using the spontaneous heating capability) *IAM* 17, 3–14.

Novak, B. (1972): Ausnutzung biochemischer Tests in der Bodenmikrobiologie. I. Verwendete Methoden (Utilisation of biochemical tests in soil microbiology. I. Used methods). *Zbl. Bakt. II. Abt.*, 127, 699–705.

Ohlrogge (1998): Berechnung der Grundwassererneuerung in Niedersachsen zur Begrenzung der Wasserentnahmen nach den Berechnungen des NLÖ. Persönliche Mitteilung der Agrarmeteorologischen Station der FAL Braunschweig an den Verfasser des Kapitels 2 (Calculation of groundwater regeneration in Lower Saxony to limit water withdrawals according to the calculation of NLÖ. Personal communication of the Agricultural-Meteorological Station of the FAL Braunschweig to the Author of Chapter 2).

Oi, S. and T. Yamamoto (1977): A streptomyces spec. effective for conversion of cyanate to thiocyanate. *J. Ferment. Technol.* 55, 560–569.

Öman, C. and P.-A. Hynning (1993): Identification of organic compounds in municipal landfill leachates. *Environ. Pollut.* 3, 265–271.

Otake, Y., T. Kobayashi, H. Asabe, N. Murakami and K. Ono (1995): Biodegradation of low-density polyethylene, polystyrene, polyvinyl chloride and urea formaldehyde resin buried under soil for over 32 years. *J. Appl. Polym. Sci.* 56, 1789–1796.

Ottow, J. C. G. (1969): Qualitative und quantitative mikrobiologische Populations-untersuchungen an unterschiedlich verlegten Böden unter Berücksichtigung der Ökologie und Physiologie eisenreduzierender Bakterien (Qualitative and quantitative microbiological population investigations in differently arranged soils under consideration of ecology and physiology of iron-reducing bacteria), Thesis. Gießen.

Pantke, M. (1996): General aspects and test methods. In: Heitz, E.; Flemming, H.-C.; Sand, W. (eds): Microbially influenced corrosion of materials – scientific and engineering aspects. Berlin: Springer-Verlag, 379–391.

Pecher, P. (1990): Untersuchungen zum Verhalten ausgewählter organischer Chlorkohlenwasserstoffe während des sequentiellen Abbau von kommunalen Abfällen. Dissertation an der Fakultät Biologie, Chemie und Geowissenschaften der Universität Bayreuth (Investigations into the behaviour of selected organic chlorohydrocarbons during sequential degradation of local wastes. Thesis at the Faculty of Biology, Chemistry and Geosciences of the University of Bayreuth).

Pestemer, W. and H. Nordmeyer (1988): Sorption von ausgewählten Pflanzenschutzmitteln an unterschiedlichen Schlauchmaterialien. Zentralblatt für Bakteriologie, Mikrobiologie u. Hygiene (Zbl. Bakt. Hyg. B.) (Sorption of selected pesticides at different tube materials. *Central Journal for Bacteriology, Microbiology and Hygiene*) 186, 375–379.

Poller, Th. (1990): Hausmüllbürtige LCKW/FCKW und deren Wirkung auf die Methangasbildung. Diss. Abfallwirtsch., TU Hamburg-Harburg.- Hamburger Berichte 2 (Municipal waste generated slightly chlorinated hydrocarbons/fluorochlorohydrocarbons and their effect on methane gas generation. Thesis. Waste Management, Technical University of Hamburg-Harburg. Hamburg Reports 2), Economica Verlag.

Pouwels, A. D., G. B. Eijkel and J. J. Boon (1989): Curie-point pyrolysis-capillary gas chromatography-high-resolution mass spectrometry of microcrystalline cellulose. *J. Anal. Appl. Pyrolysis* 14, 237–280.

Pouwels, A. D., A. Tom, G. B. Eijkel and J. J. Boon (1987): Characterisation of beech wood and its holocellulose and xylan fractions by pyrolysis-gas chromatography-mass spectrometry. *J. Anal. Appl. Pyrolysis* 11, 417–436.

Ramke, H.-G. and M. Brune (1990): Untersuchungen zur Funktionsfähigkeit von Entwässerungsschichten in Deponieabdichtungssystemen – Abschlußbericht zum Forsch.-vorh. (Investigations into the efficiency of drainage layers in landfill liner systems. Final Report.) Project No. BMFT 145 0457 3. Umweltbundesamt (Federal Environment Agency) Berlin.

Reichard, W. (1978): Einführung in die Methoden der Gewässermikrobiologie (Introduction into the methods of water microbiology), G. Fischer Stuttgart.

Reichard, W. and M. Simon (1972): Die Mettma – ein Gebirgsbach als Brauereivorfluter: Mikrobiologische Untersuchungen entlang eines Abwasser-Substratgradienten (Mettma – a mountain brook as a brewery receiving stream: Microbiological investigations along a waste water substrate gradient). *Arch. Hydrobiol./Suppl.* 42, 125–138.

Reinhard, M. and N. L. Goodman (1984): Occurrence and distribution of organic chemicals in two landfill leachate plumes. *Environ. Sci. Tech.* 12, 953–961.

Rettenberger, G. (1992): see Landesanstalt für Umweltschutz Baden-Württemberg (Baden-Württemberg State Office for Environment) (ed.) (1992), Chapter 2.

Richter, J. and Großbauer (1987): The soil as a reactor. Catena Verlag, Cremlingen.

Rolle, G., B. Orsanic, W. Obrist and B. Aerne (1970): Methoden zur Untersuchung von Abfallstoffen. EAWAG – Eidgenössische Anstalt für

Wasserversorgung, Abwasserreinigung und Gewässerschutz, Abteilung Müllforschung, Schweiz (Methods for the investigation of waste materials. EAWAG – Swiss Federal Institute for Water Supply, Waste Water Cleaning and Water Protection, Department of Waste Research, Switzerland).

Saiz-Jimenez, C. (1994): Analytical pyrolysis of humic substances: Pitfalls, limitations, and possible solutions. *Environ. Sci. Tech.* 11, 1773–1780.

Saiz-Jimenez, C., J. J. Ortega-Calvo and B. Hermosin (1994): Conventional pyrolysis: A biased technique for providing structural information on humic substances? *Die Naturwissenschaften* 81, 28–29.

Savage, G., C. Golueke and E. von Stein (1993): Landfill mining: past and present, *BioCycle* May, 58–61.

Savar, I. (2003): Communication of Dr. Irfan Savar at http://www.user. tu-berlin/arhagthi/1-1.htm.

Scheibel, H.-J., P. Harborth, E. Lang and H. Hanert (1991): Einsatz des Leuchtbakterien- und Daphnientests zur ökotoxikologischen Bewertung von Grundwasser- und Bodenreinigung bei der Altlastensanierung. GWF Wasser u. Abwasser (Use of luminous bacteria and Daphnia tests for ecotoxicologic assessment of groundwater and soil cleaning in reclamation of contaminated land. *Water and Waste Water*), 132, 441–447.

Scheijen, M. A., J. J. Boon, W. Hass and V. Heemann (1989): Characterization of tobacco lignin preparations by Curie-point pyrolysis-mass spectrometry and Curie-point pyrolysis-high-resolution gas chromatography/mass spectrometry. *J. Anal. Appl. Pyrolysis* 15, 97–120.

Schiffmann, L., Rostock University (1998): Senkung des kanzerogenen Potentials eines anaeroben Deponiekörpers durch dessen Aerobisierung (Reducing the carcinogenic potential of an anaerobic waste body using aerobisation). – In: ARGE Biopuster (1998).

Schiffmann, Fritzenschaf, Kohlpoth and Rusche (1993): Testing of known carcinogens and noncarinogens in the Syrian hamster embryo (SHE) micronucleus test in vitro; correlations with in vivo micronucleus formation and cell transformation. *Mutation Research* 319, 47–53.

Schmidt, H.-H. (1996): Grundlagen der Geotechnik (Fundamentals of geotechnics). Teubner Verlagsges., Stuttgart.

Schneider, I. (2005): Beschleunigung der anaeroben Stabilisierung eines noch in Betrieb befindlichen Deponieausschnitts durch linienförmige Infiltration von gereinigtem Sickerwasser über das horizontale Gasfassungssystem. Diss. a.d. Agrar- u. Umweltwiss. Fak d. University of Rostock (Acceleration of anaerobic stabilisation of a landfill sectional core still in operation using linear infiltration of cleaned leachate through the horizontal gas collection system. Thesis at the Faculty of Agriculture and Environmental Protection, University of Rostock).

Scholwin, F. (2003): Ein modellbasiertes Regelkonzept für biologische aerobe Abfallbehandlungsanlagen auf der Grundlage von Fuzzy Logic. – Diss. a.d. Fak. Bauingenieurwesen, Bauhaus Universität Weimar (A model-based regulation concept for biological aerobic waste treatment facilities based on fuzzy logic – Thesis at the Faculty of Civil Engineering, Bauhaus University Weimar).

Schulten, H.-R., C. Sorge-Lewin and M. Schnitzer (1997): Structure of 'unknown' soil nitrogen investigated by analytical pyrolysis. *Biol. Fertil. Soils* 24, 249–254.

Sleat, R., C. Harries, I. Viney and H. Rees (1987): Activities and distribution of key microbial groups in landfill. Process, technology and environmental impact of sanitary landfill. ISWA – Intern. Sanit. Landfill Sympos. 19th–23rd Oct. 87, Cagliari, Sardinia.

Sorge, C. (1995): Struktur der organischen Substanz in Böden und Partikelgrößenfraktionen: Pyrolyse-Gaschromatographie/Massenspektrometrie und Pyrolyse-Feldionisation Massenspektrometrie. Kiel, Universität, Inst. f. Pflanzenernährung und Bodenkunde, Diss., 177 S. (Structure of organic substances in soils and particle size fractions: pyrolysis-gas chromatography/mass spectrometry and pyrolysis-field ionisation mass spectrometry. Kiel, University, Institute for Plant Nutrition and Soil Science, Thesis).

Spencer, R. (1990): Landfill Space Reuse. *BioCycle* Feb. 30–33.

Spillmann, P. (1986a): Konstruktive Maßnahmen gegen Restbelastungen. Fachseminar 'Bodensanierung u. Grundwasserreinigung'. Zentrum f. Abfallforschung (ZAF) H. 1 (Constructional measures against residual contamination. Specialist seminar 'Soil reclamation and groundwater cleaning', Waste Research Centre), No. 1, Technical University of Braunschweig.

Spillmann, P. (ed.) (1986b): Wasser- und Stoffhaushalt von Abfalldeponien und deren Wirkung auf Gewässer. Forschungsbericht, Deutsche Forschungsgemeinschaft (Water and material balance of landfills and their effect on waters. Research Report, German Research Foundation). VCH Verlagsgesellschaft mbH, Weinheim.

Spillmann, P. (1988): Einflüsse verschiedener Deponietechniken einwohnergleicher Müll- und Klärschlammmassen auf die Nutzungsdauer von Abfalldeponien. Mitt. d. Leichtweiß-Inst. f. Wasserbau d. TU Braunschweig, H. 96 (Influence of different landfill technologies of population equivalent sewage sludge waste masses on the service life of landfills. Publications of Leichtweiß Institute for Hydraulic Engineering. Technical University of Braunschweig, No. 96).

Spillmann, P. (1989): Die Verlängerung der Nutzungsdauer von Müll- und Müll-Klärschlamm-Deponien. – Abfallwirtschaft in Forschung und Praxis (Extension of the use of waste and waste-sewage sludge landfills. – Waste Management in Research and Practice), 27, E. Schmidt-Verlag, Berlin.

Spillmann, P. (1991): Vermeidung unerwünschter Gasemissionen während der Räumung alter Abfallablagerungen oder Industriebelastungen durch gezielte Be- und Entgasung in situ. Gutachten an die Stadt Wien im Rahmen der Altlast Abbauarbeiten Donaupark (Avoidance of unwanted gas emissions during mining of old waste deposits or industrial contaminants by focused in situ degassing. Expert opinion for the City of Vienna within the reclamation of the contaminated site in Donaupark). EXPO 95, available at MA 45, Wien.

Spillmann, P. (1993a): Anforderungen an die Vorbehandlung von Deponiegut zum Aufbau langzeitstabiler Deponiekörper. Verein Deutscher Ingenieure, Gesellschaft Energietechnik, Düsseldorf, in: VDI-Berichte 1033: Technik der Restmüllbehandlung, kalte und/oder thermische Verfahren. (Provisions for the pre-treatment of landfill material to establish long-term stable waste bodies. Association of German Engineers, Society Energy Technology, Düsseldorf, in: VDI Reports 1033: Technology of residual waste treatment, cold and/or thermal methods). VDI-Verlag.

Spillmann, P. (1993b): Neues Deponieabdichtungssystem für Siedlungs- und Industrieabfälle (A new landfill liner system for municipal and industrial wastes). *Waste Magazine* 1/93.

Spillmann, P. (1995a): Stabilization of Solid Waste by Biochemical and Thermal Process-Application on Existing Landfill Sites. Intern. Sympos. on 'Resanitation of the old Nanjido Landfill-Area' of National Assembly Library Auditorium, Seoul, ed. by Intern. Inst. of Labour and Environment (ILE), Environmental Management Corp.

Spillmann, P. (1995b): Kostenneutrale Verbesserung der Identifizierung und Lokalisierung toxischer Stoffe in Altlasten durch Einsatz von Anreicherungsverfahren. In: Rettenberger u. Spillmann Hrsg.: Sicherheitstechnische Aspekte und Arbeitsschutz während der Betriebs- und Nachsorgephase von Abfalldeponien. (Cost-neutral improvement of identification and localisation of toxic substances in contaminated land using enrichment methods. In: Rettenberger and Spillmann eds.: Safety engineering aspects and occupational safety during operation and aftercare phase of landfills). Conference at the University of Rostock 06–07 Nov. 1995.

Spillmann, P. and H.-J. Collins (1978): Einfluß eines Sickerwasserkreislaufes auf den Wasserhaushalt eines rottenden, ländlichen Hausmülls. Müll und Abfall (Influence of leachate re-circulation on the water balance of decomposing, rural municipal waste. *Waste and Refuse*) 10(11), 331–339.

Spillmann, P. and H.-J. Collins (1979): Verminderung der Sickerwasserfrachten und Verlängerung der Nutzungsdauer einer Hausmülldeponie durch Nutzung aerober Abbauvorgänge. Müll und Abfall (Reduction of transported masses in leachate and extension of service life of a municipal waste landfill using aerobic degradation processes. *Waste and Refuse*), 11(3), 61–77.

Spillmann, P. and H.-J. Collins (1981): Das Kaminzug-Verfahren – eine einfache und zielsichere Belüftung als Vorraussetzung des aeroben Abbaus im Betrieb einer geordneten Mülldeponie (The chimney draught method – a simple and efficient aeration as prerequisite of aerobic degradation in operating an organized landfill). Forum Städte-Hygiene (*Town Hygiene Forum*) 32, 15–24.

Spillmann, P. and Meseck, H. (1986): Möglichkeiten zur Erkundung von Altdeponien und kontaminierten Standorten. Fachseminar 'Bodensanierung und Grundwasserreinigung' ZAF-Heft 1 (Possibilities for the investigation of old deposits and contaminated sites. Specialist seminar 'Soil reclamation and groundwater cleaning'), ZAF-No 1, Technical University of Braunschweig.

Spillmann, P., D. Ranner, M. Reisner (1992/93): Low-Emission Waste Transposition by Converting from Anaerobic to Aerobic Decomposition, in: Technical University Budapest and Florida State University: Intern. Sympos. on Environmental Contamination in Central and Eastern Europe, Budapest, 12–16 October 1992. Published in: *Soil and Environment* (1).

Spillmann, P., H. Eschkötter and G. Morscheck (1998): Neueste Erkenntnisse der mechanisch-biologischen Abfallbehandlung im Landkreis Stendal mit Druckschwallbelüftung (BIOPUSTER®-Verfahren). Tagung Abfallwirtschaft 'Restabfallbehandlung – handeln statt abwarten'. (Latest results of the mechanical-biological waste treatment in the Stendal district using pressure surge aeration (BIOPUSTER® method). Waste management conference 'Treatment of residual waste – acting instead of waiting'). Magdeburg, May 1998.

Spillmann, P., A. Nassour and H. Eschkötter (2002): Logistik der Sammlung und Trennung der Abfälle in definierte Rohstoffe. In: Loll, U. (Hrsg.) (2002): 'Mechani-sche und biologische Abfallbehandlung', Kap. 13 (Logistics of waste collection and separation into specified raw materials. In: Loll, U. (ed.) (2002): 'Mechancial and biological waste treatment', Chapter 13). ATV-Handbuch Ernst and Sohn, Berlin.

Spillmann, P., H.-J. Collins, G. Matthess and W. Schneider (eds) (1995): *Schadstoffe im Grundwasser, Bd. 2: Langzeitverhalten von Umweltchemikalien und Mikroorganismen aus Abfalldeponien im Grundwasser – Deutsche Forschungsgemeinschaft – DFG. (Contaminants in groundwater, Vol. 2: Longterm behaviour of environmental chemicals and microorganisms from landfills in groundwater – German Research Foundation).* VCH-Verlagsgesellschaft mbH, Weinheim.

Spillmann, P., T. Dörrie, A. Nassour and M. Struve (2001): Re-use of landfill areas through specific treatment. 4th Middle East Regional Conference on the Role of Environmental Awareness in Waste Management, 10–12 November, State of Kuwait, Kuwait City.

Steinbrecht, D. and Neidel, W. (2001): Verbrennung von BRAM in der stationären Wirbelschicht. Tagungsband 4. Dialog 'Abfallwirtschaft M-V'. Universität Rostock, Institut für Landschaftsbau und Abfallwirtschaft. (Incineration of BRAM in a stationary fluidised bed. In: Proceedings 4th Dialogue 'Waste management Mecklenburg-Western Pomerania'. Rostock University, Institute of Landscape Engineering and Waste Management) pp. 147–157. Rostock, June.

Straube, G. (1991): Microbial Transformation of Hexachlorocyclohexane – *Zentralbl. Mikrobiol.* 146, 327–338.

Struve, M. (1996): Ergebnisse Stickstoffversuch vom 07-11.03.1996, Bericht für die Bezirksregierung Braunschweig (Results of nitrogen tests of 07–11 March 1996, Report for the Braunschweig district government), unpublished.

Struve, M. (2003): Gefahrlose Umlagerung und Recycling der Deponie Chiriah Tel Aviv, Projektskizze (Safe relocation and recycling of the Chiriah, Tel Aviv landfill, project sketch), unpublished.

523

Struve, M. (2005a): Bericht an die Bezirksregierung Braunschweig (Report for the Braunschweig district government), 6/2005, unpublished.

Struve, M. (2005b): Untersuchungen zur mikrobiellen Kontamination von Deponieproben (Investigations into microbial contamination of landfill samples), Rostock, 04 January 2005, unpublished.

Swannel, R. P. J. (1993): Biological Treatment of HCH, Recent Developments, 2nd HCH-Forum, Oct. 1993, Magdeburg.

TAA (1991): Techn. Anleitung zur Lagerung, chemisch/physikalischen, biologischen Behandlung, Verbrennung und Ablagerung von besonders überwachungsbedürftigen Abfällen (Zweite allgemeine Verwaltungsvorschrift zum Abfallgesetz – TA Abfall) vom 12. März 1991. (Technical instructions on storage, chemical/physical, biological treatment, incineration and disposal of waste requiring special surveillance. Second General Administrative Provision to the Waste Avoidance and Waste Management Act: – TI Hazardous Waste of 12 March 1991). (GMBl. No. 8 p. 139) last amended on 21 March 1991 by amendment of united text, GMBl. No. 16 of 23 May.

TASi (1993): Technische Anleitung zur Verwertung, Behandlung und sonstigen Entsorgung von Siedlungsabfällen. Dritte Allgemeine Verwaltungsvorschrift zum Abfallgesetz – TASi vom 14. Mai 1993, Köln, (Technical Instructions on Recycling, Treatment and Disposal of Municipal Waste. Third General Administrative Provision to the Waste Avoidance and Waste Management Act: TI Municipal Waste of 14 May 1993). Bundesanzeiger (Federal Gazette).

TBG (1997): Regeln für Sicherheit und Gesundheitsschutz bei der Arbeit in kontaminierten Bereichen. BGR 128, vormals ZH 1/183. 7. Ausgabe, Tiefbau-Berufsgenossenschaft. (Safety and health protection regulations for work in contaminated areas. BGR 128, formerly ZH 1/183. 7th edition, Foundation Engineering Professional Association). Munich.

Thalmann, A. (1967): Über die mikrobielle Aktivität und ihre Beziehung zu Fruchtbarkeitsmerkmalen einiger Böden unter besonderer Berücksichtigung der Dehydrogenaseaktivität (On microbial activity and its relationship with fertility properties of some soils with special consideration of the dehydrogenase activity). Thesis. Gießen.

TrinkwV (2001): Verordnung zur Novellierung der Trinkwasserverordnung – Verordnung über die Qualität von Wasser für den menschlichen Gebrauch (TrinkwV 2001) – vom 21. Mai 2001 – Bundesgesetzblatt Teil I (Regulation for amending the Drinking Water Regulations –Regulation on the quality of water for human use (TrinkwV 2001) – of 21 May 2001, Federal Gazette Part I), No. 24, Bonn, 21 May 2001.

Tsuge, S. and H. Matsubara (1985): High-resolution pyrolysis–gas chromatography of proteins and related materials. *J. Anal. Appl. Pyrolysis* 8, 49–64.

Turk, M. (1997): Der Einfluß der maximalen Abfallstückgröße auf den Gasaustausch bei dem Kaminzugverfahren. Braunschweig, Technische Universität, Fachbereich für Bauingenieur- und Vermessungswesen (Influence of maximum waste size on gas exchange in the chimney draught method.

Technical University of Braunschweig, Faculty of Civil Engineering and Surveying). Thesis.

UBA (2000): http://www.umweltbundesamt.de. Bundesweite Übersicht zur Altlastenerfassung – Zusammenstellung des Umweltbundesamtes vom Dezember 2000 auf der Grundlage von Angaben aus den Bundesländern (Country-wide overview of contaminated land registry – compilation of the Federal Environment Agency of December 2000 based on data of the States).

Waschke, C. (1994): Entwicklung eines mikrobiologischen Verfahrens zur Vorbehandlung fester Siedlungsabfälle nach dem Alternanz-Prinzip (aerob/anaerob), Diplomarb. Inst. f. Mikrobiol., Technische Universität Braunschweig (Development of a microbiological method for pre-treatment of municipal solid wastes according to the alternation principle (aerobic// anaerobic). Final thesis. Institute of Microbiology, Technical University of Braunschweig).

Weis, M., F. H. Frimmel and G. Abbt-Braun (1995): Charakterisierung huminstoffähnlicher Substanzen aus Deponiesickerwasser (Characterisation of humic-like substances from leachate). In: Spillmann *et al.* (1995), Chapter 5.2.

WHG (2002): Gesetz zur Ordnung des Wasserhaushalts (Wasserhaushaltsgesetz – WHG) Bundesgesetzblatt Teil I (Water Management Regulation Act, Federal Water Act) – Federal Gazette, Part I) No. 59, Bonn, 19 August 2002.

WHO (1991): Environmental Health Criteria, Vol. 124 (Lindane), World Health Organisation.

WRRL (2000): Directive 2000/60/EC of the European Parliament and the Council of 23 October 2000 establishing a framework for Community action in the field of water policy, OJ L 327, 22.12.2000, p. 1.

Consultants, laboratories and construction companies who were involved with the project

CDM - SOLUTIONS
CLEAN DEVELOPMENT MECHANISM

WE HAVE THE SOLUTION FOR GREEN HOUSE GAS EMISSIONS

Aerobic material specific waste treatment combined with recycling

CDM-SOLUTIONS GmbH

CLEAN DEVELOPMENT MECHANISM

D- 38100 Braunschweig, Leopoldstraße 38

Struve@CDM-Solutions.de

BIOPUSTER®

technology

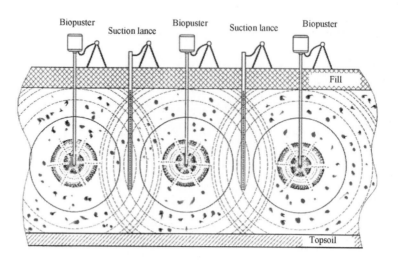

Design and aeration technology
for landfills and composting
directly from the inventor of the system

Helmholtz Center Munich, German Research Center for Environmental Health (formal GSF) Institute of Groundwater Ecology, D-85764 Neuherberg

The research in the Institute of Groundwater Ecology is interdisciplinary and ranges from hydrology, environmental tracers and stable isotope analysis to microbiology and ecology. "Clean water – healthy life" stands for our vision to contribute to the protection of one of man kind's most important resource for life, clean drinking water. We want to clarify the self-purification processes in groundwater ecosystems, the general principles why and when contaminants of anthropogenic or natural origin are removed by natural attenuation processes, and under which circumstances such resistance or resilience mechanisms of the ecosystem fail. By gaining a thorough understanding of the governing principles that drive a groundwater ecosystem and trigger self-purification processes we provide a sound basis for regulators taking decisions to protect and manage groundwater resources. In the institute are seven working groups: Hydrogeology and Geochemistry, Hydrological Modelling, Environmental Isotopes, and Contaminant Stable Isotopes, Microbial Ecology, Molecular Ecology, and Anaerobic Degradation.

The hydrology group is specialised in the assessment of hydrological systems and flow regimes using environmental isotopes. The institute has a well known environmental isotope analysis group that can, among others, analyse such tracers as ^{3}H, ^{2}H or ^{18}O. Based on isotope contents measured in the groundwater, different water flow-paths and origins can be distinguished. Using mathematical models developed in the institute the mean transit time of water (age) and transit time distribution of water in the groundwater systems are evaluated. The isotope data are further processed in the groundwater modelling group which aims to describe and predict groundwater resources, water dynamics and heterogeneity in different aquifer systems (fissured rocks, karsts, porous) under saturated and unsaturated flow conditions.

Another focal point is the assessment of microbial transformation reactions of inorganic pollutants such as nitrate or sulphate. Based on stable isotope analysis different sources of nitrate can be distinguished on a local and catchment scale and biodegradation processes can be evaluated. Analysis of the stable isotope fractionation of sulphur, nitrogen, and oxygen reveals self attenuation processes even in complex groundwater system such as karst.

A particular focus of the institute is on anaerobic microbial degradation of organic pollutants in groundwater systems. We isolate new organisms that can degrade mono- or polycyclic aromatic hydrocarbons with sulphate as the terminal electron acceptor and study the enzymatic degradation pathways. With that knowledge, we analyse the microbial degradation activity in the subsurface with compound specific analysis of stable isotope fractionation in the residual contaminant phase and compound specific metabolite analysis. The two methods which were partly developed in our laboratory can give an excellent picture on the in situ degradation activities in groundwater systems. We also investigate the interaction of microbes with solid minerals and humic acids and their role in weathering or precipitation of minerals.

The influence of aquifer heterogeneities on the degradation activity and limitations to microbial activity are investigated in the microbial ecology group. This part will be supported in due time by a molecular microbial ecology group that just has been established.

Another microbiology group is working on the removal of heavy metal contaminants with biotechnological usage of adsorption in aquatic fungi.

Waste management

Waste legislation for municipal solid waste treatment facilities places great emphasis on treatment of waste material flow and energy from waste.

Energy from waste

Regula Beratungsgesellschaft

Stromtal 81 • 38226 Salzgitter • Germany

fon +49-5341-1 33 08 or 5341-40 11 61

fax +49-5341-40 11 62

E-mail: Regula . Beratung @ t - online . de

Treatment of raw and residual materials

The purpose of modern treatment techniques is to produce the best possible separation of waste into materials suitable for recycling or the manufacture of quality saleable goods.

Treatment of secondary combustible substances

Contaminated sites

Early co-operation with the expert, planner, consultant, client, investor and local authorities makes it possible to optimise the treatment of contaminated areas which will later be built upon. This enables minimisation of costs and reduces remedial activity right from the start.

Revitalisation of formerly used and decontaminated industrial areas

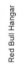

G. HINTEREGGER
& SÖHNE
Baugesellschaft m.b.H.

- FOUNDATION ENGINEERING
- BUILDING ENGINEERING and INDUSTRIAL CONSTRUCTION
- POWER PLANT CONSTRUCTION
- ROAD CONSTRUCTION
- CIVIL ENGINEERING
- TUNNELLING
- ENVIRONMENTAL TECHNOLOGY

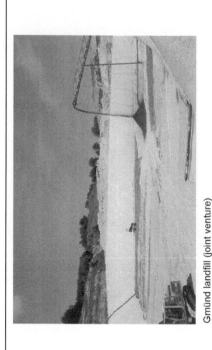

Gmünd landfill (joint venture)

Vibration piling

Reichsbrücke overhaul

Deesener Wood North

Horse race track

Red Bull Hangar

A-1220 Wien, Baranygasse 7
Phone: + 43 1 282 1537, Fax + 43 1 282 1537-13
Email: wien@hinteregger.co.at
www.hinteregger.co.at

SALZBURG   NIKLASDORF   WIEN   KLAGENFURT